Hazardous Waste Management

In memory of Robert W. Scrivner

Hazardous Waste
REDUCING

Foreword Lee B. Thomas, Jr.

Project Directors Benjamin A. Goldman
James A. Hulme
Cameron Johnson

Research Alexander C. Stewart, Jr.
Bridget Weighart

Editors Alice Tepper Marlin
Steven S. Ross

Managing the Risk

THE RISK

Council on Economic Priorities

ISLAND PRESS

Washington, D.C. □ Covelo, California

Island Press publishes, markets, and distributes the most advanced thinking on the conservation of our natural resources—books about soil, land, water, forests, wildlife, and hazardous and toxic wastes. These books are the practical tools used by public officials, business and industry leaders, natural resource managers, and concerned citizens working to solve both local and global resource problems.

Founded in 1978, Island Press reorganized in 1984 to meet the increasing demand for substantive books on all resource-related issues. Island Press publishes and distributes under its own imprint and offers these services to other nonprofit research organizations. Funding to support Island Press is provided by The Donner Foundation, The Ford Foundation, The William and Flora Hewlett Foundation, The Joyce Foundation, The Rockefeller Brothers Fund, and The Tides Foundation.

For additional information about Island Press publishing services and a catalog of current and forthcoming titles, contact: Island Press, 1718 Connecticut Avenue NW, Suite 300-B, Washington, D.C. 20009.

Library of Congress Cataloging-in-Publication Data
Goldman, Benjamin A.
Hazardous waste management.
Includes bibliographical references and index.
1. Hazardous waste management industry—United States.
2. Hazardous waste sites—United States. I. Hulme, James A. II. Johnson, Cameron.
III. Marlin, Alice Tepper. IV. Ross, Steven S. V. Council on Economic Priorities.
VI. Title.
HD9975.U52G65 1986 363.7'28 36-2748
ISBN 0-933280-30-0
ISBN 0-933280-31-9 (pbk.)

Text and cover design Rodelinde Albrecht
Copyediting Toni Murray
Proofreading Janine Hannel
Line art Irene Imfeld
Indexing Kate Gross
Production Coordination Rodelinde Albrecht
Composition Graphic Typesetting Service
Printing and binding The Maple-Vail Book Manufacturing Group
9 Manufactured in the United States of America

Grateful acknowledgment is made to the following for permission to reprint material previously published:

Table 3-2 derived from *State of the Environment: An Assessment at Mid-Decade,* © 1984, The Conservation Foundation. **Table 5-4** reprinted from the May 16, 1984 issue of *Chemical Week* by special permission, © 1984, by McGraw-Hill, Inc., New York, NY. **Figure 6-2** from *Chemical Catastrophes: Regulating Environmental Risk through Pollution Liability Insurance,* by Martin T. Katzman (Homewood, IL: Richard D. Irwin, 1985), p. 117. **Table 6-2 and Figure 6-3** reproduced with permission of Institute of Chemical Engineering, Rugby, England. Reprinted by special permission from *Chemical Engineering,* April 1, 1985, © 1985 by McGraw-Hill, Inc., New York, NY 10020. **Table 6-3 and Figure 6-5** derived from Environmental Management, Inc., "Technical Criteria for Identification and Screening of Sites for Hazardous Waste Facilities" (West Chester, PA: Delaware River Basin Commission and New Jersey Department of Environmental Protection, August 1981).

Contents

PROBLEMS AND SOLUTIONS

SIX. How to Choose: Tools for Evaluating Facilities, 99

CASE STUDIES

SEVEN. The Big Eight: Corporate Comparisons, 127

EIGHT. Picking a Winner: CEP Comparative Facility Evaluation Method, 173

Figures

Tables

Foreword

Public concern over toxic waste is only a decade old. Knowledge about what is really toxic and the effects of toxicity is growing rapidly. Disposal technology is expanding too fast for anyone to follow accurately.

Many small, medium, and large companies have been caught by these events. They have disposed of wastes to the best of current knowledge to discover later that they have a financial obligation for a cleanup. We at Vermont American Corporation have had this problem. Sometimes it is a local landfill, sometimes a Superfund site. Many companies are now concerned that even with the manifest system that supposedly assures their wastes will end up in licensed sites, they may still have a liability. Why? Because governments assume no liability even when one or another of them has approved and licensed the site.

Companies are further concerned that when they pay to clean up a site where some of their wastes were found, the wastes may be moved to another site only to cause trouble in the future—the company will have to pay again. We are among the companies so concerned. We welcome this study.

Several substantial waste disposal companies have grown up with the need. That's good. New technologies will necessitate huge capital investment. Perhaps some less hazardous wastes will have to be handled in less than the best facilities while improvements are made. But eventually, the public and waste-generating companies will demand nothing less than the best technology has to offer, at a price they can pay.

The Council on Economic Priorities (CEP) has undertaken to study and rate operating toxic waste disposal sites. CEP's motives are to help the body politic better protect our population, to help toxic waste generators determine where they might dispose of toxic wastes more safely, and to help the waste disposal companies know what constitutes a good job.

This is a tall order. The CEP study is necessarily imperfect. There are going to be honest differences of opinion as to which technologies are best. Between the time of inspection or evaluation and publication, constructive changes can take place. There can even be errors although the researchers and the members of the advisory panel were carefully chosen. I urge that instead of merely accepting this study at face value, use it as a basis for asking additional questions.

Lee B. Thomas, Jr.
Louisville, Kentucky

Note: To my knowledge I am not in any way related to Lee M. Thomas, administrator of EPA. There is no actual or potential conflict of interest.

Acknowledgments

Tackling a subject as complex as evaluating the commercial hazardous waste management industry has required the help of many people from diverse backgrounds.

The original impetus for the study was the proposal by Lee B. Thomas, Jr., chairman of the board of CEP, that the council provide generators with help in selecting hazardous waste management contractors. As chief executive officer of Vermont American, Inc., Thomas had been directly affected by the lack of available information about the quality of commercial hazardous waste managers. Along with some 500 other generators, his company had used the Seymour Recycling facility to manage wastes. Although this commercial facility was thought to be "state of the art," Indiana proposed it for the Superfund National Priorities List in October 1981. Since then, $7.7 million have been invested in cleaning up Seymour Recycling. Had better information and alternatives been available, Vermont American probably would not have used this facility for environmental and economic reasons.

This study was made possible by the generous contributions of North Shore Unitarian Veatch Program; Rockefeller Family Fund; The New York Community Trust; Arca Foundation; Ottinger Foundation; Lucy Lemann; Boehm Foundation; Pittsburgh Bridge and Iron Works Charitable Trust; and, of course, the council's board of directors, donors, and members. We are deeply grateful for their support.

Alice Tepper Marlin worked with us to structure the study and provided editorial oversight of the manuscript. Steven S. Ross wrote the initial project proposal and provided expert and knowledgeable editing. Robert A. Michaels oversaw the preliminary research of hazardous waste management in New York City and established the basis for the study.

Alexander C. Stewart and Bridget Weighart skillfully conducted research and data preparation. Nancy Bennett, Cameron Gordon, Daniel Johnson, and Tracy Weaver provided invaluable help during the later stages of the project. Carol Goodman and Kevin Heyeck assisted us during the hot summer of 1984.

Ellen Davidson accurately entered the data into our information system; Donald Jenner helped us computerize the system; and David Alsberg helped to determine the statistical accuracy of preliminary evaluations.

CEP was very fortunate to have the counsel of an active advisory panel whose members carefully reviewed our methodology and manuscript at various stages of the project. Their comments enabled us to strengthen both our methods and presentation. The findings and opinions expressed in this book, however, are solely those of the authors and CEP. Due to the controversial nature of the subject and the advisors' wide range of backgrounds and expertise, their views may differ from ours. Although CEP has taken into account the advisory panel comments, the advisors are, of course, not responsible for the final product.

The panel included: Jane Bloom; Steven Cohen; Will Collette; Dan C. Edwards; Thomas G. Eubank; Richard C. Fortuna; Jay M. Gould; Jeffrey A. Klein; John B. Leuzarder; B. Charles Malloy; Patrick McVeigh; William E. Oatess; Carol O'Cleireacain; Richard A. Peluso; Michael B. Picker; Emily S. Plishner; Robert B. Pojasek; W. B. Rossnagel; Eric Scherzer; Howard Singer; Timothy P. Slauson; Frank J. Sudol; Lee B. Thomas, Jr.; Jeffery Tryens; Carey L. Weiss; Rae Zimmerman; and others who provided substantial comments but preferred to remain anonymous.

Ruth Ruttenberg provided an overall evaluation of our OSHA data and presented CEP's findings to Congress. Marshall Beil generously contributed his time and keen legal advice by reviewing every page of the manuscript. Barbara Dean of Island Press made excellent editorial suggestions.

Island Press speedily edited and produced the book with the assistance of Rodelinde Albrecht, Toni Murray, Janine Hannel, Irene Imfeld, and Kate Gross. We are truly grateful for their efforts.

Special thanks to the individuals who live near the facilities CEP visited. Without their insights and hospitality, we would never have been able to experience the truly distressing problems felt by those with waste sites in their backyards. We hope our efforts will help them in their struggles.

Special thanks also to the owners and operators of these sites, who spent many hours showing us their facilities and reviewing our data, as well as the many EPA, state, and local officials who provided us with the large amounts of information necessary for our evaluations.

Benjamin A. Goldman
James A. Hulme
Cameron Johnson

COUNCIL ON ECONOMIC PRIORITIES

About CEP

STATEMENT OF PURPOSE

The Council on Economic Priorities is a public service research organization, dedicated to accurate and impartial analysis of some of the most vital issues facing our country today. CEP is nonaligned, independent, and nonprofit.

Founded in 1969, CEP believes that a well-informed community can express its convictions and play a far more effective and influential role in molding public policy when it is acting upon carefully conducted, documented, and factual research.

A clear and primary goal of the council is to enhance corporate performance as it affects society in the critically important areas of: energy, the costs and consequences of military spending, political influence, fair employment practices, and environmental impact. CEP acts as a clearinghouse for corporate activities and monitors the role of government as it relates to these social issues.

Credibility is absolutely essential to the work of the council. Our achievements can be measured by the frequency with which Congressional committees and administrative agencies invite CEP staff to testify. The national press reports regularly and extensively on the council's work. A broad spectrum of corporations, libraries, and government agencies are institutional subscribers. We receive support from individuals and foundations who share our concerns and beliefs.

The remarkable impact of CEP studies on policy makers, the media, and government is a result of careful and unbiased research. But the council is far more than a research and reporting organization. Once a completed study points to a course of action that we believe to be correct, we are willing to adopt that point of view and press vigorously for its wide acceptance.

Our country is faced with issues of unimaginable magnitude and complexity. The role of the Council on Economic Priorities is to assist in uncovering the facts behind these issues and to bring them to the attention of both the public and the government in such a way that they can be effectively resolved in a logical, humane, and public-spirited manner.

For your convenience, a membership application envelope is included in this book.

BOARD AND STAFF

Abbreviations

ACES: Associated Chemical and Environmental Services
ADPC&E: Arkansas Dept. of Pollution Control and Ecology
AMEC: American Ecology, Inc.
ATSDR: Agency for Toxic Substances and Disease Registry
BEHP bis: 2-ethylhexyl phthalate
BFI: Browning–Ferris Industries, Inc.
BTU: British Thermal Unit
CAA: Clean Air Act
CADIR: California Dept. of Industrial Relations
CADOHS: California Dept. of Health Services
CALOSHA: CADIR's Division of Occupational Safety and Health
CBO: Congressional Budget Office
CECOS: CECOS International, Inc.
CECOS NY: CECOS International, Inc.'s facility in Niagara Falls, New York
CECOS OH: CECOS International, Inc.'s facility in Williamsburg, Ohio
CEP: Council on Economic Priorities
CERCLA: Comprehensive Environmental Response, Compensation and Liability Act of 1980 (42 USC 9601)
CERCLIS: CERCLA Information System
CFR: Code of Federal Regulations
CWA: Clean Water Act
CWM: Chemical Waste Management
CWM IL: Chemical Waste Management's Facility in Chicago; also known as Chicago Incinerator
CWM LA: Chemical Waste Management's facility in Lake Charles, Louisiana
DISC: Developers International Services Corp.
ECC: Environmental Control Commission
EERU: Environmental Emergency Response Unit Contract
EIL: Environmental Impairment Liability
ENSCO: Environmental Systems Co., Inc.
ENSCO AR: Environmental Systems Co.'s facility in El Dorado, Arkansas
EP: Extraction Procedure
EPC: Environmental Protection Corp.
ESI: Envirosafe Services, Inc.
ESII ID: Envirosafe Services of Idaho, Inc.'s facility in Grandview, Idaho (Site B)
FEI: Fondessy Enterprises, Inc.
FIFRA: Federal Insecticide, Fungicide and Rodenticide Act
FML: Flexible Membrane Liner
FOIA: Freedom of Information Act of 1966 (5 USC 552)
GAO: Government Accounting Office
HAZAN: Hazard Analysis
HAZOP: Hazard and Operability Study
HMIS: Hazardous Materials Information System
HRI: Hydrocarbon Recyclers, Inc.
HRS: Hazard Ranking System
HSWA: The Hazardous and Solid Waste Amendments of 1984
HWDMS: Hazardous Waste Data Management System
IEPA: Illinois Environmental Protection Agency

IT: International Technology Corp.
IT CA: International Technology Corp.'s facility in Benicia, California
IU: IU International Corp.
KO51: Primary sludge produced by oil-refining processes
LADEQ: Louisiana Dept. of Environmental Quality
LADNR: Louisiana Dept. of Natural Resources
MMT: Million metric tons
NA: Not Applicable
NAL: National Analytical Laboratories
NAS: National Academy of Sciences
NCI: National Cancer Institute
NE: Not Evaluated
NECO: Nuclear Engineering Co.
NEIC: National Enforcement Investigations Center
NIPDWS: National Interim Primary Drinking Water Standards
NPL: National Priorities List
NRC: Nuclear Regulatory Commission
NYSDEC: New York State Dept. of Environmental Conservation
OAT: Office of Appropriate Technology
OEPA: Ohio Environmental Protection Agency
OKI: Ohio-Kentucky-Indiana Regional Council of Governments
OSW: Office of Solid Waste, U.S. EPA
OTA: Office of Technology Assessment, U.S. Congress
Part A: Part A of a RCRA permit application
Part B: Part B of a RCRA permit application
PCB: Polychlorinated biphenyl
PA/SI: Preliminary Assessment/Site Investigation
POTW: Publicly Owned Treatment Works
PRP: Potentially Responsible Party
QC/QA: Quality Control/Quality Assurance
RCRA: Resource Conservation and Recovery Act of 1976 (42 USC 6901)
RES: Rollins Environmental Services, Inc.
RES (LA): Rollins Environmental Services' facility in Baton Rouge, Louisiana
RI/FS: Remedial Investigation/Feasibility Study
SDWA: Safe Drinking Water Act
SFBWQCB: San Francisco Bay Region Water Quality Control Board
SIA: Surface Impoundment Assessment
SIC: Standard Industrial Classification
TACB: Texas Air Control Board
TCE: Trichloroethylene
TDS: Total Dissolved Solids
TDWR: Texas Dept. of Water Resources
TECO TX: Texas Ecologists' facility in Robstown, Texas
TOC: Total Organic Carbon
TOX: Total Organic Halides
TSDF: Treatment, Storage, and Disposal Facility
UAREP: Universities Associated for Research and Education in Pathology, Inc.
U.S. DOC: United States Dept. of Commerce
U.S. DOE: United States Dept. of Energy
U.S. DOJ: United States Dept. of Justice
U.S. DOL: United States Dept. of Labor
U.S. DOT: United States Dept. of Transportation
U.S. EPA: United States Environmental Protection Agency
U.S. OSHA: United States Occupational Safety and Health Administration
U.S. SEC: United States Securities and Exchange Commission
UIC: Underground Injection Control
USC: United States Code
USEC: U.S. Ecology, Inc.
USGS: United States Geological Survey
USHWC: Utah Solid and Hazardous Waste Commission
USPCI: U.S. Pollution Control, Inc.
USPCI UT: U.S. Pollution Control, Inc.'s facility in Clive, Utah
VOC: Volatile Organic Compounds
WET Model: Waste, Environment, and Technology Model
WMI: Waste Management, Inc.

CEP FINDINGS

CHAPTER ONE

Busting the Shell Game: Where CEP Looked and What CEP Found

The fact is that industrial societies generate hazardous wastes. Another fact is that plenty of room for improvement remains in the industrial production processes and practices that reduce waste and reuse waste byproducts. The public must recognize the need for hazardous waste management facilities and work with industry and the regulators to ensure that these facilities are properly located, that they are optimally managed, and that they use the most appropriate technologies.

Moving hazardous wastes is not a final solution. In fact the policy of the United States Environmental Protection Agency (U.S. EPA) in most cases is to keep wastes where they are and to stabilize leaking sites rather than to move the wastes elsewhere. Nevertheless, beset by political pressures to get rid of toxic wastes quickly, many public officials still implement out-of-state, out-of-mind policies. Proper management of hazardous wastes means minimizing the risks to all of society, not shifting risks to other sectors.

This study by the Council on Economic Priorities (CEP) tracks a nationwide toxic "shell game" and provides practical help to arrest its continuation. The CEP study does some of the work that U.S. EPA should have been doing all along and points the way to an adequate method for screening operating hazardous waste facilities. By taking the time to look through hundreds of files at regional U.S. EPA and state offices and by evaluating them with an intensive comparative approach, CEP discovered much that is going unnoticed by state and local regulators.

The CEP study focuses on the eight publicly held commercial U.S. hazardous waste management companies that own and operate treatment and disposal facilities. The table on page 4 lists the companies in alphabetical order.

CEP estimates that these eight companies own one-fifth of the commercial hazardous waste management industry's active facilities and garner almost one-half of the industry's revenues. These eight major waste firms, if managed improperly, could adversely affect the health of many Americans. The hazardous waste management industry receives wastes from over 5000 industrial plants, and the number is growing by leaps and bounds.

Parent Company	Principal Hazardous-Waste Management Subsidiary(ies)
American Ecology, Inc.	U.S. Ecology, Inc.
Browning–Ferris Industries, Inc.	CECOS International, Inc.
Environmental Systems Co., Inc.	ENSCO, Inc.
International Technology Corp.	IT Corp.
IU International Corp.	Envirosafe Services, Inc.
Rollins Environmental Services, Inc.	RES (LA), RES (NJ), RES of LA, and RES (TX)
USPCI, Inc.	U.S. Pollution Control, Inc.
Waste Management, Inc.	Chemical Waste Management

Congress has strengthened the laws under which active hazardous waste facilities are regulated. Nevertheless, after visiting a number of active treatment, storage, and disposal facilities, CEP found that not only do generators not know which facilities are the best, but that U.S. EPA has not always selected the best facilities to receive wastes removed from Superfund sites (closed sites that endanger public health and the environment). Other facilities were better managed, better located, and better at using more advanced technologies than the facilities U.S. EPA selected. In fact, of the ten facilities CEP evaluated in detail, U.S. EPA chose the one that performed worst—CECOS International, Inc. in Williamsburg, Ohio—to receive Superfund wastes in more instances than any of the other nine facilities. Data from a house subcommittee survey indicate that almost half (46 percent) of the operating hazardous waste facilities U.S. EPA chose to receive wastes removed from Superfund sites (see chapter 3) may have contaminated ground water. Some of the chosen facilities may even be partially responsible for a share of the wastes they are being paid to clean up (see chapter 4). If the decision making of U.S. EPA has resulted in poor site selection, is it any wonder that companies in the private sector have also had difficulty selecting appropriate options for waste disposal?

A CHRONICLE OF ACHIEVEMENTS AND GROWING PAINS

CEP identifies a number of the factors that led to the gap that now exists between sound practices and many of those that are currently used in the hazardous waste management industry. Many factors arise from the fact that the industry is relatively new. It has expanded rapidly in response to environmental regulations promulgated over the past 5 years.

As the industry matures, however, the trend is toward improvement. Federal and state governments are developing guidelines for proper hazardous waste management; industry compliance, in turn, is also improving—but not always fast enough to respond to tightened regulations.

CEP's study discusses federal and state regulations that currently affect hazardous waste management, presents a guide to the technologies used to manage these wastes, and offers a profile of the strengths and weaknesses of specific companies. CEP's goal is to promote the proper management of hazardous wastes by highlighting regulatory strengths and deficiencies, by naming companies and facilities with the best performance records, by defining the most appropriate technologies, and by developing strategies for selecting hazardous waste management contractors.

THE REGULATORS: STATES AND REGIONAL EPAs

Unavailability of accurate, consistent, and thorough information about hazardous waste management companies and facilities poses a major obstacle to effective regulation. CEP finds that the regulators' abilities to gather and process data are deficient, and so are their abilities to communicate findings inside and outside government. CEP ratings reveal that Illinois, Michigan, Nevada, and Texas are the best at communicating information to the public; California, Idaho, New York, and Ohio are worst.

In terms of inspection and enforcement performance during fiscal year 1984, CEP ranks Maryland's hazardous waste regulatory program as the best. On average, Maryland inspected each major facility ten times and took seven enforcement actions against every serious violator. The scores of New Jersey and California, two states that produce large amounts of waste, are also excellent in this regard. Hawaii and Wyoming, on the other hand, receive the lowest marks because neither state inspected any major facilities during the fiscal year. The records of Louisiana and New York are also among the poorest in terms of site inspection and enforcement of regulations.

The performance of the ten U.S. EPA regional offices, which manage much of the information on the waste industry, also varies: Region 3 (the Mid-Atlantic States) does the best; Region 1 (New England), on the other hand, places last (see chapter 10).

EVALUATION METHODS

Appropriate methods for evaluating facilities and companies are hard to come by. CEP searched the literature of many relevant disciplines; sources ranged from chemical engineering models to actuarial models. The object was to find an easy-to-use, consistent, and relatively accurate method for comparing the risks and benefits of existing hazardous waste management facilities. CEP selected U.S. EPA's Hazard Ranking System (HRS) as a starting point for developing an analytical tool to compare facilities on the bases of site, management, and technology (see chapters 6 and 8).

Events have supported the relevance of CEP screening procedures. The facility of Rollins Environmental Services (LA) in Baton Rouge, Louisiana, scores poorly in terms of potential hazards posed by fugitive hazardous waste emissions and worker exposure, for example. Since our evaluation the state temporarily shut down the incinerator for those very reasons. CECOS International in Williamsburg, Ohio, scores poorly in terms of potential hazards associated with ground- and surface-water contamination; since then, Ohio closed the RCRA landfill because of inadequate ground water monitoring.

SUMMARY OF THE EVALUATIONS

The Companies

CEP evaluated the eight major commercial hazardous waste management firms and the facilities they control. CEP considered a number of significant criteria, including financial viability, compliance with environmental and occupational safety regulations, ground water monitoring, potential hazardous waste liabilities, technology, corporate and public relations, and our in-depth evaluations of selected facilities.

CEP's examination of each company as a whole reveals that the performance record of USPCI, a relatively little-known Oklahoma-based company, is the best of the eight companies. The examination rates California-based International Technology Corp. (IT), the second largest hazardous waste management firm, Environmental Systems Company and Rollins Environmental Services, two leading incineration firms, as above average. CEP concludes that the performances of Browning–Ferris Industries and Waste Management, the largest firm, reflect the average performance for the eight companies.

CEP rates American Ecology (AMEC), which began as a nuclear waste management firm, as a company with a below-average performance record. Conglomerate IU International, parent company of Envirosafe Services, receives the least desirable grades of the eight hazardous waste firms.

The leaders of the commercial hazardous waste management industry are working to overcome significant problems. If the performance record of these eight firms reflects the industry as a whole, CEP believes that four of the Big Eight are performing above the industry average, two are close to the norm, and two are below average.

The Facilities

The CEP study assigns low overall scores to CECOS International in Williamsburg, Ohio, Envirosafe Services of Idaho in Grandview, Idaho, and Texas Ecologists in Robstown, Texas. On the other hand U.S. Pollution Control's facility in Utah, Chemical Waste Management in Louisiana and Illinois, and CECOS International in New York receive high marks in their geographic regions.

Unfortunately, only one of the facilities in the study performs significantly better than the regulatory minimum. CEP believes the locations, the management, and the technology of all the facilities need to improve. The study does belie the popular belief that hazardous waste facilities are not making a sustained effort to improve, however. The facilities are improving, and the improvements are substantial.

WHO NEEDS INFORMATION ABOUT HAZARDOUS WASTE MANAGEMENT?

The Generators of Hazardous Waste. In the past the lack of comparative evaluations forced some generators of hazardous waste to rely on hearsay and the lowest bid to select waste management firms. Few generators have the resources to conduct in-depth environmental audits of the waste management facilities they use. The thousands of small- and medium-size companies that use commercial hazardous waste disposal firms now or will need to use them as more stringent laws come into effect will find that this study is essential. The study is a consumers' guide that will help companies minimize the significant environmental and legal risks posed by hazardous waste mismanagement.

Policy Makers. This book advises policy makers about regulatory deficiencies. By using a modified version of U.S. EPA's own methodology, the HRS, CEP has acted as an independent user and tester of the HRS. CEP concludes that the HRS overstates certain risks (air pollution, for example) and underestimates others (including direct contact and ground water contamination). CEP's critique of regulations and the ranking system have broad-reaching implications for policy making and implementation.

Community and National Organizations. Hundreds of community and national organizations focus on the hazardous waste management industry. CEP contacted many such organizations in the process of gathering and sharing information. These groups all desire a detailed and objective view of what is really happening to hazardous wastes. In the past a few theoretical and general works had been published, but CEP's study is one of the first extensive real-world investigations.

State and Local Governments. State and local governments face the need to select sites and contractors for hazardous waste facilities. CEP's critique should help them by pointing out the better companies, technologies, and facility evaluation methods.

Academics. Academics can use this book to verify their ideas with actual statistics. Students will find it a useful introduction to the multifaceted field of environmental management.

Investors. The waste management industry is rapidly changing and growing in response to increased regulation. Investors will find CEP's analysis of the long-term financial viability of these major publicly owned companies useful in identifying investment opportunities. This study could provide the basis for many appreciating investments.

THE BOTTOM LINE

An actively engaged public that commands prompt government response to citizen inquiries and complaints will be an important factor in improving the performance of hazardous waste management companies. Better enforcement of state and federal regulations will also help. Government regulation alone cannot solve the problems of environmental degradation, however. Companies must support the general intent of the law before they will comply with it.

CEP believes the market economy largely determines whether industry executes the intent of the law. CEP's study and others like it will inform companies about proper hazardous waste management. The response of informed companies and investors to the bottom line may ultimately be the catalyst that improves the standard of hazardous waste management in the United States. If generators begin to send their wastes only to those facilities that they believe pose the least liability, the marketplace could upgrade waste management practices. Commercial waste disposers will be forced either to improve their operations or to face stagnation and bankruptcy.

CHAPTER TWO

Recommendations: A Starting Point for Individual Perspectives

The subject of hazardous waste management comprises a bewildering mix of perspectives. Anyone who undertakes the task of learning about the developments, technology, and state of the industry may have trouble finding a starting point that reflects individual priorities. This following chapter is an overview of the most important recommendations that CEP reached while evaluating the hazardous waste management industry. These recommendations and the accompanying references to later chapters should help you find the information that is most relevant to your perspective. A review of the recommendations may also help readers who are unfamiliar with the industry to gain an overview of current concerns.

RECOMMENDATIONS FOR INDUSTRY AND LOCAL ORGANIZATIONS

For Generators

1. Be smart shoppers. The marginally higher short-term expense of the best waste management strategies and contractors is ultimately less costly than the liabilities for mismanaged wastes, as chapters 3 and 4 describe.
2. Consider wastes as carefully as raw materials; understand the physical and chemical properties of the waste.
3. Review waste-producing processes from beginning to end. At every step ask whether the process can be modified, controlled, or eliminated to decrease waste generation.
4. Implement the steps that chapter 5 describes to reduce waste generation, the hazard of wastes, and—possibly—overall costs:
   - process modification, raw material substitution, waste segregation, and source separation
   - reuse by recovery and recycling or waste exchange

- treatment for hazard elimination or reduction
- treatment for volume reduction
- land emplacement of residual waste in a secure facility.

5. After implementing every possible means of elimination, modification, and volume reduction, choose a hazardous waste management plan that balances long-term benefits and risks.
6. Use all available information, the expertise of trade associations, and, if necessary, the services of consultants to determine which hazardous waste contractor(s) and facility(ies) to use. Begin with chapters 7 and 9, which evaluate eight hazardous waste management firms and ten RCRA facilities. Chapter 9 cites the most and least desirable of ten facilities in three broad geographical areas.
7. Use chapter 6 as a starting point for reference to documents and analytical methods relevant to selecting a waste facility.
8. Join with other generators of similar wastes to obtain technical evaluations of facilities at reduced cost.
9. Find out if a contractor's financial problems affect operations, maintenance, and quality of operating personnel. Chapter 7 presents financial data and investigatory techniques.
10. Review the adequacy of the commercial waste manager's financial protection in case of accidents or other liabilities. Federal requirements are specific, and how well a company meets those requirements is an indication of financial stability and reliability.
11. CEP believes that the site of USPCI's Grassy Mountain facility in Clive, Utah, is the best of the ten facilities evaluated. If wastes can be treated with the processes available at the facility or can only be landfilled, and if transportation does not pose undue costs or risks, we recommend using this site.

For Host Communities

1. Ask for information about prospective hazardous waste management contractors. Information is the key to accurate evaluation. Examine the contractor's history, finances, regulatory compliance, and processes. Chapter 7 shows you how to start.
2. Find out how well the company complies with U.S. EPA regulations. Has U.S. EPA or a state agency inspected the company's facilities? Are there outstanding enforcement actions? What requirements have not been met? How large have the penalties been? Have the penalties been paid?
3. Find out whether the company has received citations for poor compliance from U.S. OSHA or U.S. DOT. A good safety record may indicate a company's attitude toward safety in general and toward its responsi-

bility in dealing with hazardous wastes. Beware, however, that reports of accidents and spills can be an indication of awareness and compliance and that the lack of reports does not necessarily indicate the lack of accidents.

4. Review company finances. Get the company's annual, 10-K, and 10-Q reports to U.S. SEC. Use the consolidated balance sheet, the statement of earnings, and the statement of changes in financial position to make a basic analysis. Obtain the assistance of an accountant if necessary. A smaller developing company may have short-term financial shortcomings that do not affect its technical capability.
5. To compare companies in a consistent manner, use the financial overview method described in chapter 7 to prepare a summary similar to table 7-2.

For Generators and the Public

1. Understand the regulatory status of a facility. *Interim status* does not mean that a facility is safe; it means that U.S. EPA has granted interim permission to operate until U.S. EPA or the state can fully evaluate the facility.
2. Use CEP's methods as described in chapter 8 to select hazardous waste contractors, sites, and technologies. U.S. EPA's requirement that all RCRA facilities submit exposure assessment reports will give interested parties access to much of the information needed to perform a facility evaluation.
3. Find out how well a facility complies with ground water-monitoring requirements. Know if monitoring has been evaluated by U.S. EPA, if well systems and record keeping are adequate, if waste leakage has occurred, and even if there has been a significant increase in any of the monitoring test parameters. If wastes have leaked, have corrective actions been taken, are corrective actions planned, and is the corrective action on schedule?
4. Review Part B of the RCRA permit application for a site-specific operating plan. Evaluate the plan for its application to the process or waste in question.
5. Ask for operating reports. The most sophisticated technology will be useless if the contractor applies it poorly or inconsistently. Poorly maintained equipment will not operate as specified, and it will not be effective in treating or destroying hazardous wastes. A good operating plan should be understandable to an informed citizen; many government permit writers also have limited expertise in the applicable fields. Common sense combined with well-planned questions can highlight problem areas that you need to check more closely.

RECOMMENDATIONS FOR CONGRESS AND REGULATORS

For Superfund and RCRA Administrators

1. Move wastes from leaking sites only to sites where no leaks have been found and where leaks can be prevented or contained, as chapters 3 and 4 describe.
2. Provide standards for remediation measures that reduce long-term risks. Enforce these standards despite marginally increased short-term costs if the standards reduce long-term risks.
3. Recognize that RCRA is this country's preventive medicine against future hazardous waste mismanagement; RCRA stipulations deserve as much attention as Superfund requirements as chapters 3 and 4 detail. Focus equal effort on risk assessment and management—the key environmental policies of the future—as on remedial action.
4. Orient policy and regulation toward the goal of helping the most environmentally successful technologies become the most profitable. Use financial requirements to bring the long-term costs of less sophisticated technologies, such as land emplacement, to bear in the present; make the development and use of improved technologies worth the cost for generators and disposal firms.
5. As chapter 4 recommends, maximize the impact of regulatory oversight by investigating corporations as a whole.
6. Damage to ground water has potential costs of tremendous magnitude (see chapter 4). Require facilities to comply fully with ground water monitoring requirements and to adhere strictly to schedules for bringing monitoring programs into compliance. Close facilities that fail to comply with monitoring regulations, and close any facility that fails to meet a compliance schedule for any reason within its control.
7. Strictly enforce existing requirements for financial instruments. Close facilities that do not have adequate financial instruments to pay the costs of corrective action or to compensate victims for damages.
8. Provide economic incentives for industry to set up its own insurance pool or fund. A proven incentive is to allow companies to deduct payments into a pool as business expenses just as they would be allowed to deduct premium payments to an insurance company.
9. If other means to cover the costs of mismanaged hazardous waste are unavailable, ensure that Congress and the public realize that CERCLA or the federal treasury are the only other sources of revenue that will pay for damages at hazardous waste sites. Provide estimates for these potential costs.
10. The CECOS facility in Williamsburg, Ohio, ranked lowest in the CEP evaluation outlined in chapter 9. CEP concluded that the facility's location is unsuitable for the land emplacement of hazardous waste

and recommends that the facility not be allowed to reopen, especially if it has not complied with ground water-monitoring requirements. The site may be suitable for treatment technologies.

Risk Assessment

1. Develop a mathematical model similar to the HRS for screening remedial action and removal options with consistent and cost-effective comparative analyses (see chapter 6). The National Contingency Plan requires that all contractors that perform a Remedial Investigation/Feasibility Study consider every possible treatment and destruction alternative prior to removal and landfill burial. Calculate cost-effectiveness with a long-term perspective.
2. Adopt compatible and consistent policy and regulatory procedures so that the public can compare the risks posed by active facilities relative to those by closed sites.
3. Modify the HRS to reflect developing RCRA regulations more closely. Relevant standards will affect treatment requirements, liner systems, and location.
4. Modify the HRS to account for migration from a contaminated aquifer to an aquifer of concern by changing containment, depth, and permeability scoring criteria.
5. Within the legislated 3 years, provide adequate funds for the development of air-monitoring standards and standards for the prevention of air releases. Incorporate containment factors in HRS air-route scoring as soon as these standards are developed.

Improve Information Management

1. Accurate and consistent information about hazardous waste is critical for the public, regulators and industry and must remain in federal hands. The complexity and national scope of information about hazardous waste necessitate that the federal government retain responsibility for managing information. State and local governments can discover problems and act sooner than federal agencies, and they can tailor solutions to the needs of specific locations, but the fact remains that wastes cross state lines. Waste management is, therefore, an interstate industry.
2. Recognize that public participation and support is the key to a successful RCRA program. Place as much emphasis on providing accurate information to local citizens and responding to their concerns as to industry and its requests.

3. The wide range in performance among states and regional U.S. EPA offices that chapter 10 describes indicates the need for better communication within government and for improving the performance of some agencies. Improve the coordination within state agencies, among states, between states and federal government, among overlapping U.S. EPA programs (especially RCRA, CERCLA, and TSCA), among regional federal offices, and between the federal agencies that have some jurisdiction over hazardous waste regulation (U.S. DOT, U.S. DOL, U.S. DOJ, and NRC).
4. Recognize that information is too valuable to underutilize. Millions of dollars have already been spent on U.S. EPA's data systems; mismanagement of data is a waste of taxpayers' money, and it can result in significant environmental damage. Train information managers properly and consistently.
5. Integrate the federal government's waste data systems to facilitate national, regional, and local research that can detect precarious trends before they become catastrophes. Lack of available information has been one of the most serious obstacles to research that could confirm or deny correlations between hazardous waste and health impacts. The U.S. DOC's census database and the National Cancer Institute's cancer mortality database are prime examples of systems that could, in conjunction with U.S. EPA's hazardous waste data, locate risk areas. Maximizing compatibility is the first step toward this goal.
6. State agencies should budget money specifically for public education and information management—the need to respond to public inquiries about hazardous waste is particularly critical. Every state should establish information hotlines.
7. U.S. EPA's CERCLA System (CERCLIS) should track:
   - potentially responsible parties and their responsibility for each waste site
   - the remedial-action contractor and the dollar value of the contract
   - the selected remedial action and removal action and the cost
   - the RCRA recipient of CERCLA wastes that were moved pursuant to removal action and contract value.

This database could aid U.S. EPA in making consistent and cost-effective decisions during the Remedial Investigation/Feasibility Study (RI/FS) process at Superfund sites by providing data on past activities. Overuse of certain contractors would be easier to detect, and potential conflict of interest in selecting potentially responsible parties for remedial work could be avoided. Even though U.S. EPA argues that all Superfund sites are unique, there are aspects of the review process that U.S. EPA should treat consistently. Chapter 10 details tracking requirements and deficiencies.

PROBLEMS AND SOLUTIONS

CHAPTER THREE

Setting the Stage: Some Technical and Legal Background

WHAT IS HAZARDOUS WASTE?

Waste is any material that is no longer wanted and has no current or perceived future value. Hazardous waste is waste that has physical, chemical, or biological characteristics that can place human health at risk or adversely affect the natural environment.

Hazardous wastes can be solids, liquids, or gases. They are the byproducts of a wide variety of manufacturing and service industries—everything from hairdressing to the production of drugs, from sign painting to steelmaking. In short, consumer demand results in processes that generate hazardous wastes. Table 3-1 shows some common consumer products and the types of

TABLE 3-1. *Common Products and Potential Hazardous Wastes Generated during Manufacturing*

Products	Potential Hazardous Waste Byproducts
Plastics	Organic chlorine compounds, organic sludges
Pesticides	Organic chlorine compounds, organic phosphate compounds, organic sludges
Medicines	Organic solvents and residues, heavy metals such as mercury and zinc
Electrical equipment	Cyanides, heavy metal sludges, caustics, solvents, acids
Paints	Heavy metals, pigments, solvents, organic residues
Oil, gasoline, and other petroleum products	Oil, phenols and other organic compounds, heavy metals, ammonia salts, acids, alkalis
Metals	Heavy metals, fluorides, cyanides, acid and alkaline cleaners, solvents, pigments, abrasives, plating salts, oils, phenols
Leather	Heavy metals, organic solvents
Textiles	Heavy metals, dyes, organic chlorine compounds, solvents

DERIVED FROM: California Hazardous Waste Management Council, "Draft Hazardous Waste Management Plan" (January 1984) and U.S. EPA's general brochures about hazardous waste.

NOTE: Processes for making these products vary; not all manufacturers produce the same waste.

wastes their manufacture produces. Commercial dry cleaners and automobile repair shops, for example, are both sources of hazardous wastes because they use solvents that can be toxic. In addition, those who use pesticides on gardens and lawns or oven, septic tank, or cesspool cleaners are all likely to add, though minutely, to the overall hazardous waste disposal problem.

Many materials that one might consider as wastes are not currently regulated by law. Also, the definition of the term *hazardous* differs in accordance with the federal law that controls a particular substance. Some of the controlling laws are the Clean Water Act (CWA), the Clean Air Act (CAA), the Toxic Substances Control Act of 1976 (TSCA), the Safe Drinking Water Act (SDWA), the Resource Conservation and Recovery Act (RCRA), the Comprehensive Environmental Response, Compensation and Liability Act of 1980 (CERCLA, more commonly known as Superfund), and the Federal Insecticide, Fungicide and Rodenticide Act (FIFRA). Table 3-2 lists other applicable federal statutes.

The term *toxic* describes materials that have deleterious biological effects, such as poisoning, on humans and other animal life. Toxic wastes are only part of the total universe of wastes that fall under the classification *hazardous*. The rubric *hazardous* includes materials that have the physical and chemical properties of ignitability, corrosivity, and reactivity—properties that can lead to harmful effects not only to living organisms but also to inanimate objects, including buildings and sewage treatment plants.

Table 3-3 outlines U.S. EPA's definition of hazardous waste characteristics. Table 3-4 specifies the concentration of various substances (expressed in milligrams per liter) that U.S. EPA defines as toxic. These concentrations are known as Extraction Procedure (EP) limits. At present, RCRA defines EPs for only a very small portion of the many different toxic wastes that society generates. The 1984 amendments to RCRA require a review of EP toxicity, however, and this review is likely to result in significant changes.

THE REASONS FOR PUBLIC CONCERN

The general public's concern about hazardous waste is well-founded; all too often contaminated ground water, irritating and noxious gases, and discolored vegetation and surface water are the results of a mismanaged waste site. Almost every state has waste sites—some operating, some abandoned—that could endanger the environment or human health. A report prepared in 1984 by Fred C. Hart Associates (see chapter 4) relates many incidents of environmental damage to contaminated aquifers. Unfortunately, because the contamination of underground water occurs away from human eyes, it is not obvious. Detection usually requires a number of monitoring wells. Ground water in many areas is not analyzed until a problem has already been detected or is strongly suspected.

TABLE 3-2. *Federal Pollution-Control Statutes, 1970–1984*

Law	Date Enacted
Statutes Focusing on Nonagricultural "Conventional" Pollutants in Regard to Hazardous Substances	
Clean Air Act Amendments of 1970 and in years following	1970
Resource Recovery Act—amends the Solid Waste Disposal Act	1970
Water Pollution Control Act Amendments	1972
Ocean Dumping Act—a title of the Marine Protection Research and Sanctuaries Act	1972
Statutes Focusing on Toxics or Hazardous Substances	
Federal Environmental Pesticide Control Act—amends the Federal Insecticide, Fungicide and Rodenticide Act	1972
Safe Drinking Water Act—amends the Public Health Service Act	1974
Resource Conservation and Recovery Act—amends the Resource Recovery Act, which focuses on hazardous wastes	1976
Toxic Substances Control Act	1976
Clean Air Act Amendments of 1977	1977
Clean Water Act of 1977—amends the Water Pollution Control Act	1977
Federal Insecticide, Fungicide and Rodenticide Act Amendments of 1978	1978
Comprehensive Environmental Response, Compensation, and Liability Act (Superfund)	1980
Hazardous and Solid Waste Amendments—amend the Resource Conservation and Recovery Act	1984
Statutes Focusing on Consumer and Worker Health and Safety in Regard to Hazardous Substances	
Federal Railroad Safety Act—provides tightened controls on shipment by rail of hazardous substances	1970
Occupational Safety and Health Act—regulates worker exposure to toxic chemicals in the workplace	1970
Consumer Product Safety Act—relates to the manufacture and sale of hazardous or toxic consumer products (except cosmetics, drugs, and tobacco)	1972
Hazardous Materials Transportation Act—provides general guidelines and regulations to control the transport of toxic substances	1975
Federal Mine Safety and Health Act—addresses the problem of human exposure to toxic substances and harmful physical agents in mines	1977

DERIVED FROM: Conservation Foundation, *State of the Environment: An Assessment at Mid-Decade* (Washington, DC: 1984).

Moreover, waste sites have traditionally been located near industrial areas in or near major centers of population. Such proximity to population centers greatly adds to the potential danger from fire and explosion or from contamination of ground water, surface water, or air.

Public concern over the effects of contamination is aimed at both short- and long-term hazards.

TABLE 3-3. *Characteristics of Hazardous Wastes as Defined by U.S. EPA*

Ignitability
U.S. EPA considers wastes with the following characteristics of ignitability to be hazardous: ■ liquids with a flash point (the temperature at which the vapor easily ignites in air) less than 140°F (The only exceptions are aqueous alcohol solutions containing 24% by volume or less of alcohol.) ■ materials that are not liquids and are capable, under standard temperature and pressure, of causing a fire by means of friction, absorption of the moisture of spontaneous chemical changes ■ materials that burn so vigorously and persistently when ignited that they create a hazard ■ ignitable compressed gases ■ oxidizers.
Corrosivity
U.S. EPA considers wastes with the following characteristics of corrosivity to be hazardous: ■ aqueous wastes with a pH less than or equal to 2 or greater than or equal to 12.5 ■ liquid wastes that corrode steel at a rate equal to or greater than 0.25 inch per year at a test temperature of 130°F.
Reactivity
U.S. EPA considers wastes with the following characteristics of reactivity to be hazardous: ■ materials that are normally unstable and readily undergo violent change without detonating ■ materials that react violently with water ■ materials that form potentially explosive mixtures with water ■ materials that, when mixed with water, will generate toxic gases, vapors, or fumes in quantities sufficient to endanger human health or the environment ■ cyanide- or sulfide-bearing materials that, when exposed to a pH between 2 and 12.5, can generate sufficient quantities of toxic gases, vapors, or fumes to present a danger ■ materials capable of detonation or explosive reaction if subject to a strong initiating source or if heated under confinement ■ materials that are readily capable of detonation or explosive decomposition or reaction at standard temperature and pressure ■ forbidden explosives of Class A (primers, dynamite, and other high-energy explosives) and Class B (propellants and other low-energy explosives).
Toxicity
U.S. EPA defines as toxic those wastes that qualify under the terms of U.S. EPA's Extraction Procedure, a test that identifies hazardous concentrations of a constituent in ground water, and the national Interim Primary Drinking Water Standards (NIPDWS). The following steps summarize the determination procedures: ■ Constituents are extracted from the wastes in a manner designed to simulate the leaching action that occurs in landfills. ■ This extract is analyzed to determine whether it possesses any toxic contaminants identified in NIPDWS. ■ If the extract contains any of the contaminants in concentrations 100 times greater than that specified in NIPDWS, the waste is considered hazardous. ■ If the original solid-waste stream contains less than 0.5% solid matter, then technicians analyze the original solid-waste stream rather than the leachate.

SOURCE: *United States Code of Federal Regulations,* Title 40, Part 261, Subpart C, Section 261.24 (1983).

TABLE 3-4. *Extraction Procedure Toxicity Limits*

Contaminant	Maximum Concentration (mg/1)[a]
Arsenic	5.0
Barium	100.0
Cadmium	1.0
Chromium	5.0
Lead	5.0
Mercury	0.2
Selenium	1.0
Silver	5.0
Endrin	0.02
Lindane	0.4
Methoxychlor	10.4
Toxaphene	0.5
2,4-D (2,4-dichlorophenoxyacetic acid)	10.0
2,4,5-TP silvex (2,4,5-trichlorophenoxypropionic acid)	1.0

SOURCE: *United States Code of Federal Regulations,* Title 40, Part 261, Subpart C, Section 261.24 (1983).

a. The standards listed here, for the metals in particular, are general. For several of these metals the toxicity of some organic forms, most notably arsenic and mercury, can be over 1000 times as high.

- Short-term hazards include acute exposures to toxic substances via ingestion, inhalation, or skin absorption; corrosivity; and fire and explosion.
- Long-term hazards are posed by chronic exposures at doses too low to cause immediate illness. Long-term effects include cancer and birth defects.

A number of health studies have attempted to specify the results of chronic exposure, but direct cause-and-effect relationships have been difficult to establish. Often monitoring and contact periods has been too short to determine whether or not the incidences of cancer, birth defects, and miscarriages are higher near waste sites. Much of the evidence in reported cases is still controversial. Love Canal, for instance, first gained attention because children received chemical burns from playing in schoolgrounds at the site. The burns provided visible proof of the danger of acute exposure. But those who have tried to link other harmful effects with chronic, off-site, low-level exposure have rarely succeeded.

An extensive review of existing literature conducted in 1984 by Universities Associated for Research and Education in Pathology, Inc. (UAREP) concluded that few scientific reports document a cause-and-effect relationship between hazardous waste disposal and health effects. The UAREP review, sponsored by the Chemical Manufacturers Association, found only one report, a study of a waste site in Perham, Minnesota, that "offered overwhelmingly convincing evidence that serious health effects had resulted from improper disposal of waste chemicals."[1] UAREP's overall findings, however, could well be due to seriously deficient data and not to an insignificant relationship between hazardous waste and human health.

Tables 3-5A and 3-5B summarize the effects on the health of humans or laboratory animals that scientists have associated at a level of statistical significance with hazardous wastes.

THE TWO KEY HAZARDOUS WASTE LAWS

The Resource Conservation and Recovery Act

The Resource Conservation and Recovery Act (RCRA) passed in 1976 contains within subtitle C the key to federal and state oversight of hazardous waste management. The law was designed to control existing and future generators, transporters, treaters, storers, and disposers of hazardous wastes. Information submitted in compliance with this act's reporting requirements provided much of the data in this study.

Regulations pursuant to RCRA set standards for generators of hazardous wastes. The regulations define notification procedures, require the acquisition of an identification number for each site, outline manifest or waste-tracking requirements, and list training and reporting requirements (see *United States Code of Federal Regulations* title 40, part 262 (40 CFR 262)).

RCRA regulations require transporters to obtain a U.S. EPA identification number and comply with all U.S. EPA and U.S. Department of Transportation (U.S. DOT) manifesting (tracking) and labeling requirements (40 CFR 263). Standards for the design and operation of treatment, storage, and disposal facilities (TSDFs) are described in detail in 40 CFR 264 and 265. Regulations specify that the owner or operator must provide a projection of maintenance requirements and environmental impact for current and future operations as well as during closure and for a period of at least 30 years after closure.

RCRA stipulates civil and, since 1984, criminal penalties of up to $50,000 per day and 2 years in jail. U.S. EPA can seek an injunction against any person handling hazardous waste in a manner that would create an imminent and substantial danger to public health or the environment. Citizens (including companies) have the right to sue any company or the U.S. EPA for violations.

State Regulatory Programs. As Congress intended most of the states have assumed responsibility for implementing at least part of the hazardous waste management regulatory program. As of October 21, 1985, 27 states and the District of Columbia had received full final authorization to implement the RCRA program (see table 3-6). All authorized states must have hazardous

TABLE 3-5A. *Acute Effects of Certain Hazardous Wastes*

Type of Waste	Nervous System Damage	Gastrointestinal System Damage	Neurological System Damage	Respiratory System Damage	Skin Damage	Death
Pesticide wastes						
Halogenated organic pesticides	+		+	+		+
Methyl bromide			+			
Halogenated organic phenoxy herbicides					+	
2, 4-D	+					
Organophosphorous pesticides	+		+	+		+
Organonitrogen herbicides (Paraquat and Diquat)		+				+
Carbamate insecticides	+		+	+		+
Dimethyldithiocarbanate fungicide compounds				+		
Aluminum phosphide		+				
Rotenone		+		+		
Polychlorinated biphenyls		*			+	
Cyanide wastes	+		+	+		+
Toxic metals						
Zinc, copper, selenium, chromium, nickel		+		+	+	
Arsenic		+			+	+
Organic lead compounds	+	+	+			
Mercury	+	+	+			+
Cadmium		+		+		+
Halogenated organics	+		+		+	
Nonhalogenated volatile organics			+	+		

DERIVED FROM: Toxic Waste Assessment Group, Governor's Office of Appropriate Technology, *Alternatives to the Land Disposal of Hazardous Wastes, An Assessment for California* (1981).

\+ Statistically verifiable effects on human beings.

* Statistically verifiable effects on laboratory animals.

TABLE 3-5B. *Chronic Effects of Certain Hazardous Wastes*

Type of Waste	Carcinogenic Effects	Mutagenic Effects	Teratogenic Effects	Damage to Reproductive System
Pesticide wastes				
Halogenated organic pesticides	*	*	*	+
Methyl bromide				
Halogenated organic phenoxy herbicides	*	*	*	*
2, 4-D				
Organophosphorous pesticides	*	*	*	
Organonitrogen herbicides (Paraquat and Diquat)	*	*	*	
Carbamate insecticides				
Dimethyldithiocarbanate fungicide compounds				
Aluminum phosphide				
Rotenone				
Polychlorinated biphenyls	*		*	
Cyanide wastes				
Toxic metals				
Zinc, copper, selenium, chromium, nickel	+			
Arsenic				
Organic lead compounds				
Mercury			+	
Cadmium	+			
Halogenated organics	+	+		
Nonhalogenated volatile organics	*	*		

DERIVED FROM: Toxic Waste Assessment Group, Governor's Office of Appropriate Technology, *Alternatives to the Land Disposal of Hazardous Wastes, An Assessment for California* (1981).

\+ Statistically verifiable effects on human beings.

* Statistically verifiable effects on laboratory animals.

TABLE 3-6. *States Granted Final Authorization for Pre-HSWA Program as of October 21, 1985*

State	Date Authorized	Federal Register Page Number
Delaware	June 22 1984	23837 (June 8)
Mississippi	June 27 1984	24377 (June 13)
Montana	July 25 1984	28245 (July 11)
Georgia	August 21 1984	31417 (August 7)
North Dakota	October 19 1984	39328 (October 5)
Utah	October 24 1984	39683 (October 10)
Colorado	November 2 1984	41036 (October 19)
South Dakota	November 2 1984	41038 (October 19)
Virginia	December 18 1984	47391 (December 4)
Texas	December 26 1984	48300 (December 12)
North Carolina	December 31 1984	48694 (December 14)
New Hampshire	January 3 1985	49092 (December 18)
Oklahoma	January 10 1985	50362 (December 27)
Vermont	January 21 1985	775 (January 7)
Arkansas	January 25 1985	1513 (January 11)
New Mexico	January 25 1985	1515 (January 11)
Kentucky	January 31 1985	2550 (January 17)
Tennessee	February 5 1985	2820 (January 22)
Massachusetts	February 7 1985	3344 (January 24)
Nebraska	February 7 1985	3348 (January 24)
Louisiana	February 7 1985	3348 (January 24)
Maryland	February 11 1985	3511 (January 25)
Minnesota	February 11 1985	3756 (January 28)
Florida	February 12 1985	3908 (January 29)
New Jersey	February 21 1985	5260 (February 7)
District of Columbia	March 22 1985	9427 (March 8)
Kansas	October 17 1985	40377 (October 3)
Nevada	November 1 1985	42181 (October 18)

SOURCE: U.S. EPA, Office of Solid Waste, State Programs Branch (November 1985).

waste programs and regulations that are at least "substantially equivalent" to their federal RCRA counterparts. Some states have more stringent and numerous requirements than U.S. EPA.[2] Others have fallen behind in implementing minimum requirements; as a result U.S. EPA revoked some authorities.

Land Disposal Bans. The Hazardous and Solid Waste Amendments of 1984 (HSWA) go a long way toward discouraging land disposal methods such as landfills and deep-well injection. Outright bans on such methods are a possibility under the law.

Deep-well injection is usually used to dispose of dilute wastes. Dilute wastes comprise two-thirds of the total volume of hazardous wastes generated annually in the United States. (Much of the volume is water, however.) Over 99 percent of hazardous wastes are eventually placed in the ground into deep wells, surface impoundments (pits, ponds, and lagoons) and landfills. The debate over land emplacement usually includes concern for the future of deep and shallow aquifers as sources of drinking water and the contamination of aquifers with hazardous wastes.

In most parts of the country where deep-well injection is used, wastes are injected into aquifers more than 2000 feet below the surface of the ground. Drinking water usually comes from much shallower aquifers. If pressure forces the wastes upward, if the casing of the well leaks, if fractures develop in the subsurface, or if subsurface layers prove to be impermeable, aquifers nearer the surface can become contaminated.[3] Similarly, hazardous leachate (leaking rainwater contaminated by wastes) from surface impoundments and landfills can percolate downward and contaminate underground water sources.

A 1981 study prepared by the Governor's Office of Appropriate Technology in California sought alternatives to the land disposal of the 1.3 million tons of hazardous waste that were put underground annually in the state. The report identified many categories of high-priority wastes that could be recycled, treated, or destroyed.[4]

Three years later section 201 of the RCRA reauthorization law mandated that the hazardous wastes on the California list be destroyed or treated nationally, not buried in landfills. In addition, U.S. EPA is on a 5½-year schedule to prohibit direct land emplacement of all the other untreated hazardous wastes that RCRA lists unless findings indicate that they are harmless to human health and the environment. If U.S. EPA takes no action, land disposal is automatically banned. U.S. EPA must also decide whether to ban the deep-well injection of many types of hazardous waste.

The 1984 amendments also prohibited landfill disposal of bulk liquids as of May 1985 because EPA had not developed alternative regulations by that date. This amendment outlawed solidification, the most common treatment applied to liquid wastes destined for landfills. Landfills can no longer accept liquid wastes treated with absorbant materials that alter the physical state of the wastes regardless of whether the generator or the waste contractor added the materials.[5] The rationale behind this ruling is that absorbants are "sponges." Under pressures that normally occur within a landfill, the absorbant may release the hazardous liquids. The goal of the ban is to provide an incentive to reduce the generation of hazardous liquid wastes. The ban requires the use of chemical, biological, or thermal treatment processes and requires that all liquid wastes are chemically stabilized before the landfill accepts them. In the short term, however, it is likely that most liquid wastes destined for a landfill may now be injected into a deep well. As discussed the increased use of deep-well injection may not enhance safety.

Minimum Standards for Technology. Revised design and performance standards also aim at reducing the risks associated with land emplacement. The new standards required that by May 9, 1985, all interim-status landfill, surface impoundment, waste-pile units, and lateral expansions that first received wastes after November 8, 1984 have double linings and leachate collection systems between them. Landfills must also have a collection system above the liners.

U.S. EPA identified two generic liner designs that meet the minimum performance standards. The top synthetic or flexible membrane liner (FML) should prevent migration of any hazardous constituents during the entire time that the facility remains in operation and during the 30-year postclosure monitoring period. The second liner required by the new standards may be of clay or a synthetic-and-clay composite.[6] Standards for existing interim-status units, in comparison, allow for 3 feet of recompacted clay with a saturated hydraulic conductivity of 1×10^{-7} centimeters per second. Assuming a porosity of 0.495, this clay has a breakthrough time of less than 2 years according to accepted modeling procedures.

Most interim-status surface impoundments that first received wastes prior to November 8, 1984 must be retrofitted with double liners and leachate collectors by November 8, 1988. These facilities must also satisfy the ground water-monitoring standards that apply to new impoundments.

Air Emission and Monitoring Standards. A little-publicized provision of the 1984 amendments requires that U.S. EPA develop standards for air emissions and air monitoring at RCRA facilities. U.S. EPA is developing a computer model to determine the magnitude of emissions from RCRA facilities and their potential environmental impact. U.S. EPA will commission contractor studies, site visits, sampling, and validation of testing procedures.[7]

Location Standards. In October 1984 U.S. EPA released a guidance manual for permit writers that outlines the phase I criteria for location acceptability and the existing applicable regulations. This manual is the first of five publications, which will also include:

- "Review of State Siting Criteria for Hazardous Waste Treatment, Storage, and Disposal"
- "Permit Writers' Guidance Manual for the Location of Hazardous Waste Land Storage and Disposal Facilities. Phase II: Vulnerable Hydrogeology Guidance Criteria and Location Analysis Methods"
- "Permit Writers' Guidance Manual for the Location of Hazardous Waste Land Storage and Disposal Facilities. Phase II: Case Studies"
- "Permit Writers' Guidance Manual for the Location of Land Treatment Facilities."

"Phase I" describes the criteria for determining location acceptability.

1. *Site Characterization.* The inherent geologic, hydrogeologic, and pedologic features of the site should be fully characterized.

2. *High-Hazard and Unstable Terrains.* The acceptable site provides a stable foundation for the containment structure. It is not subject to natural or human-induced events that would damage the structure, which should not require ongoing maintenance because of terrain instability. The following may constitute high-hazard terrain:
   - flood-prone areas
   - seismic impact zones
   - volcanic impact zones
   - landslide-susceptible areas
   - subsidence-prone areas
   - weak and unstable soils.
3. *Ability to Monitor at Location.* Potential contamination of any ground water pathway can be monitored, or the facility is not located in a recharge zone.
4. *Protected Lands.* The site should not be on federally protected land. Protected lands include:
   - archeological and historic places
   - endangered and threatened species and critical habitat
   - parks, monuments, and scenic rivers
   - coastal areas
   - wildlife refuges
   - significant agricultural lands
   - wetlands.[8]

Permit Application Requirements. RCRA requires that a permit application for a TSDF be divided into two parts. Part A lists the location of the facility, the names and addresses of owners and operators, and the processes the facility will use. In this section the applicant estimates the types and quantities of wastes that the facility will handle. Part A does not necessarily show what the facility is currently handling but what is planned. Most states have authority to collect Part A data.

Part B requires a detailed set of technical and managerial documents. A facility that uses more than one treatment and disposal technology must complete one Part B for each technology. The preparation of Part B data can easily take a year. The facility's management usually hires environmental consultants to prepare these documents.

U.S. EPA does not provide a specific form for Part B because the detailed information required is site-specific and several different formats may be appropriate. U.S. EPA permit writers do, however, use checklists and standard formats to determine if an application is complete (see the checklists in chapter 6).

Exposure Assessments. As of August 1985 regulations require that all permit applications for landfills and surface impoundments be accompanied by an assessment of the potential for and the effects of a release of hazardous substances. A release could be the result of spilling, leaking, pumping, pouring, emitting, emptying, discharging, injecting, escaping, leaching, dumping, disposing, or any other normal or accidental operation at a facility. Regulations also required applicants to assess the risks of transporting hazardous wastes to or from the facility.

If either U.S. EPA or an authorized state agency concludes that the exposure assessment indicates a substantial health risk, 1984 RCRA amendments require that the Agency for Toxic Substances and Disease Registry (ATSDR) conduct a health assessment. U.S. EPA's RCRA appropriations fund these assessments unless findings indicate that a facility is contributing to a release of hazardous substances, in which case the funds can be recovered by EPA from the operator/owner under section 107 of CERCLA. Members of the public can submit evidence of release or associated health risks to U.S. EPA and ATSDR.[9]

Corrective-Action Provisions. All facilities seeking a RCRA operating or closure permit must now identify all continuous and prior releases, take measures to clean them up, and demonstrate financial capability for carrying out the corrective action.

Facilities that stopped receiving wastes prior to July 26, 1982 are exempt from the corrective-action requirements. U.S. EPA, however, plans to take RCRA and CERCLA enforcement actions on these exempt facilities even though it cannot require corrective action as part of the permitting process. Furthermore, the corrective-action provisions also apply to all inactive or nonhazardous units at an operating RCRA facility—a closed solid-waste landfill at the site of an active RCRA incinerator, for example.[10]

The Comprehensive Environmental Response, Compensation and Liability Act

In 1980 Congress passed CERCLA, better known as Superfund. The 5-year fund set up by the law was to collect $1.6 billion (88 percent through an industry tax and 12 percent directly from the federal treasury).

The best-known portion of CERCLA deals with the identification and response plans for uncontrolled hazardous waste sites. More than 700 sites are currently on the National Priorities List (NPL). The list includes those sites nationwide that are most in need of some type of response action. A very small number of these sites have been totally "cleaned up." Many, however, have been investigated, inspected, and have progressed through at least one step in the remedial-action program.

Also under CERCLA, generators must report spills and other releases that are greater than the minimum reportable quantity of the released chemical. Violations are criminal offenses with penalties of one year in jail and fines up to $10,000.

The Off-Site Disposal Policy. On January 28, 1983 U.S. EPA released a guidance entitled "Requirements for Selecting an Off-Site Option in a Superfund Response Action." The guidance stated that off-site treatment, storage, and disposal of CERCLA materials must take place at facilities that have RCRA permits or interim status. The facilities must have been inspected within 12 months of the response action and could have no deficiencies that may have resulted in unsound management of waste or future release. Furthermore, off-site responses had to be cost-effective in comparison with other options, and wastes requiring transportation off-site had to be properly manifested in accordance with 40 CFR part 262 and any applicable state requirements.[11] The bias was toward stabilizing and containing wastes at Superfund sites—not toward moving the wastes elsewhere.

"Procedures for Planning and Implementing Off-Site Response Actions," proposed May 6, 1985, modified the older guidance in a number of ways.

- The new procedures extend coverage beyond CERCLA remedial and removal actions to all CERCLA and RCRA enforcement actions intended to remedy situations of immediate hazard (CERCLA section 106 and RCRA section 7003).
- The procedures favor treatment, reuse, or recycling over land disposal for all responses except when an on-scene coordinator determines that such methods are not feasible during an emergency removal action.
- The modifications require that cost-effectiveness be evaluated from a long-term perspective.
- The new guidance maintains that the off-site facility must have been inspected within 6 months prior to a remedial action.
- If a facility has significant Class 1 violations or other conditions that prevent satisfactory operation, the new policy prohibits use of the facility unless the owner/operator agrees to a consent order or decree that provides full correction, and the unit to be used does not contribute in any significant way to the adverse conditions at the facility.
- Land-disposal technology must meet minimum RCRA standards for existing single-lined units and new or expanded double-lined units.[12]

Effective implementation of U.S. EPA's policy for off-site disposal depends on a rigorous inspection and enforcement program that has not been achievable so far because of inadequate resources. The policy does not favor the use of double-lined landfill units over existing single-lined or even unlined units. Moreover, the new policy is less stringent than the original policy in that it allows the use of facilities that have significant RCRA violations if the managements agree to correct them. The original policy required that management correct violations before U.S. EPA awarded a contract or subcontract.

The Hazard Ranking System. To achieve the goal of CERCLA—to identify and clean up the country's most dangerous hazardous waste sites—U.S. EPA uses the Hazard Ranking System (HRS). HRS was developed by Mitre Corp. and other organizations to score waste sites on the basis of their potential to bring about environmental damage and harm to human health by way of:

- ground water
- surface water
- air
- fire and explosion
- direct contact.

CEP used a modification of HRS (see chapters 8 and 9) to evaluate the facilities in this study. The strengths and weaknesses of the HRS are discussed in chapter 6.

Who Pays for Cleanup? Under ideal circumstances and RCRA rules, one of the financial mechanisms—surety bonds, trust funds, corporate financial test, insurance, and so forth—would assure society that sufficient monies would be available for 30 years beyond facility closure to pay for any necessary remedial actions. Under CERCLA previous or present owners and operators of a site help pay for removal and remedial actions. If the previous firm is out of business or, as is frequently the situation, bankrupt, U.S. EPA has usually been able to seek and obtain funds from companies that sent the waste for treatment or disposal.

Strong court precedents have interpreted strict, joint, and several liability under CERCLA for any or all contributors to an abandoned waste site. That means that a minor waste contributor may end up paying for the entire cleanup if other contributors cannot be found.

The cost of cleanups at Superfund sites has averaged from below what a company has to contribute for removal, additional treatment, or redisposal.

Even if a facility is operating, it may be on the verge of bankruptcy. A U.S. EPA task force began investigating compliance with financial instrument requirements early in 1985. The extent of noncompliance may not be known for years. U.S. EPA has proposed a number of strategies that allow facilities to operate without adequate insurance. If any of these strategies is adopted, the potential liabilities associated with sending wastes off-site will increase.

SELF-REGULATION

It may seem unfair to generators, but their liability is not limited by U.S. EPA inaction. Congress did not allow this by accident. Congress wants to enlist generators in the enforcement effort.

Even if a site is in compliance with federal requirements—an unlikely event for a number of years, because many regulations will not even be promulgated—the site may not represent the best option available for a potential customer. Generators need some method of comparative analysis to reduce exposure to liabilities and to protect human health and the environment. Those willing to pay top dollar for first-class waste management ought to be able to determine where to buy it.

Generators should seek indemnification or release from future liabilities. But if the waste-management company has insufficient assets or financial assurance to correct possible future problems, the waste generator will be held responsible no matter what releases the company has secured. Let the buyer beware. As of September 1985 all generators of hazardous waste must certify in their manifests that they have reduced the amount and toxicity of every waste they generate to the maximum degree that is economically practicable. Waste generators rather than commercial waste managers are the targets of this requirement. This is one example of the goal to enlist generators in the enforcement effort.

Environmental Auditing

As part of the settlement agreements negotiated between U.S. EPA and companies that have violated RCRA regulations, the agency can require that noncompliers implement internal auditing programs. The first case in which U.S. EPA applied this new policy was in an agreement with Chemical Waste Management that resulted from violations at its Emelle, Alabama facility.

Elements of an effective auditing program include:

- explicit top-management support for environmental auditing and commitment to follow-up on audit findings
- an environmental auditing function that is independent of audited activities
- adequate team staffing and auditor training
- a process that collects, analyzes, interprets, and documents information sufficient to achieve audit objectives.[14]

U.S. EPA encourages all regulated entities to develop an environmental auditing program. Not only can effective environmental auditing ensure compliance with regulatory requirements, but it also makes sound business sense. Auditing establishes protocols and checklists that enable a firm to manage pollution control affirmatively, not just react to crises. Auditing allows a company to minimize potential liabilities and potential environmental damage.

Risk Management

Risk management is a technique that can lead managers to enlightened self-regulation. Risk management seeks to mitigate identifiable risks by developing an understanding of all potential conditions. Formal risk management contains qualitative and quantitative steps.

The first step, qualitative risk identification, is similar to what many waste management firms regard as an environmental audit. Next, quantitative risk characterization determines what the estimated consequences of any exposure might be. Risk management can be an effective tool for evaluating long-term environmental consequences and liabilities, for example.

U.S. EPA has proposed guidelines for conducting health-risk assessments that underlie the general qualitative and quantitative steps mentioned above.[15] Figure 3-1 outlines the general steps needed to make management decisions about health risks.

An insurance carrier usually requires an Environmental Risk Assessment Report before a TSDF can acquire an environmental insurance policy. Recently, Environmental Impairment Liability (EIL) insurance has been difficult to obtain. Premiums rose sharply, deductibles were large, and few insurance carriers were even active in the market. The prospects of short- and long-term liabilities, both for on-site and off-site management of hazardous waste, became much too uncertain for insurers to handle.

THE REALITIES OF REGULATION: INDEPENDENT ASSESSMENT IS IN THE GENERATOR'S INTEREST

Congress and U.S. EPA are to be commended for legislating and attempting to implement the many provisions and programs that aim to assess and reduce the risks that hazardous waste management facilities pose. These federal initiatives, in and of themselves, are not adequate to assure generators that all legally operating RCRA facilities are safe, however. For a number of reasons, a generator should perform an independent assessment of waste management facilities.

Depending on U.S. EPA to provide an appropriate analysis may be asking too much of the beleaguered agency. As a result of the Hazardous and Solid Wastes Amendments of 1984, U.S. EPA and the states will be swamped with increased responsibilities without concomitant increases in personnel and fiscal resources. The result: Many of the exemplary provisions initiated by the RCRA reauthorization will take more time to complete than current schedules allow. They may, in fact, never be completed.

U.S. EPA work groups have been established in 37 areas to develop guidance documents, establish policies, and address technical issues. Even Jack McGraw, former acting administrator of the RCRA–CERCLA office, warned:

> [In] spite of our best efforts, these documents will take a long time to develop and we expect that many will have to be used in draft. . . . In implementing the amendments, regions will have to act with whatever advice is available at the time from headquarters and from regional counsels.[16]

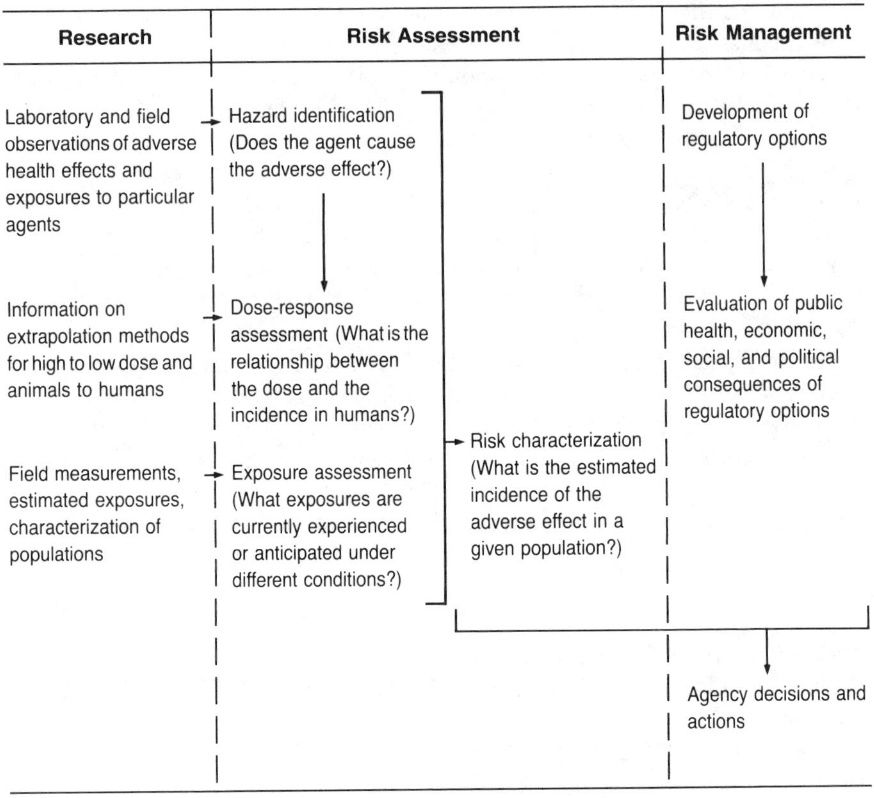

FIGURE 3-1. *General Steps for Managing Health Risks*

SOURCE: National Research Council, *Risk Assessment in the Federal Government: Managing the Process* (Washington DC: National Academy Press, 1983).

Similar delays are manifest in the permitting process. As U.S. EPA's Sylvia Lowrance admits, "we never thought it would take this long."[17] The new deadlines require only that facilities submit Part B of their applications by the dates listed in table 3-7. Facilities will not be allowed to operate under interim status if they do not comply with this schedule. On the other hand no deadlines oblige EPA to process the applications in a hurry. At U.S. EPA's current rate, it could easily take 10 more years to make final permitting determinations on all existing TSDFs.[18]

Problems with U.S. EPA's ground water-monitoring task force typify this scheduling problem. In late 1984, U.S. EPA established a task force to assess the extent of noncompliance with ground water-monitoring requirements at TSDFs. The task force consists of two groups. One analyzes guidance documents to identify and solve regulation problems. The second conducts full-

TABLE 3-7. *Hazardous and Solid Waste Amendments of 1984 Permit Application Deadlines*

Facility Type	Interim Status Terminates	Unless Part B Submitted
Land disposal	October 1985	October 1985
Incinerators	October 1989	October 1986
Other facilities	October 1992	October 1988

SOURCE: U.S. EPA, "Summary of 1984 Amendments to Resource Conservation and Recovery Act" (10/9/84).

scale evaluations of ground water-monitoring systems and programs at TSDFs. Of the 59 off-site hazardous waste management facilities that are subject to ground water monitoring, the proposed number of facilities that the task force is scheduled to investigate has fallen repeatedly. In the most recent drop the number fell from 18 to 6 during 1985. At this rate it could take over 9 years to complete full-scale reviews of all subject off-site facilities, let alone on-site plants. This raises the specter of capacity shortages; what will happen if U.S. EPA prohibits uninspected facilities from operating?

If internal instability continues to plague the task force, reviews could take longer than 9 years. Several key task force members have already left. After the first four facilities were selected for review, William Myers resigned; he claimed that his role had become "ceremonial" and that the task force's efforts as a whole were "a front." Myers, who had been the task force's technical coordinator, told *Hazardous Waste News* that his principal reason for resigning was that he did not believe the panel would be allowed to do its job. He maintained that the task force was created for political reasons to convey merely the appearance of action. Furthermore, Myers claimed that task force director Alfred Lindsay (who has also since resigned because of reportedly unrelated health reasons) requested that all internal comments be presented orally or in informal documents that would not be accessible through the Freedom of Information Act (FOIA). Myers maintained that Lindsay did not want outsiders to know what the task force was or was not doing. Lindsay denied this charge.[19]

Whether or not Myers's allegations are valid, a number of structural problems limit the task force's effectiveness. The task force will share its obligations with the U.S. EPA's National Enforcement Investigations Center (NEIC) and the regional offices. Different management approaches and overlapping responsibilities have resulted in a number of turf squabbles. Some see the NEIC as "slick and glossy"—an organization with political clout but not much substance. Most of NEIC's work is kept in strict confidentiality, a factor that limits public inquiry. The regions, in turn, often display a parochial attitude. Region 4 (the Southeast) refused to give the task force access to the Pine Wood, South Carolina, facility until former enforcement chief Gene Lucero and former acting solid waste and emergency response chief Jack McGraw personally arrived on the scene.[20]

Even with cooperation, the experts often disagree on conclusions regarding ground water monitoring. Differences develop between U.S. EPA regions and headquarters and between U.S. EPA and the states. Consensus-building or swift action under these circumstances is difficult to achieve.

U.S. EPA expects that it will not be able to meet six of the most important HSWA deadlines. The agency has already missed four others. Delays will range from several months to a year. One example: Two-thirds of all hazardous waste facilities subject to ground water-monitoring requirements did not meet the November 8, 1985 deadline for certification of compliance with monitoring and financial responsibility requirements. In response, U.S. EPA first decided to consider facilities that are meeting extended compliance schedules under administrative orders as having "met" the deadline. Fearing Congressional disapproval, however, U.S. EPA will now use a stricter interpretation of HSWA: If a site is not in compliance, it will close. How will U.S. EPA verify the certification validity of the more than 1500 facilities?

In the meantime, what if a third party seeks compensation for bodily injury or property damage resulting from releases of hazardous waste from an inadequately insured facility? Three scenarios may occur:

1. The facility's owner will pay out of his or her own pocket.
2. If the owner has inadequate funds, the waste generator(s) will pay.
3. If the generators either cannot be identified or if they do not have adequate funds, the victim will not be compensated.

At this point noncomplying active facilities may cross the thin line and become become Superfund sites.

In the meantime, regulators have limited authority to act on certain problems. Until the standards for location acceptability are promulgated, for example, the Phase I guidance recommends that permit writers "strongly discourage" granting permits to facilities at sensitive locations, even if no legal basis exists that would deny a permit. If the facility already exists at an inappropriate site, all a permit writer can do is to warn the site owners that permit renewal may be denied on the basis of site characteristics regulations that may exist in the future.

In short, during the next years of rapid change in regulatory requirements and industrial compliance, generators of hazardous waste will have to decide—with limited input from U.S. EPA input—which facilities are good bets and which are not.

CHAPTER FOUR
A NEW INDUSTRY:
Commercial Hazardous Waste Management

Commercial hazardous waste management is unusual among industries: Its growth and emerging character are largely a product of environmental regulation. Many industries have fought regulation by claiming that governmental encroachment upon free enterprise undermines competitiveness. In contrast hazardous waste management companies often lobby to protest weakening environmental regulation. Their business is transporting, storing, treating, and disposing of industrial wastes; they stand to gain from regulation.

Unfortunately, most people learn about hazardous waste management only after a toxic leak makes front-page news. Successes are rarely publicized. Establishing a site for a new waste disposal facility is nearly impossible today because of public wariness of the potential hazards. The commercial hazardous waste management industry has succeeded in establishing only half a dozen or so new disposal sites since 1980. As current facilities for destroying, neutralizing, or disposing of hazardous wastes reach capacity, and as liability insurance gets more expensive, firms that rely on commercial disposers are seeing disposal prices skyrocket. Without affordable alternatives to environmentally questionable disposal methods, problems caused by hazardous waste mismanagement will continue for years to come.

Regulators fear that rising costs and the lack of new facilities may promote illegal "midnight dumping." It is far cheaper for a tanker full of liquid wastes to pull up to an abandoned gas station, drop a hose into a toilet that connects with a sewer system, and "let 'er pour," than to wait in line and pay for testing and proper management. Cheaper, that is, so long as no one gets caught.

CEP believes the percentage of hazardous wastes that are dumped illegally is very small. U.S. EPA reports that most hazardous wastes are disposed of at the site where they are generated. Most companies want to manage their wastes properly. Yet whether generators engage in midnight dumping or use government-approved facilities, the courts and the public may hold them liable for mismanagement.

FINE TUNING THE MARKET FORCES

The business of hazardous waste management encompasses as many conflicting self-interest groups as any other sector of the marketplace. Disposal firms may try to avoid capital investment in safer technology. Regulators want to enforce safety but have to battle for the appropriations to do so. CEP believes that most generators want to have the best contractor handle their wastes, not only to protect public health and the environment from unnecessary damage, but also to reduce their potential liabilities. No company would send its wastes to a possible future Superfund site intentionally. But generators need adequate information to make their selection. Figure 4-1 just begins to

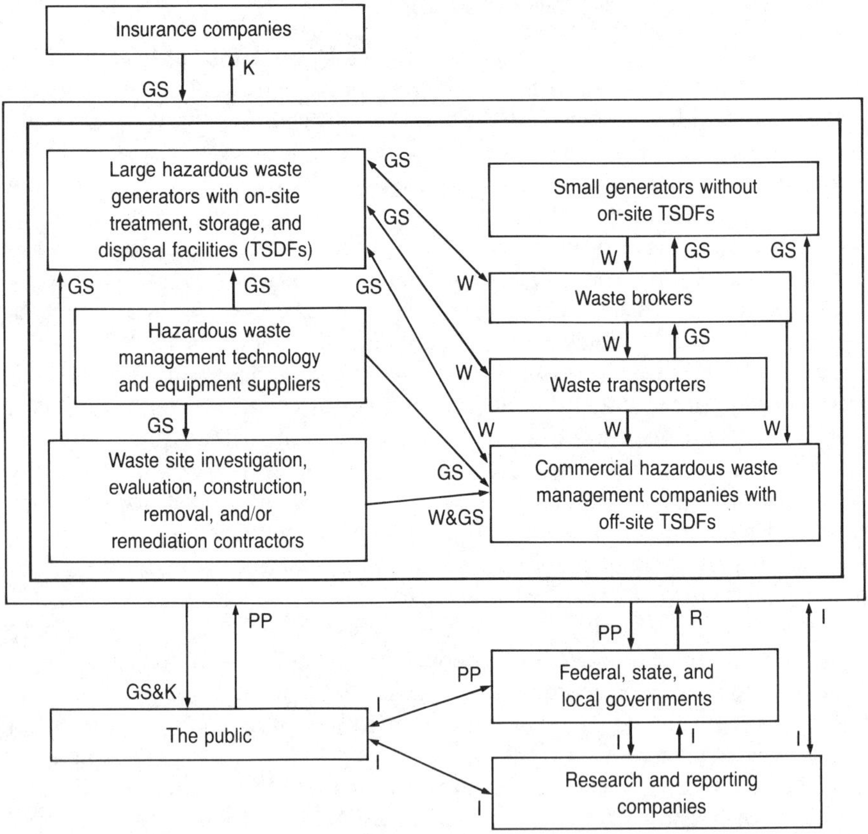

Principal Flows: GS = goods and/or services; I = information;
K = risks; PP = political pressure; R = regulation;
W = hazardous waste

FIGURE 4-1. *Major Interrelations among Key Parties Affecting and Affected by Hazardous Waste Management*

outline the groups that affect and are affected by hazardous waste management.

The lack of information often widens the gaps between the groups. Many state governments (and recall that state governments will eventually be responsible for regulating hazardous waste) are presently unable to respond to even rudimentary requests for information about waste facilities. What's more, although wastes routinely travel across state lines, states cannot easily share and compare information. State agencies usually lack the money and the personnel to keep up with developments in their own state let alone inquire about what is happening elsewhere. Independent analyses such as this one will serve as important communication bridges between the many forces of the hazardous waste marketplace.

The firms CEP evaluated are already managing over half of all the waste handled away from generating sites. We hope that government agencies, waste generators, and trade associations will further refine our methodology and expand our study.

THE GROWTH OF COMMERCIAL HAZARDOUS WASTE MANAGEMENT

By 1980 several states had passed regulations that governed hazardous waste. But the requirements were neither uniform nor comprehensive. Some states had initiated a waste-tracking system and required the segregation of hazardous and nonhazardous wastes in different types of landfills. Because the key regulatory provisions of RCRA became effective only in that year, a national profile of the commercial hazardous waste management industry did not exist.

In late 1980 U.S. EPA ordered an assessment of the amount of wastes generated and the capacity of the commercial hazardous waste management industry. The agency intended the report "to serve as a baseline against which the future effects of RCRA on the hazardous waste management industry and hazardous waste generators can be evaluated".[1]

As of June 1980 U.S. EPA knew of commercial hazardous waste treatment and disposal facilities operated by approximately 90 separate firms. These facilities treated or disposed of approximately 7.2 million wet metric tons annually. Not included in the U.S. EPA study were the many solvent recovery firms and oil re-refiners.[2]

U.S. EPA reports that from 1971 to 1980 revenues of the U.S. hazardous waste management industry had grown at a rate of approximately 20 percent per year—from $60 million to $315 million. Several estimates now place the entire U.S. market revenues for 1984 at $1 billion (see chapter 7).[3] In 1980 over half of the 127 commercial facilities were in three U.S. EPA regions—Region 5 (with headquarters in Chicago), Region 6 (with headquarters in Dallas), and Region 9 (with headquarters in San Francisco).[4]

Since 1980 much public and media attention has focused on commercial managers of hazardous waste—and with good reason. U.S. EPA estimates that over 80 percent of all generators regulated in 1981 relied either entirely or partially on commercial hazardous waste management firms to dispose of their wastes (see figure 4-2). Yet U.S. EPA also estimates that over 90 percent of the total volume of hazardous wastes generated in the U.S. was handled on site by the generators themselves. Table 4-1 shows the varied reliance of a dozen major industry groups on the commercial waste sector in 1983.

WASTE GENERATION, MANAGEMENT, AND REGULATION: WHAT THE LITERATURE SHOWS

The Westat Survey

Although more recent data exist for specific sites, the most comprehensive information on hazardous waste management practices was gathered for U.S. EPA by Westat, Inc. and published by the agency in April 1984. Westat con-

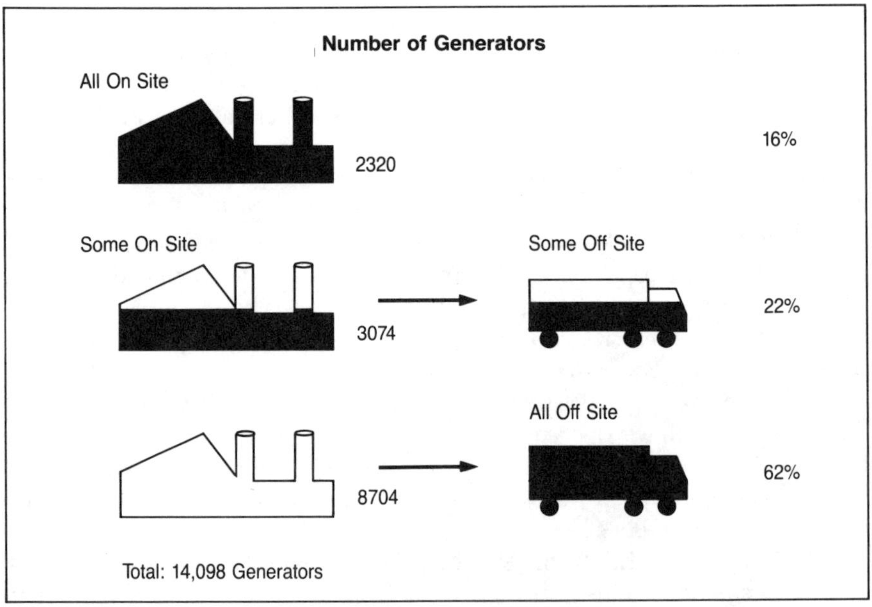

FIGURE 4-2. *Number of Generators Shipping Hazardous Waste Off Site in 1981*

SOURCE: Westat, Inc., National Survey of Hazardous Waste Generators and Treatment, Storage, and Disposal Facilities Regulated under RCRA in 1981 (Washington, DC: U.S. EPA, 1984).

TABLE 4-1. *On-Site and Off-Site Waste Flows Managed by Major Industry Groups in 1983 (in Millions of Metric Tons)*

	Flow		Percent of Total[a]	
Major Industry	*On Site*	*Off Site*	*On Site*	*Off Site*
Chemicals and allied products	125.9	1.2	99	1
Primary metals	47.3	0.3	98	2
Petroleum and coal products	31.0	0.3	99	1
Fabricated metal products	24.8	0.6	98	2
Rubber and plastic products	11.5	3.0	80	20
Miscellaneous manufacturing	4.5	0.9	83	17
Nonelectrical machinery	4.3	0.6	88	12
Transportation equipment	2.5	0.5	83	17
Motor freight transportation	0.2	2.0	11	89
Electrical and electronic machinery	1.7	0.2	90	10
Wood preserving	1.7	[b]	100	0
Drum Reconditioners	[b]	[b]	11	89
Total	255.0	9.6	96[c]	4[c]

SOURCE: Congressional Budget Office, *Hazardous Waste Management: Recent Changes and Policy Alternatives* (1985).

a. Percents calculated before rounding
b. Less than 0.1 million metric tons
c. Weighted average

ducted a national survey of hazardous waste generators and TSDFs regulated under RCRA in 1982 and 1983. All data in the survey were reported for the year 1981.[5] The Westat survey proved to be an indictment of EPA's poor record keeping.

The Westat study indicated industry in United States generated 264 million metric tons (71.3 billion gallons) of hazardous wastes in 1981. The new total of 264 million metric tons is more than one ton of waste for every person in the United States. The annual production could cover the entire island of Manhattan under a one-story-high hazardous waste pile.

The primary purpose of the Westat survey was not to estimate the quantity of hazardous waste generated in 1981 nor the total number of generators. Among the wastes and facilities excluded from regulation under RCRA (and therefore from the survey) were wastes that passed through a sewer system to publicly owned treatment works (POTW), permitted discharges of industrial waste water, irrigation runoff, and many wastes from ore processing and other sources.

The survey also found that in 1982 some 55,739 facilities had notified EPA they handled hazardous wastes. After exclusions, however, there were only about 14,000 facilities that actually handled RCRA-regulated wastes. And many of these, in fact, had notified U.S. EPA just in case they might be covered by new regulations.

Westat used Standard Industrial Classification (SIC) codes to arrange data about hazardous waste generation in 1981 according to the industry that produced it.

Industry		*Percent of Total Waste Generation*
Chemical (SIC 28)		68%
Metal-Related Industries (SIC 33–37)		22%
Machinery, except Electrical (SIC 35)	10%	
Transportation Equipment (SIC 37)	6%	
Primary Metals (SIC 33) Fabricated Metal Products (SIC 34) Electrical and Electronics Machinery and Supplies (SIC 36)	6%	
Petroleum, including Petroleum Refining (SIC 29)		3%
Motor Freight Transportation and Warehousing Industry (SIC 42)		3%
All Others		4%
Total		100%

Westat revealed that surface impoundment in pits, ponds, and lagoons (one of the least secure technologies) was used to store the largest quantity of hazardous wastes (see table 4-2). Also, about as much waste was treated in surface impoundments as by all other treatment methods combined. Incinerators treated only a very small portion (see table 4-3). Underground injection wells "disposed" of the largest volume of hazardous waste (see table 4-4). (Although injecting wastes deep into the ground is actually a form of long-term storage, U.S. EPA calls any ultimate disposition of waste a "disposal" method.) Table 4-5 summarizes these data and the number of facilities involved.

Discrepancies in U.S. EPA Data. The Westat figure of 264 million tons of hazardous waste generated in one year was six times greater than U.S. EPA's initial estimate. The database from the national survey was of such poor quality, however, that the report estimated only the regional data. Distribution of the actual waste quantities involved was not estimated (see figure 4-3). Four years later, U.S. EPA is still unable to determine the actual quantities of waste that these plants generate.[6]

The wide range of plant sizes and process variations accounts for the enormous disparity in waste volumes and waste hazard produced by different

TABLE 4-2. *Quantities of Hazardous Waste Stored in 1981 (Total of 135 Million Metric Tons)*

Storage Type	Total Quantity in Million Metric Tons[a]	Percentage of Total[b]
Surface impoundments	52.0	70.2
Tanks	19.0	25.6
Waste piles	1.5	2.0
Containers	0.6	0.8
Other	1.0	1.4
Total	135.0	100.0

SOURCE: Westat, Inc., *National Survey of Hazardous Waste Generators and Treatment, Storage, and Disposal Facilities Regulated under RCRA in 1981* (Washington, DC: U.S. EPA, 1984).

a. Subtotals are underestimated due to uncertainties in Westat's generator questionnaire; therefore, they do not equal the U.S. EPA estimated total when added.

b. Percentages are of the additive total of subtotals (74.1 million metric tons), not of U.S. EPA estimated total.

TABLE 4-3. *Quantities of Hazardous Waste Treated in 1981 (Total of 176 Million Metric Tons)*

Treatment Type	Total Quantity in Million Metric Tons[a]	Percentage of Total[b]
Surface impoundments	62.0	55.0
Tanks	32.0	29.4
Incinerators	1.7	1.5
All other treatment types[c]	17.0	15.1
Total	176.0	100.0

SOURCE: Westat, Inc., *National Survey of Hazardous Waste Generators and Treatment, Storage, and Disposal Facilities Regulated under RCRA in 1981* (Washington, DC: U.S. EPA, 1984).

a. Subtotals are underestimated due to uncertainties in Westat's generator questionnaire and therefore do not equal the U.S. EPA estimated total when added. Additionally, total is overstated by 18.6 million metric tons due to failure to exclude RCRA-exempt wastewater discharge.

b. Percentages are of the additive total of subtotals (112.7 million metric tons), not of U.S. EPA estimated total.

c. Other treatment types included open burning, explosion, treatment in waste piles, and treatment in containers.

plants—even plants within the same industry. So U.S. EPA's figures tell very little. Although over eighty percent of all wastes are produced by fewer than one percent of all generators (see figure 4-4), these high-volume wastes may not be the most toxic or the most hazardous because most are dilute aqueous solutions rather than concentrated hazardous compounds.

U.S. EPA required that by late 1984 all states submit reports of the quantities and types of wastes that are generated and disposed of in their territories. The reports are due every 2 years. Because there is no requirement that the states use compatible data systems, however, U.S. EPA must rely on paper

TABLE 4-4. *Quantities of Hazardous Waste Disposed of in 1981 (Total of 55 Million Metric Tons)*

Disposal Type	Total Quantity in Million Metric Tons	Percentage of Total
Underground	32.0	58.7
Surface impoundments	19.0	34.9
Landfills	3.0	5.5
Land application	0.4	0.7
Other	0.1	0.2
Total	54.5	100.0

SOURCE: Westat, Inc., *National Survey of Hazardous Waste Generators and Treatment, Storage, and Disposal Facilities Regulated under RCRA in 1981* (Washington, DC: U.S. EPA, 1984).

TABLE 4-5. *Total Quantities of Hazardous Waste Treated, Stored, and Disposed of in 1981, and the Number of Facilities Engaged in Each Process*

Process	Total Quantity Managed, by Process (Million Metric Tons)[a]	Number of Facilities Engaged in Each Process[b]
Treatment	176	1495
Storage	135	4299
Disposal	55	430
Total	264	4818

SOURCE: Westat, Inc., *National Survey of Hazardous Waste Generators and Treatment, Storage, and Disposal Facilities Regulated under RCRA in 1981* (Washington, DC: U.S. EPA, 1984).

a. Total of 265 million metric tons is overstated by 18.5 million metric tons due to failure to exclude RCRA-exempt waste water discharges. Treatment + Storage + Disposal exceeds total amount managed due to multiple processing of some wastes.

b. Treatment + Storage + Disposal exceeds 4818 total facilities due to multiple processing at facilities. As of 6/30/85, U.S. EPA estimates that there are 5037 TSDFs in the United States.

copies—usually computer printouts—to conduct its analyses. As a result, any data manipulation will be far less efficient and precise than had EPA required state information to be standardized, computerized, and submitted on tape. In addition the new waste statistics are likely to be published several years after the year they describe. Data published in 1985, for instance, may detail the quantities of waste that facilities handled in 1983.

Researchers can perform only simple calculations with the paper-based data, and the resulting research can be no more reliable than the quality of the data, which is suspect. The variety of reporting formats that generators use, for example, undermines data accuracy. The generator of a specific waste could use the company's own internal code to identify its waste in the last biennial report; the firm hired to manage the waste, however, could use an

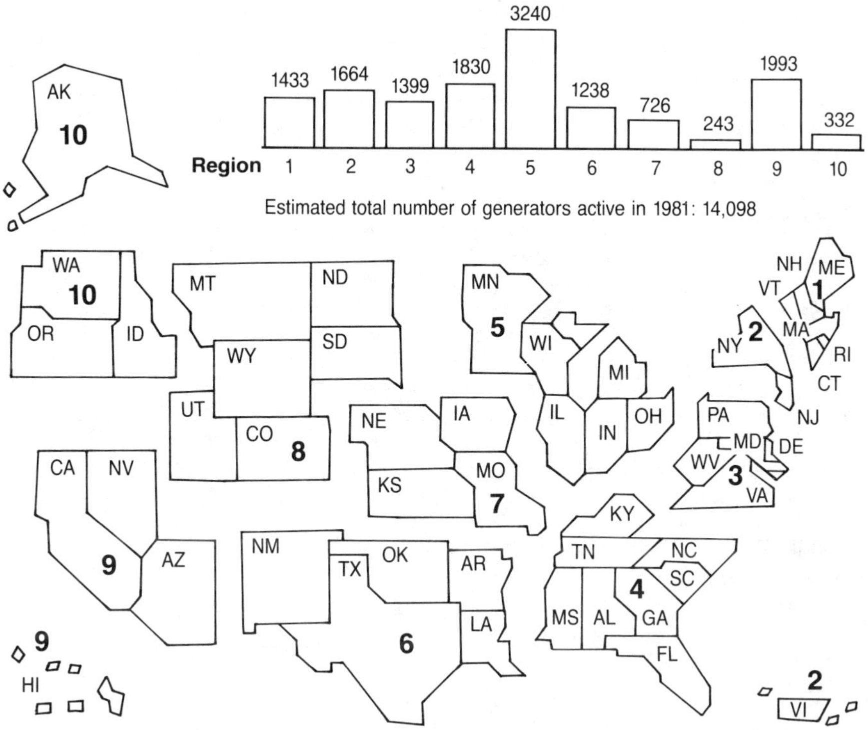

FIGURE 4-3. *Regional Distribution of Hazardous Waste Generators in 1981 (Numbers on Map Indicate U.S. EPA Region)*

SOURCE: Westat, Inc., National Survey of Hazardous Waste Generators and Treatment, Storage, and Disposal Facilities Regulated under RCRA in 1981 (Washington, DC: U.S. EPA, 1984).

entirely different code to identify the same waste stream upon receipt. Thus, for the purposes of estimating capacity shortages or identifying illegal activities, this data is extremely limited. Regulators cannot use it to track specific wastes from their points of origin to the process used to dispose of it.[7]

The Westat survey of hazardous waste generation and disposal in 1981 estimates that almost one-tenth of the waste—more than 20 million metric tons—was produced at industrial plants that did not have on-site disposal capabilities. Most of this material, then, would presumably have been shipped to commercial hazardous waste management facilities for storage, treatment, and disposal (see figure 4-5). But the Westat survey reports that the commercial hazardous waste management industry received only 3.0 million metric tons of hazardous wastes that year from generators without TSDFs. Much of the "missing" waste could simply be due to statistical "noise" or measurement error. Even if half of the discrepancy is due to error, a minimum of 6.35

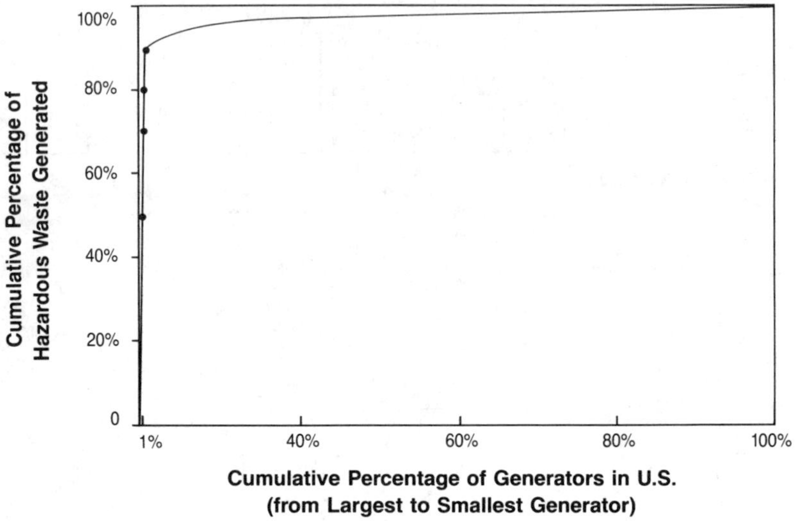

FIGURE 4-4. *Concentration of Hazardous Waste Generation in the United States in 1981*

SOURCE: Westat, Inc., National Survey of Hazardous Waste Generators and Treatment, Storage, and Disposal Facilities Regulated under RCRA in 1981: Preliminary Findings (Washington, DC: U.S. EPA, 1983).

million tons would still be unaccounted for. That's a small part of the total, but hardly insignificant. Was that much disposed of illegally? In short, U.S. EPA cannot presently determine whether or not commercial disposal capacities are adequate to handle the amounts of hazardous waste requiring off-site management.

The 1981 survey estimates that more than half of the wastes managed by the commercial industry during 1981 came from small-quantity generators. At the time, federal law did not require these small generators to bring their wastes to regulated commercial facilities. Many states did, but not nearly all. Now federal law requires all plants that generate more than 100 kilograms of waste a month (220 pounds, or about half a drum) to send their wastes to legally registered facilities.

U.S. EPA estimates that the new regulations concerning smaller quantities will affect at least 175,000 companies. This newly regulated class of customers is expected to create a substantial increase in service demand. The added volume of waste will be relatively small, perhaps 0.8 million tons according to a U.S. EPA survey, but the need for transport and testing will increase tremendously. Small shipments could overwhelm the ability of disposal laboratories to test the waste composition. The excess demand could force prices up. Without strict or adequate enforcement, higher prices could provide a strong incentive for noncompliance.

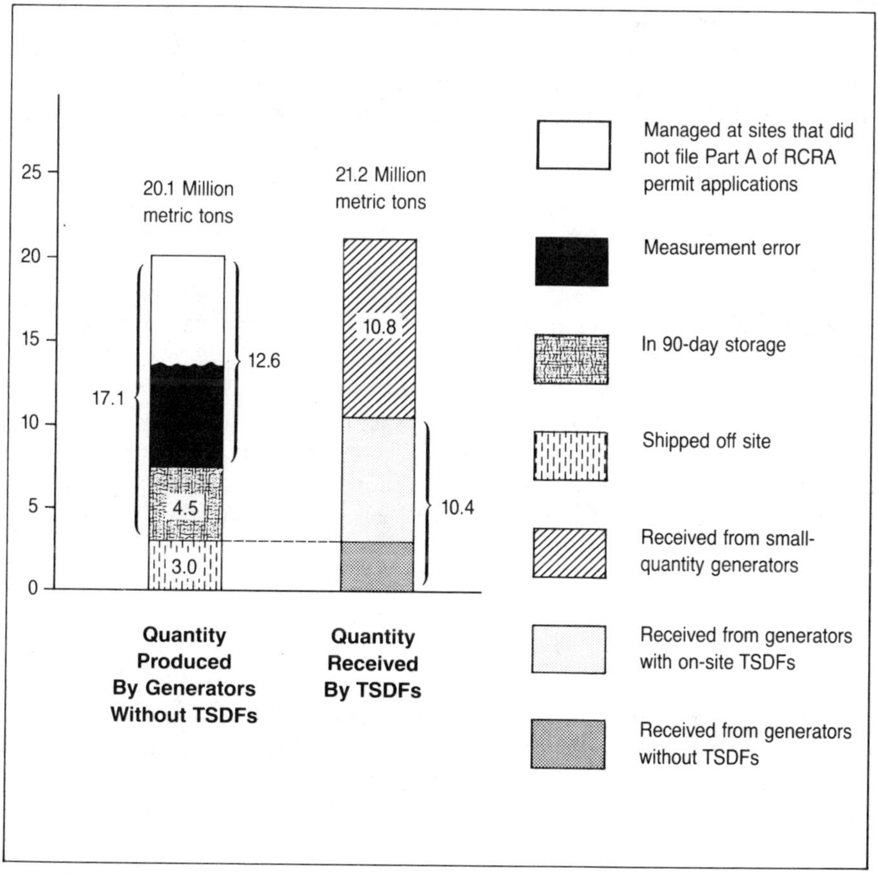

FIGURE 4-5. *Quantities of Wastes Produced by Generators without TSDFs, and Quantities Received from Off Site by TSDFs in 1981*

DERIVED FROM: Westat, Inc., National Survey of Hazardous Waste Generators and Treatment, Storage, and Disposal Facilities Regulated under RCRA in 1981 (Washington, DC: U.S. EPA, 1984).

The Hart Report

We've all read about the improper management of hazardous wastes. A report commissioned by U.S. EPA and released in 1984 summarizes many of the damage cases on file at U.S. EPA or at its data-analysis contractors.[8] The report, which was written by Fred C. Hart Associates, examined a total of 927 sites, many of which contained multiple facilities. The sites selected for this report were suspected of mismanagement and therefore do not reflect all sites, many of which are properly managed. Nevertheless, the report shows that the potential liabilities are great.

The Hart report identified documented or suspected contamination (pollutants in greater concentration than in the background environment) at 834 sites, or 90 percent of the sites evaluated. At 555 sites (60 percent of the sites) the report documented contamination along the following routes of exposure:

Ground Water	32%
Soil	31%
Surface Water	29%
Air	8%

Documented or suspected contamination-related damage was identified in at least 544 (58 percent) of the sites.

At most of the sites where contamination occurred, the Hart report made positive identification of chemicals (although often at low concentrations). Table 4-6 lists the ten chemical categories that appeared most often among the 1347 positive identifications. The large number of contamination incidents involving metals and volatiles is likely to be because of their widespread use.

NEIC Reports

Creation of the National Enforcement Investigations Center (NEIC) in Denver, Colorado, is an encouraging attempt on the part of U.S. EPA to overcome the myopia resulting from decentralized enforcement. NEIC investigates enforcement issues that concern more than one state. NEIC is trying to find out what happened to the wastes that the Westat survey could not account for.

TABLE 4-6. *The Ten Chemicals Documented Most Frequently in Selected Contamination Incidents*

Chemical Category	Total Positive Identifications	Percentage of Total
Metals	316	23
Volatile halogenated organics (mainly solvents)	233	17
Volatile nonhalogenated organics	159	12
Acid-extractable compounds	127	9
Base-neutral extractables	117	9
Pesticides	85	6
Polychlorinated biphenyls (PCBs)	82	6
Inorganics	60	5
Cyanide	40	3
Acids	37	3

SOURCE: Fred C. Hart Associates, *Assessment of Hazardous Waste Mismanagement Damage Case Histories* (Washington, DC: U.S. EPA, 1984).

NOTE: Sampled sites were not randomly selected.

Most enforcement actions taken by U.S. EPA or state hazardous waste agencies focus on individual facilities. Many times, however, the responsible parties are corporations that own and operate waste facilities in many states. One NEIC strategy to stop improper waste handling across state lines is to look at many facilities owned by one parent corporation. According to Richard Mays, EPA's senior enforcement counsel:

> We're starting at the top and working down. It's important for [companies] to know this [because] eventually we're going to get to them. If we can raise the consciousness of top corporate management toward environmental compliance, then maybe we can raise the level of compliance.[9]

NEIC's first "top-down" investigation in late 1983 focused on CWM, probably the largest commercial hazardous waste management firm in the world. NEIC reportedly found 120 violations at 14 facilities across the country. Violations at just one of these sites resulted in a $1.05 million fine.[10]

Because of the sensitive nature of NEIC investigations, they are usually highly confidential. Except for leaks to the press, the public rarely hears NEIC findings. The limited number of reviews that NEIC undertakes and the confidentiality of the reports minimize the impact of NEIC's program in the short term.

Although confidentiality is important during an investigation (to avoid tipping off the targets), CEP believes that NEIC's final reports should be made public. Publicity over such investigations could be helpful in two ways: (1) by educating the public and (2) by promoting awareness of the rigor of the federal enforcement program. Both of these results work toward dissuading other bad actors from committing illegal acts.

The Sanjour Memo Highlighted a Regulatory Oversight

Logic indicates that if a company places wastes in the same facilities as U.S. EPA does, then the company will be as safe as possible. Not so. Contrary to popular opinion, wastes that EPA has removed from Superfund sites and disposed of at active RCRA facilities do not pose greater danger than those received directly from an industrial generator. In fact, Superfund wastes are likely to be more dilute than wastes coming directly from generators because Superfund wastes often include large volumes of contaminated soils. Still, CEP is convinced that EPA should have evaluated and chosen the best sites; EPA is a role model for industry. EPA has not always made the best choices, however. Bidding for Superfund waste contracts is highly price-competitive. Potential Superfund contractors may be tempted to cut corners to make up for lost profits as they try to underbid their competitors.

This unfortunate regulatory oversight was first brought to public attention during the spring of 1984, when Cass Peterson of the *Washington Post* obtained an internal memorandum from the congressional Office of Technology Assessment (OTA) dated May 17, 1984 that indicated some of the facilities that received Superfund wastes were leaking.[11] The *Post* quotes Lee M. Thomas as acknowledging that "probably quite a bit" of the Superfund wastes were being sent to "substandard" disposal sites. At the time, Lee Thomas was head of U.S. EPA's hazardous waste program. He is now the overall EPA Administrator.

By the end of 1984, U.S. EPA had directed and disbursed funds for "emergency" cleanups at 459 Superfund national priority sites but had completed cleanup activities at only six (see figure 4-6).[12] U.S. EPA's William Sanjour, who wrote the internal OTA memorandum while he was "on loan" to the congressional office, questioned why cleanup activities were considered "emergency" responses when EPA had been aware for almost a decade that many of the sites needed remedial action. "Why," Sanjour asks in the memo, "were [these wastes] not incinerated or otherwise treated rather than moved to another dump?"[13]

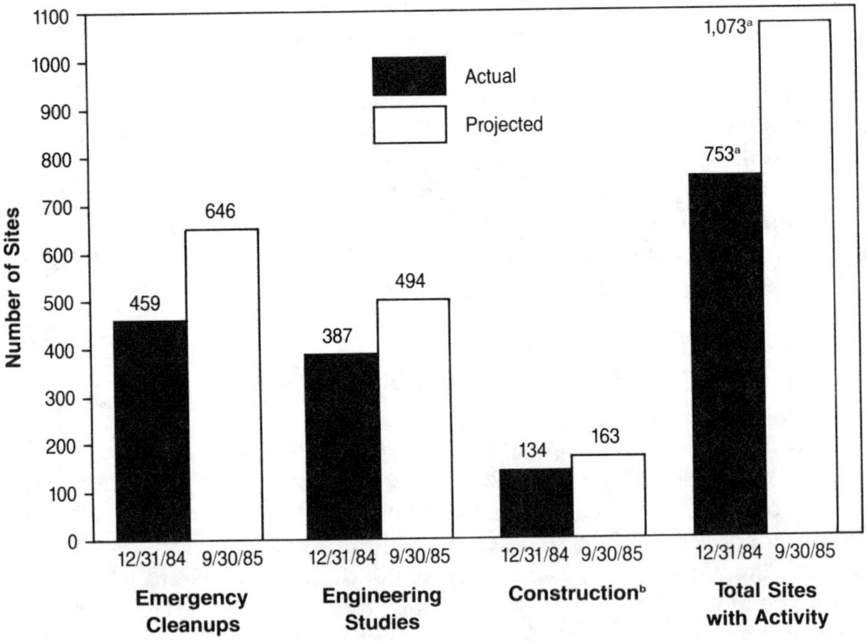

ª Some sites underwent more than one type of activity.

ᵇ The category includes commitments for construction by responsible parties as part of administrative or judicial enforcement action

FIGURE 4-6. *Total Superfund Sites with Cleanup Activity*

SOURCE: Lee M. Thomas, Attachment to Statement of February 22, 1985.

Tracking CERCLA Wastes

Paucity of information makes it difficult to determine whether the recipient sites are "dumps"—that is, sites where hazardous wastes were deposited without adequate precautions to minimize risks—or proper facilities. William Sanjour, for example, referred to two commercial newsletters (*Hazardous Waste Report* and *Hazardous Materials Intelligence Report*) rather than U.S. EPA documents to document which RCRA-regulated hazardous waste facilities received the Superfund wastes.

After that, EPA set up a database, "Disposition of CERCLA Wastes Data Base," to track the RCRA facilities that received these materials. But according to David Fege, a U.S. EPA staff member who asked that the system be set up, the database is not being updated as more sites are added to the Superfund NPL, nor have there been any quality-assurance checks of the information that has been entered.[14]

CEP made a FOIA request to acquire "Disposition of CERCLA Wastes Data Base." The data indicate that as of October 1984, 102 RCRA facilities received materials removed from Superfund sites. All of these RCRA facilities were interim-status facilities (they did not have final permits).

Thirty-eight, or over one-third, of these recipient facilities are owned by the eight publicly held commercial hazardous waste management companies that CEP evaluated in detail.[15] In all, U.S. EPA data indicate that the 38 facilities owned by the eight commercial companies have received Superfund wastes in over 200 instances (wastes from one Superfund site were in some instances brought to more than one RCRA facility). The contract cost of cleaning up only 17 of these Superfund sites totals close to $50 million according to the Sanjour memo.

The majority of the recipient facilities owned by these eight companies does not appear to consist of dumps. Although more than half of the recipient facilities owned by the eight companies include landfills, more than half of the facilities also have treatment processes available on site. In fact, one-quarter of the recipient facilities have incinerators and 40 percent can perform chemical treatment. See columns C–H in table 4-8A and table 4-9 to gain additional perspective on facilities that accept Superfund wastes.

How much of these Superfund wastes was incinerated or otherwise treated? U.S. EPA data offer no answers.

In all, Sanjour's memo alleges potential ground water problems at eight facilities owned by Browning–Ferris, Rollins Environmental Services, or Waste Management. All of the eight had received wastes from Superfund sites. Based on his own FOIA inquiries while at OTA, Sanjour's memo alleges that three facilities belonging to Chemical Waste Management—CID Landfill south of Chicago; ESL Landfill in Joliet, Illinois; and five deep wells in Vickery, Ohio—were leaking. Sanjour also alleged that the CECOS landfill in Niagara Falls, New York, and the Rollins facility in Baton Rouge, Louisiana, were leaking.

Furthermore, because of inadequate ground water-monitoring systems, U.S. EPA did not have the data to show that there were *not* leaks at Waste Management's landfills in Emelle, Alabama, and Model City, New York, or at CECOS's landfill in Williamsburg, Ohio. The data shortage existed in spite of the fact that, according to U.S. EPA's database, each of these three facilities had received wastes from more Superfund sites than any other RCRA facility in the country.

After the *Post* made Sanjour's allegations public, other U.S. EPA officials publicly discredited his memo. Waste Management claims that two of the three sites mentioned were not leaking; Browning–Ferris denies that its site leaked; and while Rollins (LA) admits to leakage, the company emphasizes that the contamination is contained within the site's boundaries in unused aquifers and is being removed and treated under state-approved plans.

The Thomas Memos

Two months after the *Post* disclosed the Sanjour memo, Lee M. Thomas, who was U.S. EPA's assistant administrator in charge of the federal hazardous waste program at the time, sent his own memoranda to the administrators of U.S. EPA's ten regional offices. The documents highlighted RCRA facilities with particularly disturbing deficiencies relating to ground water monitoring.[16]

Thomas's memos of August 1984 revealed that almost one-quarter of all the RCRA facilities that had received Superfund wastes were deficient. One site had no ground water-monitoring wells at all, and 13 had inadequate monitoring systems, sampling, analysis, programs, or reporting procedures. Eight facilities had either provided conflicting information that indicated inadequacies or had not been adequately evaluated by U.S. EPA.

The Thomas memos provided the data in tables 4-7A and 4-7B and 4-8A and 4-8B that indicate the incidence of deficiencies at sites that received CERCLA wastes. The sequence of tables compares the data for the CERCLA waste-recipient sites with the incidence of deficiencies at all major facilities. The data reveal that 10 percent of the sites that received CERCLA wastes had either inadequate or no monitoring wells compared to 17 percent for all major facilities. Tables 4-7A and 4-7B also provide data on two other categories: an incomplete or unevaluated waiver and insufficient information. A general comparison of all four categories indicates that the performance of major facilities owned by the major commercial hazardous waste management companies—the Big Eight—was as good and perhaps slightly better than the average for all major facilities. CEP data show that 22 percent of the Big Eight's major facilities were deficient compared to 25 percent for all majors.

TABLE 4-7A. *Compliance of RCRA Facilities That Have Received CERCLA Wastes, with Certain Ground-Water Monitoring and Financial Instrument Requirements as of August 1984*

	Deficient Sites by Type of Deficiency								
Parameter	*a*	*b*	*c*	*d*	*e*	*f*	*g*	*h*	*i*
Totals for CERCLA recipient facilities (102 total)	1	9	5	9	1	3	5	4	3
Percent of total 102 recipient facilities	1	9	5	9	1	3	5	4	3
Percent e + i deficient recipients (insufficient information) of total recipients						4/102 (× 100) = 4%			
Percent a + b deficient recipients (inadequate or no wells) of total recipients						10/102 (× 100) = 10%			
Percent total deficient recipients of total recipients						22/102 (× 100) = 22%			

a. Facilities with no ground-water monitoring well system in place
b. Facilities in detection or assessment with well system evaluated and inadequate
c. Facilities in detection or assessment with ground-water sampling and analysis program evaluated and inadequate
d. Facilities in detection or assessment with ground-water monitoring records evaluated and inadequate
e. Facilities claiming waiver but not evaluated, or facilities with waiver incomplete
f. Facilities in assessment with inadequate or no report
g. Facilities with land-disposal activities that did not indicate that they were subject to ground-water monitoring
h. Facilities with land-disposal activities that checked "not applicable" for non-sudden liability insurance
i. Facilities with insufficient information for analysis

DERIVED FROM: U.S. EPA, "Disposition of CERCLA Wastes Data Base" (Washington, DC: 10/17/84) and Lee M. Thomas, "Internal Memoranda to Regional Administrators Regarding RCRA Major Facility Status Sheets" (8/84).

The Congressional Ground-Water Monitoring Survey

On April 29, 1985 the Subcommittee on Oversight and Investigations of the House Committee on Energy and Commerce released a report called "Groundwater Monitoring Survey."[17] The subcommittee found that at least 1246 facilities were subject to ground water-monitoring requirements and that 39 CERCLA-waste recipients were subject to the monitoring requirements (see table 4-10). The survey revealed that as of April 1985, more than half of all the facilities that were required to monitor for ground water contamination had inadequate monitoring wells, no wells, or well systems whose status U.S. EPA and the states could not verify. Moreover, the data indicated possible ground water contamination in 45 percent of the facilities. Facilities in the

TABLE 4-7B. *Compliance of All Major RCRA Facilities, with Certain Ground-Water Monitoring and Financial Instrument Requirements as of August 1984*

	Deficient Sites by Type of Deficiency								
Parameter	*a*	*b*	*c*	*d*	*e*	*f*	*g*	*h*	*i*
Totals for major facilities (972 subject, 1749 total)	85	205	*	*	41	*	*	*	114
Percent of total 1749 major facilities (6% are CERCLA recipients)	5	12	n/a	n/a	2	n/a	n/a	n/a	7
Percent e + i deficient majors (insufficient information) of total majors	155/1749 (× 100) = 9%								
Percent a + b deficient majors (inadequate or no wells) of total majors	290/1749 (× 100) = 17%								

a. Facilities with no ground-water monitoring well system in place
b. Facilities in detection or assessment with well system evaluated and inadequate
c. Facilities in detection or assessment with ground-water sampling and analysis program evaluated and inadequate
d. Facilities in detection or assessment with ground-water monitoring records evaluated and inadequate
e. Facilities claiming waiver but not evaluated, or facilities with waiver incomplete
f. Facilities in assessment with inadequate or no report
g. Facilities with land-disposal activities that did not indicate that they were subject to ground-water monitoring
h. Facilities with land-disposal activities that checked "not applicable" for non-sudden liability insurance
i. Facilities with insufficient information for analysis

DERIVED FROM: Lee M. Thomas, "Internal Memoranda to Regional Administrators Regarding RCRA Major Facility Status Sheets" (8/84).

* Information not available. n/a = not applicable.

group of possible contaminators had either monitoring wells that showed a statistically significant increase in indicator parameters, an ongoing program to assess contamination, or confirmation of a release. Representative John Dingell, a Democrat from Michigan who is chairman of the Energy Committee and the Oversight Subcommittee, called the survey findings "shocking."

The results of "Groundwater Monitoring Survey" were consistent with the data from the Lee Thomas memos; the facilities that have received wastes from Superfund sites performed only marginally better than all facilities. The survey showed that 46 percent of CERCLA-waste recipients had adequate monitoring wells whereas 41 percent of all subject RCRA facilities had adequate monitoring wells (see table 4-10). The fact that the survey also identified the number of CERCLA-waste recipients that are subject to ground water monitoring allows for analysis that overcomes the statistical uncertainty that plagued U.S. EPA's databases.

Although the data were preliminary, they indicated that almost half (46 percent) of RCRA facilities U.S. EPA chose to landfill wastes removed from

TABLE 4-8A. *Statistics for CERCLA Recipient Facilities Owned by the Big Eight*

	Number of Sites																		
	Status		Processes						CERCLA		Deficiencies								
Parameter	*A*	*B*	*C*	*D*	*E*	*F*	*G*	*H*	*I*	*J*	*a*	*b*	*c*	*d*	*e*	*f*	*g*	*h*	*i*
Totals for recipient facilities (38 total)	33	32	4	15	20	9	4	21	218*	4	0	3	12	4	1	1	2	2	3
Percent of total 38 recipient facilities	87	84	11	40	53	24	11	55	n/a	11	0	8	5	11	3	3	5	5	8

Percent e + i deficient recipients (insufficient information) of total recipients	4/38 (× 100) = 11%
Percent a + b deficient recipients (inadequate or no wells) of total recipients	3/38 (× 100) = 8%
Percent total deficient recipients of total recipients	12/38 (× 100) = 32%

A. Active facilities
B. Major facilities
C. Facilities providing biological treatment
D. Facilities providing chemical treatment
E. Facilities providing physical treatment
F. Incinerator
G. Injection well
H. Landfill
I. Number of CERCLA sites from which wastes have been received
J. Number of CERCLA sites for which facilities are potentially responsible
a. Facilities with no ground-water monitoring well system in place
b. Facilities in detection or assessment with well system evaluated and inadequate
c. Facilities in detection or assessment with ground-water sampling and analysis program evaluated and inadequate
d. Facilities in detection or assessment with ground-water monitoring records evaluated and inadequate
e. Facilities claiming waiver but not evaluated, or facilities with waiver incomplete
f. Facilities in assessment with inadequate or no report
g. Facilities with land-disposal activities that did not indicate that they were subject to ground-water monitoring
h. Facilities with land-disposal activities that checked "not applicable" for non-sudden liability insurance
i. Facilities with insufficient information for analysis

DERIVED FROM: U.S. EPA, "Disposition of CERCLA Wastes Data Base" (Washington, DC: 10/17/84); U.S. EPA, "Listing of Potentially Responsible Parties at CERCLA Sites" (Washington, DC: 1984); Lee M. Thomas, "Internal Memoranda to Regional Administrators regarding RCRA Major Facility Status Sheets" (8/84); and Environmental Information, Ltd., *Industrial and Hazardous Waste Management Firms, 1985* (Minnesota: 1984).

* Because wastes from certain Superfund sites were brought to more than one RCRA facility, this figure represents the number of instances CERCLA wastes were received.

n/a = not applicable.

TABLE 4-8B. *Statistics for Major Facilities Owned by the Big Eight*

	Number of Sites										
	CERCLA		Deficiencies								
Parameter	*I*	*J*	*a*	*b*	*c*	*d*	*e*	*f*	*g*	*h*	*i*
Totals for major facilities (78 total)	218	11	0	9	7	8	1	3	8	5	7
Percent of total 78 major facilities (41% are CERCLA recipients)	n/a	14	0	12	9	10	1	4	10	6	9
Percent e + i deficient majors (insufficient information) of total majors	8/78 (× 100) = 10%										
Percent a + b deficient majors (inadequate or no wells) of total majors	9/78 (× 100) = 12%										
Percent total deficient majors of total majors	31/78 (× 100) = 40%										

I. Number of CERCLA sites from which wastes have been received
J. Number of CERCLA sites for which facilities are potentially responsible
a. Facilities with no ground-water monitoring well system in place
b. Facilities in detection or assessment with well system evaluated and inadequate
c. Facilities in detection or assessment with ground-water sampling and analysis program evaluated and inadequate
d. Facilities in detection or assessment with ground-water monitoring records evaluated and inadequate
e. Facilities claiming waiver but not evaluated, or facilities with waiver incomplete
f. Facilities in assessment with inadequate or no report
g. Facilities with land-disposal activities that did not indicate that they were subject to ground-water monitoring
h. Facilities with land-disposal activities that checked "not applicable" for non-sudden liability insurance
i. Facilities with insufficient information for analysis

DERIVED FROM: U.S. EPA, "Disposition of CERCLA Wastes Data Base" (Washington, DC: 10/17/84); U.S. EPA, "Listing of Potentially Responsible Parties at CERCLA Sites" (Washington, DC: 1984); and Lee M. Thomas, "Internal Memoranda to Regional Administrators regarding RCRA Major Facility Status Sheets" (8/84).

n/a = not applicable.

Superfund sites may have contaminated ground water. The data indicated that U.S. EPA, overall, chose RCRA facilities for Superfund wastes deposition that are marginally better than the national average. These recipient facilities should be not marginally but significantly better.

ARE THE DATA ON THE BIG EIGHT CONSISTENT?

The Sanjour memo, the Thomas memos, and the House survey encompass different data and different methods. From these documents one can arrive at a similar conclusion about the performance of the eight major hazardous waste management companies, however. Based on evaluations of these data, CEP found that the 53 facilities subject to ground water monitoring owned by the Big Eight performed somewhat better than the average for all subject sites. The congressional data also indicate that these companies' facilities, which include 24 of the 39 CERCLA-waste recipients, performed slightly better than the average for all subject sites that received CERCLA wastes.

Harm to public health and the environment can, of course, occur via routes other than ground water contamination. Releases of contaminants to the air and to surface water, inadequate precautions to prevent direct contact, and the danger of fire and explosion are all potential dangers.

Financial Accountability

U.S. EPA asserts that "it is likely that most land disposal facilities will leak eventually, causing hazardous releases into the environment." Therefore, the ability of facility owners to comply with financial instrument requirements is the key to whether the RCRA facility will become a CERCLA site.[18]

Thomas's memos indicate that approximately 5 percent of the land emplacement facilities that received Superfund wastes were not subject to the financial instrument requirements regarding non-sudden occurrences. This is a contradiction since the law subjects all land emplacement facilities to these requirements.

Of the 468 facilities that U.S. EPA and the states evaluated, they found the non-sudden occurrences instruments of 10 percent to be inadequate. Assuming a similar rate of noncompliance for those facilities that have yet to be evaluated, close to 100 major facilities subject to financial requirements may be inadequately insured (see table 4-11).

According to the Thomas memos, the compliance record of major facilities with RCRA's four other financial-instrument requirements is similar to that for non-sudden occurrences coverage. One-third of the evaluated major facilities had inadequate insurance for sudden occurrences. U.S. EPA and the states had evaluated only half of all major facilities' closure plans and found that more than one-third of those were deficient. Forty-four percent of the facilities had made cost estimates of closure that were inadequate. Twenty-five percent did not even have the proper financial instruments—a trust fund, a bond, a financial test, insurance, or some other instrument—to pay for closure. Bear in mind that these statistics refer only to the sites that the states identified as "major." Thousands of nonmajor active facilities that are not monitored via the status sheets must also meet the requirements for closure assurance and sudden occurrences insurance.

U.S. EPA estimates that within 10 years, 130 land disposal facilities and 475 storage and treatment facilities—a total of 605 facilities—will close because of financial problems. In fact, the more stringent the government's financial instrument requirements, the greater the likelihood that U.S. EPA has underestimated the future rate of business failure.[19]

TABLE 4-9. *Facilities Owned by Eight Companies That Have Received Wastes Removed from Superfund Sites as of October 17, 1984*

Facility Name	City	State	Status A	Status B
American Ecology, Inc.				
U.S. Ecology, Inc. Chemsite	Beatty	NV	1	1
Browning–Ferris Industries, Inc.				
CECOS International	Williamsburg	OH	1	1
CECOS International	Niagara Falls	NY	1	1
CECOS International	Cincinnati	OH	1	1
BFI Chemical Services	Livingston	LA	1	1
BFI Solley Road	Glen Burnie	MD	0	1
BFI Chemical Services	Westlake	LA	1	1
BFI		MS	0	0
BFI Chemical Services	Odessa	TX	0	1
BFI Chemical Services	Chester	VA	0	0
Browning–Ferris Industries, Inc., Total			5	6
Environmental Systems Co., Inc.				
ENSCO	El Dorado	AR	1	1
International Technology Corp.				
IT Corp. Benicia	Benicia	CA	1	1
IT Corp. Vine Hill	Martinez	CA	1	1
IU International Corp.				
Fondessy Enterprises	Oregon	OH	1	1
Envirosafe Serv. of ID	Grandview	ID	1	1
IU Conversion Systems	Honey Brook	PA	1	1
Rollins Environmental Services				
Rollins Env. Serv. (NJ)	Bridgeport	NJ	1	1
Rollins Env. Serv. (LA)	Baton Rouge	LA	1	1
Rollins Env. Serv. (TX)	Deer Park	TX	1	1
USPCI, Inc.				
USPCI Lone Mountain	Waynoka	OK	1	1

A. Active facility
B. Major facility
C. Facility providing biological treatment
D. Facility providing chemical treatment
E. Facility providing physical treatment
F. Incinerator
G. Injection well
H. Landfill
I. Number of CERCLA sites from which wastes have been received
J. Number of CERCLA sites for which facility is potentially responsible
a. Facility with no ground-water monitoring well system in place
b. Facility in detection or assessment with well system evaluated and inadequate

NOTE: Companies are listed in alphabetical order; facilities are listed in order of wastes received, from highest to lowest number of superfund sites.

Processes						CERCLA		Deficiencies								
C	*D*	*E*	*F*	*G*	*H*	*I*	*J*	*a*	*b*	*c*	*d*	*e*	*f*	*g*	*h*	*i*
					H	3	0									
					H	19	0									
C	D	E			H	11	0									
		E				10	0									
	D	E			H	4	0							g	h	
					H	3	0									i
	D	E		G	H	1	0						f			
						1	0									
		E		G		1	0									
						1	0									
1	3	5		2	5	50							1	1	1	1
			F			9	0									i
					H	2	0									
C	D	E	F			2	0									
					H	14	0									
	D	E			H	4	0					e				
						1	0									
			F			8	0								h	
	D	E	F		H	6	0									
C	D	E	F		H	4	0									
	D	E			H	2	0		b		d					

c. Facility in detection or assessment with ground-water sampling and analysis program evaluated and inadequate

d. Facility in detection or assessment with ground-water monitoring records evaluated and inadequate or not present

e. Facility claiming waiver but not evaluated, or facility with waiver incomplete

f. Facility in assessment with inadequate or no report

g. Facility with land-disposal activities that did not indicate that it was subject to ground-water monitoring

h. Facility with land-disposal activities that checked "not applicable" for non-sudden liability insurance

i. Facility with insufficient information for analysis

(continued)

Table 4-9 *(continued)*

Facility Name	City	State	Status A	Status B
Waste Management, Inc.				
Chemical Waste Management	Emelle	AL	1	0
SCA Chemical Waste Services	Model City	NY	1	1
CWM-Conservation Services	Port Arthur	TX	1	0
SCA Incinerator	Chicago	IL	1	1
WM of IL/CID-CWM of IL	Calumet City	IL	1	1
Trade Waste Incineration	Sauget	IL	1	1
Chem-Security Systems	Arlington	OR	1	1
Solvent Resource Recovery	West Carrollton	OH	1	1
Evergreen Landfill	Northwood	OH	1	0
CWM-Kettleman Hills	Kettleman City	CA	1	1
CWM-Joliet	Elwood	IL	1	1
Indiana Waste Systems	Wheeler	IN	1	1
SCA Chemical Services	Newark	NJ	1	1
Ohio Liquid Disposal	Vickery	OH	1	1
G.R.O.W.S. Inc. Landfill	Morrisville	PA	0	1
CWM of Corpus Christi	Corpus Christi	TX	1	0
Waste Management, Inc., Total			17	13

A. Active facility
B. Major facility
C. Facility providing biological treatment
D. Facility providing chemical treatment
E. Facility providing physical treatment
F. Incinerator
G. Injection well
H. Landfill
I. Number of CERCLA sites from which wastes have been received
J. Number of CERCLA sites for which facility is potentially responsible
a. Facility with no ground-water monitoring well system in place
b. Facility in detection or assessment with well system evaluated and inadequate
c. Facility in detection or assessment with ground-water sampling and analysis program evaluated and inadequate

NOTE: Companies are listed in alphabetical order; facilities are listed in order of wastes received, from highest to lowest number of Superfund sites.

Conflict of Interest?

U.S. EPA data reveal that among the eight commercial waste management firms, BFI and WMI were paid to receive wastes from seven Superfund sites for which they themselves may be partially responsible (see table 4-12). In addition to questions regarding the technical suitability of a RCRA facility for receiving Superfund wastes, U.S. EPA should not allow a disposal firm to win the contract to clean up a situation it helped to create. Unfortunately, U.S. EPA has yet to ascertain which firms are responsible for creating the problems at most Superfund sites.

Processes						CERCLA		Deficiencies								
C	*D*	*E*	*F*	*G*	*H*	*I*	*J*	*a*	*b*	*c*	*d*	*e*	*f*	*g*	*h*	*i*
	D	E	F		H	38	0									
C	D	E			H	18	0									i
	D					12	1									
			F			7	0									
	D	E			H	6	0		b		d					
			F			6	0									
		E			H	6	0									
		E				3	0									
					H	2	0									
	D	E			H	1	0							g		
		E			H	1	0		b	c	d					
					H	1	0			c	d					
	D	E				1	0									
		E		G		1	0									
					H	1	0									
				G		1	1*									
1	7	10	4	2	9	113	3		2	2	3			1		1

d. Facility in detection or assessment with ground-water monitoring records evaluated and inadequate or not present

e. Facility claiming waiver but not evaluated, or facility with waiver incomplete

f. Facility in assessment with inadequate or no report

g. Facility with land-disposal activities that did not indicate that it was subject to ground-water monitoring

h. Facility with land-disposal activities that checked "not applicable" for non-sudden liability insurance

i. Facility with insufficient information for analysis

DERIVED FROM: U.S. EPA, "Disposition of CERCLA Wastes Data Base" (Washington, DC: 84); U.S. EPA, "Listing of Potentially Responsible Parties at CERCLA Sites" (Washington, DC: 1984); Lee M. Thomas, "Internal Memoranda to Regional Administrators Regarding RCRA Major Facility Status Sheets" (8/84); and Environmental Information, Ltd., *Industrial and Hazardous Waste Management Firms, 1985* (Minnesota: 1984).

* Best guess

A case that involves WMI, one of the Big Eight, provides an example of the circuitousness that seems to feature in many accounts of Superfund disposal. WMI removed 60,000 drums of wastes from the Seymour Recycling site in Seymour, Indiana, as one of 24 parties in a $7.7 million consent decree. U.S. EPA data indicate that Waste Management redisposed of the materials at three Waste Management facilities. At least one of the facilities—the Ohio Liquid Disposal deep-well operation in Vickery, Ohio—is known to have leaked. According to the *Cincinnati Enquirer*, Waste Management also brought the wastes to Solvent Resource Recovery in West Carrollton, Ohio.[20] Solvent Resource Recovery is not identified as a recipient by U.S. EPA data and was not included in the original consent agreement.

TABLE 4-10. *Summary of RCRA Facility Compliance with Ground Water-Monitoring Requirements as of April 1985*

The Number of Facilities . . .	All RCRA Facilities		Facilities That Received CERCLA Wastes		Facilities Owned by the Big Eight	
. . . subject to ground water monitoring	1246	(100%)	39	(100%)	53	(100%)
. . . with an adequate well system	508	(41%)	18	(46%)	26	(49%)
. . . with an inadequate well system	317	(25%)	14	(36%)	23	(43%)
. . . with no well system	188	(15%)	0	(0%)	1	(2%)
. . . with well-system status unknown	209	(17%)	7	(18%)	3	(6%)
. . . with an inadequate or nonexistent well system or a well system of unknown status	714	(57%)	21	(54%)	27	(51%)
. . . claiming waiver	24	(2%)	0	(0%)	6	(11%)
. . . with indications of ground water contamination	559	(45%)	18	(46%)	25	(47%)

DERIVED FROM: Subcommittee on Oversight and Investigations of the Committee on Energy and Commerce of the United States House of Representatives, "Groundwater Monitoring Survey" (Washington, DC: Government Printing Office, April 1985).

TABLE 4-11. *Compliance of Major RCRA Facilities with Financial-Instrument Requirements as of August 1984*

Financial Requirement	Number of Major Facilities Subject	Evaluated (Percent of Those Subject)	Found Inadequate (Percent of Those Evaluated)
Insurance			
Sudden occurrences	1747	339 (19%)	117 (34%)
Non-sudden occurrences	971	468 (48%)	45 (10%)
Assurance			
Closure plan	1747	878 (50%)	325 (37%)
Closure cost estimate	1747	532 (30%)	233 (44%)
Closure assur. instrument	1747	634 (36%)	158 (25%)
Post-closure plan	971	468 (48%)	72 (15%)
Post-closure cost estimate	971	501 (52%)	48 (10%)
Post-closure assur. instrument	971	545 (56%)	37 (7%)

DERIVED FROM: Lee M. Thomas, "Internal Memoranda to Regional Administrators Regarding RCRA Major Facility Status Sheets" (August 1984).

TABLE 4-12. *CERCLA Site Wastes Moved to RCRA Facilities Owned by Potentially Responsible Parties*

CERCLA Sites[a]	RCRA Facility[b]	Type of Action	Date
Browning–Ferris Industries			
Chem-Dyne of Ohio	CECOS International Cincinnati, OH	?	3/82
	CECOS International Williamsburg, OH	?	3/82
Summit National of Ohio	CECOS International Williamsburg, OH	RP	?
Bio-Ecology Systems Disposal Site of Texas	CECOS International Livingston, LA	RD	9/26/83
	CECOS International Odessa, TX	RD	9/26/83
Waste Management			
Seymour Recycling of Indiana	Chemical Waste Management Emelle, AL	RP	?
	Ohio Liquid Disposal Vickery, OH[c]	RP	?
	Evergreen Landfill Northwood, OH	RP	?
Silresium Chemical Corporation of Massachusetts	SCA Chemical Services Model City, NY	?	?
Kin-Buc Landfill of New Jersey	SCA Incinerator Chicago, IL	?	11/83
	SCA Chemical Services Model City, NY	?	11/83
Barker Chemical of Texas	Chemical Waste Management Corpus Christi, TX	RV	10/24/83

RD: CERCLA site in remedial design phase
RP: Responsible parties contracted removal
RV: EPA removal action
?: Unknown

DERIVED FROM: U.S. EPA, "Disposition of CERCLA Wastes Data Base" (Washington, DC: 10/17/84) and U.S. EPA "Listing of Potentially Responsible Parties at CERCLA Sites" (Washington, DC: 9/13/84).

a. The CERCLA sites listed here are sites for which U.S. EPA's database has identified the company as a potentially responsible party.
b. The RCRA facilities listed here are owned by the potentially responsible party to which CERCLA wastes were sent.
c. Waste Management claims that this site has not received wastes from Seymour.

The use of these contractors may have resulted from pragmatic decision making and not from a conflict of interest. Whatever the case, CEP believes that U.S. EPA should avoid rewarding the companies that may be partially responsible for the cleanup problems with contracts to do the work. The appearance of a conflict of interest has resulted in headlines about sweetheart deals, toxic merry-go-rounds, and shell games. U.S. EPA should also provide accurate data to facilitate more accurate reporting. Waste Management, for example, claims that Ohio Liquid Disposal did not receive wastes removed from the Seymour site, even though EPA data says it did.[21] According to a report of the counsel hired by Waste Management:

> Seymour is another example of conduct which was perfectly proper, had informal agency approval, but which did not have written documentation demonstrating such approval. There was nothing wrong with conduct; just the failure to have *written* approval exposed the Company [Waste Management, Inc.] to charges of noncompliance.[22]

Leaks Are Disconcerting

On October 24, 1984 U.S. EPA's Office of Solid Waste released a draft study of the quality of ground water monitoring at hazardous waste disposal sites.[23] The study revealed (1) that the total number of disposal facilities that need to monitor for ground water contamination is unknown, (2) that over 80 percent of sites that were undergoing a contamination detection and assessment program at the time were not complying with regulations, and (3) that regulatory agencies are underfunded, understaffed, or simply unwilling to perform their duties.

U.S. EPA should be commended for releasing a study that is so self-critical. But its findings merely document what earlier reports by the General Accounting Office, Congress's Office of Technology Assessment, and the National Research Council of the National Academy of Sciences had already indicated.[24] Action has been slow to correct the situation.

Five days after the U.S. EPA ground water study was released, findings of a confidential agency draft report unexpectedly showed up on the front page of the *New York Times.*[25] This confidential report alleged that toxic liquids and gases could be seeping out of the CECOS facility in Niagara Falls, New York, and entering neighborhood cellars.[26]

The situation that the *Times* reported remains extremely complex. The possibly defective facility held large quantities of toxic wastes removed from Love Canal and 11 other Superfund sites.[27] U.S. EPA could not be sure the facility was at fault, however, because the fractured bedrock beneath the facility and an adjacent site that is known to be leaking made adequate monitoring very difficult. If the CECOS facility is leaking, millions of taxpayer dollars have been spent moving hazardous wastes from abandoned sites to an active facility that may pose similar problems.

The confidential 1984 report on the CECOS facility was the product of the first in-depth agency review of the hydrogeologic conditions at a facility that received Superfund wastes. William Myers, who at the time was the lead technical person at U.S. EPA's Ground Water Task Force (see chapter 3), wrote the report. Although it is encouraging that this review took place, U.S. EPA should have had the resources to perform such analyses in every case, before spending taxpayer dollars on any of the 102 commercial facilities that accept Superfund wastes.

The report did not prove that the CECOS site was leaking. The most troubling aspect of the CECOS case was that the report alleged circumstantial evidence indicating leakage and that appropriate monitoring and corrective measures had not been undertaken. According to U.S. EPA policies, the *facility* had the responsibility to show that a statistically significant increase of contaminants in ground water was not due to a leak. The CECOS site had not adequately explained the statistical variance; therefore, the site should not have been chosen to receive Superfund wastes. Nevertheless, the facility received Superfund wastes as recently as April 1984. Just three of the more than a dozen Superfund cleanups that involved the CECOS facility in Niagara Falls procured more than $30 million worth of federal contracts.[28]

Regional U.S. EPA and state personnel immediately disputed the paper's position. Nor did U.S. EPA headquarters approve it. After the *New York Times* published findings of the confidential report, U.S. EPA disposed of 4000 cubic yards of contaminated dirt from Re-solve, Inc. of Dartmouth, Massachusetts—a Superfund site—at CECOS. CECOS had already received the cleanup contract.[29] No other nearby facility was available at the time, and time was short. The Re-solve wastes had to be removed before spring; otherwise, rain and melting snow might have carried toxic chemicals into a river bordering the site. For these reasons U.S. EPA disposed of dirt tainted with polychlorinated biphenyls (PCBs, chemicals that may cause cancer) in the CECOS facility—the same facility that, according to William Myers's confidential report, showed convincing circumstantial evidence of leakage. The *Los Angeles Times* quoted James C. Colman, a hazardous waste official in Massachusetts: "There just aren't many places to put this stuff. What's going to happen is that we're just not going to take the stuff away from the sites anymore. We'll cap it, try to neutralize it as much as possible and hope it doesn't leak."[30]

BUILDING PUBLIC CONFIDENCE

The public's confidence in the ability of government and industry to assure protection from hazardous wastes has declined as more information concerning the inherent long-term inadequacy of burying hazardous wastes comes to light. To maintain public faith in the regulatory process, officials must ensure that facilities use the best technologies and management practices and make sure that these technologies and practices do not pose significant risks to health and the environment. But references to state-of-the-art technologies mean nothing to local residents if the best "art" is a forgotten landfill in their neighborhood. The problems at Love Canal did not arise solely from faulty technology—the layout of the site was not significantly different from that of landfills used today. The problems resulted from inadequate enforcement of safeguards. Officials allowed the construction of a school and other facilities on top of the clay "cap" that covered the site. The construction pierced the cap, rainwater mixed with the wastes, and human beings suffered the consequences. If the regulatory process is to deserve the confidence of the public, vigilant public officials will play as strong a part as technology and management.

Hazardous waste managers should recognize the impact that public confidence has on the waste management industry. The public plays an increasingly important role in determining sites and technologies. In the end public confidence in the hazardous waste industry and the regulatory program will do as much to determine the industry's success or failure as hazardous waste managers.

CHAPTER FIVE

Hazardous Waste Management: Strategies and Technologies

The suitability of hazardous waste management technologies varies with the types of wastes and the specific site. This chapter presents a general discussion of various technologies. The chapter also includes a summary of the most significant emerging control technologies that have arisen because of legislative prodding.

MATCHING MANAGEMENT TECHNOLOGIES TO WASTES

Figure 5-1 shows a preference hierarchy for the management of hazardous wastes. According to this hierarchy, producing less waste in the first place by using different raw materials and alternative production processes is the best choice. However, this is not always possible. Land emplacement is the last resort (see the discussion of land disposal bans in chapter 3). Table 5-1 outlines alternatives to land disposal for six major types of wastes.

WASTE ELIMINATION AND REDUCTION STRATEGIES

During the past decade eliminating or reducing the amounts of hazardous wastes that require management has become more widely accepted for financial and environmental reasons. The goal is to change the manufacturing process by substituting raw materials, by altering the manner in which materials are combined, by modifying the processing equipment, or by installing instruments that gauge and control the chemical processes more closely.

Industries employ four methods to reduce the quantity of wastes generated:

1. source segregation and separation
2. process modification
3. raw material substitution
4. recovery and recycling.

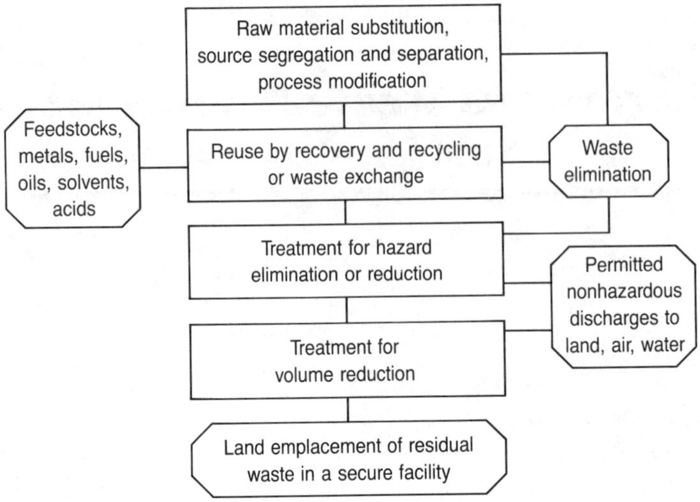

FIGURE 5-1. *Options for Managing Hazardous Waste (Strategies Listed in the Order of Most Preferable to Least Preferable)*

How effective are these methods? Waste reduction can be significant—in fact, waste production can sometimes drop to zero, especially in new plants that are designed and constructed with maximum waste reduction in mind. Changes in existing operations, however, are more difficult to make, especially in older plants. Estimating potential waste volume reductions for industry as a whole is very difficult, because most companies consider their process technologies trade secrets.[1]

Source Segregation and Separation

Companies can keep nonhazardous wastes from becoming hazardous by segregating dangerous materials. A plant may store wastes according to physical and chemical properties and according to the method of ultimate treatment or disposal. For example, many plants segregate the aqueous waste solutions that can be treated effectively at a conventional biological waste-water treatment plant. (Such plants can decompose many of the 129 priority pollutants listed in the CWA regulations.)[2] Some companies also categorize waste by the degree and type of hazard. Concentrated hazardous wastes that have been well segregated are possible candidates for one of the several operational waste exchanges or for intracompany recycling.

Process Modification

Process modifications can increase production efficiency, improve or change products, and reduce costs. Such modifications include:

- changing process conditions such as temperature or pressure
- modifying the process flow or design
- using different raw materials
- using new equipment
- using completely different processes.

Although waste reduction is often not one of the primary goals of a modification, the volume of hazardous wastes and the cost of managing them are often reduced.

Modifying a single part in a process does not usually have a large impact, but making several changes can reduce costs and waste significantly. Because modifications to existing facilities can be expensive, they are most likely to be implemented when a new production area is added. Process modifications are usually plant- or process-specific. They are difficult to transfer to another industry or to another segment of the same industry.

Raw Material or Product Substitution

Plants that employ raw material or product substitution replace hazardous waste-generating raw materials or products with a different material. The new raw materials or products reduce or eliminate the generation of hazardous wastes.

Industry has developed substitutes for two important materials, for example: polychlorinated biphenyls (PCBs) and asbestos. PCBs were used as insulating and cooling fluids in electrical equipment, especially transformers. As the toxic effects of PCBs became known, restriction on their use catalyzed the search for substitutes. Most transformers are now oil filled or air cooled. In the case of asbestos, makers of insulation and pipes who previously used the mineral as a fire retardant are now using clays, iron, and synthetics.

Substitution with a less hazardous raw material or product does not always eliminate the generation of hazardous waste. Iron pipes might pose less of a health hazard than do asbestos-concrete pipes, for instance. But the various process waste waters, sludges, roasting liquids, and pickling liquors that result from the production of iron can also be quite hazardous. Nevertheless, regulatory restrictions on the use of specific materials such as asbestos often bring about innovative substitutes.

TABLE 5-1. *Appropriate Treatment Technologies for Specific Hazardous Wastes*

	Physical Treatment						Chemical Treatment							Bio	Thermal Treatment		
Waste Type	*A*	*B*	*C*	*D*	*E*	*F*	*G*	*H*	*I*	*J*	*K*	*L*	*M*	*N*	*O*	*P*	*Q*
Pesticides																	
Pure											K	L	M		O	P	
Organic											K	L	M				Q
Aqueous						F					K	L	M				
Residues															O	P	
Contaminated solids																	Q
Contaminated solvents															O		
PCBs																	
Liquids												L					Q
Solids and Sludges																P	
Cyanide										J	K						

Heavy Metals																	
Acid and alkaline solutions			C				G	H		J							
Metal finishing and plating sludges		B	C					H		J							
Tetraethyl lead sludges				D						J					O	P	
Paint sludges and residues										J					O	P	
Organic solvents	A	B	C							J						P	
Halogenated Organics																	
Solvents											K	L			O	P	Q
Aqueous solutions				D		F					K	L		N			
Solids and sludges		B	C								K	L			O	P	Q
Nonhalogenated Volatile Organics																	
Solvents and liquids	A								I		K			N	O	P	Q
Aqueous solutions	A					F			I		K			N			
Sludges	A	B	C						I		K	L			O	P	Q

A: sedimentation
B: filtration
C: evaporation
D: distillation
E: flotation
F: absorption
G: neutralization
H: precipitation
I: stabilization
J: solidification
K: oxidation
L: dechlorination
M: ultraviolet photolysis
N: biological treatment
O: liquid injection
P: rotary kiln
Q: cement kiln

DERIVED FROM: Toxic Waste Assessment Group, Governor's Office of Appropriate Technology, *Alternatives to the Land Disposal of Hazardous Wastes, An Assessment for California* (1981).

Recovery and Recycling

Recovery and recycling of hazardous materials can bring a strong economic return. Although these practices are not new to industry, recent changes in law, liability, and compensation are bringing about more widespread use of these techniques in regard to hazardous and nonhazardous substances.

Recovery involves separating one or more substances from a mixture; the concentrated substances can then be reused later in the process or burned as a source of process heat. Recycling refers to selling the recovered material as a product or reusing it on site. Ferric chloride, a component of acidic metal-treating wastes, can be used as a settling agent in other processes, for example. Another example: The foaming agents that remove metallic ores from the crushed stone in which they were embedded can be recycled. The results are cost savings and the reduced potential for toxic discharge.

Recovery of raw materials or byproducts is usually done with standard treatment technologies.

- *Physical separation* makes use of differences in particle size and density. Techniques for physical separations include gravity settling, filtration, flotation, flocculation, and centrifugation.
- *Component separation* differentiates components by electrical charge, boiling point, freezing point, solubility, and particle size. Distillation, evaporation, ion exchange, ultrafiltration, reverse osmosis, electrolysis, carbon adsorption, and solvent extraction are some of the current techniques for component separation.
- *Chemical reaction* results in the removal of specific chemical constituents. Examples include precipitation, electrodialysis, and oxidation-reduction reactions.

Table 5-2 compares these technologies.[3]

Physical separation techniques are the most common, least expensive, and usually the safest. Component separation is also relatively easy but tends to be more expensive. The specialty chemical and pharmaceutical industries are the primary users of component separation techniques. Component separation is only now being applied to hazardous wastes and many processes are still in the demonstration stage. Chemical reactions are the most expensive and pose the greatest potential hazard.

A key goal of RCRA is to promote recovery and recycling of solid wastes to the greatest extent possible by accentuating certain economic forces in the marketplace. Recovered materials must compete with virgin stock, and they can compete more effectively if their physical or chemical properties allow them to be separated from a mixture cheaply. Table 5-3 gives some common examples of commercially applied recovery technologies. There are three major ways of recovering and recycling materials to reduce or eliminate the generation of hazardous waste.

In-Plant Recycling. Manufacturers or waste generators employ in-plant recycling if they can recycle raw materials, byproducts, process streams or waste streams on site. Many companies have found that waste materials or byproducts from one process can serve as a raw material for another process. For example, waste acids can neutralize caustic wastes to form a much less dangerous (sometimes even harmless) solution.

Off-Site Recovery. Many generators of small quantities of hazardous wastes do not own the equipment to recover or recycle the small amounts of wastes they create. If the physical or chemical properties of the wastes can be recovered, there are a number of recycling companies to which small generators can turn.[4] Recycling companies often collect specific waste types such as heavy metals, contaminated solvents, and inorganic salts from a number of generators. Recovered, often purified materials are often sold to the original generator as well as to other companies.

Waste Exchanges. Waste exchanges are "clearinghouses" that exchange wastes between generators and users. What one company considers a waste, another company uses as a raw material. The use of exchanges is growing. Almost two dozen formal wastes exchanges exist.[5] Some of these clearinghouses are simply waste information exchanges, newsletters or bulletins that list available materials. Waste materials exchanges actually transfer wastes through brokers who find markets for them and earn fees or commissions for their services (see table 5-3).

In 1976 a U.S. EPA-sponsored study estimated that about seven million tons of industrial wastes per year have potential value for exchange and reuse in the United States.[6] Since that study was done, indications are that the market for waste exchanges is somewhat larger, but that the movement toward such exchanges has not matured and that the technique is underused.[7]

TREATMENT TECHNOLOGIES

The treatment technologies described here "break down" hazardous waste into less hazardous (and nonhazardous) components or can segregate parts of the waste stream to reduce the overall hazard. Most treatment methods make waste less dangerous or more concentrated, but may not destroy waste completely. A given technology may be very suitable for one site but unsuitable for another. Many of the techniques for physically or chemically separating in-plant, in-process (listed in table 5-4) also can be considered treatment technologies.

TABLE 5-2. *Description of Technologies Currently Used for Recovery of Materials*

Technology/Description	Stage of Development	Economics
Physical Separation		
Gravity setting: Tanks, ponds provide hold-up time allowing solids to settle; grease skimmed to overflow to another vessel	Commonly used in waste-water treatment	Relatively inexpensive; dependent on particle size and setting rate
Filtration: Collection devices such as screens, cloth, or other; liquid passes and solids are retained on porous media	Commonly used	Labor intensive; relatively inexpensive; energy required for pumping
Flotation: Air bubbled through liquid to collect finely divided solids that rise to the surface with the bubbles	Commercial application	Relatively inexpensive
Flocculation: Agent added to aggregate solids together that are easily settled	Commercial practice	Relatively inexpensive
Centrifugation: Spinning of liquids and centrifugal force causes separation by different densities	Practiced commercially for small-scale systems	Competitive with filtration
Component Separation		
Distillation: Successfully boiling off materials at different temperatures (based on different boiling points)	Commercial practice	Energy intensive
Evaporation: Solvent recovery by boil-ing off the solvent	Commercial practice in many industries	Energy intensive

Types of Waste Streams	Separation Efficiency[a]	Industrial Applications
Slurries with separate phase solids, such as metal hydroxide	Limited to solids (large particles) that settle quickly (less than 2 hours)	Industrial wastewater treatment first step
Aqueous solutions with finely divided solids; gelatinous sludge	Good for relatively large particles	Tannery water
Aqueous solutions with finely divided solids	Good for finely divided solids	Refinery (oil/water mixtures); paper waste; mineral industry
Aqueous solutions with finely divided solids	Good for finely divided solids	Refinery; paper waste; mine industry
Liquid/liquid or liquid/solid separation, i.e., oil/water; resins; pigments from lacquers	Fairly high (90%)	Paints
Organic liquids	Very high separations achievable (99+% concentrations) of several components	Solvent separations; chemical and petroleum industry
Organic/inorganic aqueous streams; slurries, sludges, i.e., caustic soda	Very high separations of single, evaporated component achievable	Rinse waters from metal-plating waste

a. Good implies 50 to 80 percent efficiency, fairly high implies 80 percent, and very high implies 90 percent.

(continued)

TABLE 5-2 *(continued)*

Technology/Description	Stage of Development	Economics
Component Separation (continued)		
Ion exchange: Waste stream passed through resin bed, ionic materials selectively removed by resins similar to resin adsorption. Ionic exchange materials must be regenerated	Not common for HW	Relatively high costs
Ultrafiltration: Separation of molecules by size using membrane	Some commercial application	Relatively high costs
Reverse osmosis: Separation of dissolved materials from liquid through a membrane	Not common; growing number of applications as secondary treatment process such as metal-plating and pharmaceuticals	Relatively high costs
Electrolysis: Separation of positively/negatively charged materials by application of electric current	Commercial technology; not applied to recovery of hazardous materials	Dependent on concentrations
Carbon/resin absorption: Dissolved materials selectively absorbed in carbon or resins. Absorbents must be regenerated	Proven for thermal regeneration of carbon; less practical for recovery of adsorbate	Relatively costly thermal regeneration; energy intensive
Solvent extraction: Solvent used to selectively dissolve solid or extract liquid from waste	Commonly used in industrial processing	Relatively high costs for solvent
Chemical Transformation		
Precipitation: Chemical reaction causes formation of solids which settle	Common	Relatively high costs

Types of Waste Streams	Separation Efficiency[a]	Industrial Applications
Heavy metals aqueous solutions; cyanide removed	Fairly high	Metal-plating solutions
Heavy metal aqueous solutions	Fairly high	Metal-coating applications
Heavy metals; organics; inorganic aqueous solutions	Good for concentrations less than 300 ppm	Not used industrially
Heavy metals; ions from aqueous solutions; copper recovery	Good	Metal plating
Organics/inorganics from aqueous solutions with low concentrations, i.e., phenols	Good, overall effectiveness dependent on regeneration method	Phenolics
Organic liquids, phenols, acids	Fairly high loss of solvent may contribute to hazardous waste problem	Recovery of dyes
Lime slurries	Good	Metal-plating wastewater treatment

a. Good implies 50 to 80 percent efficiency, fairly high implies 80 percent, and very high implies 90 percent.

(continued)

TABLE 5-2 *(continued)*

Technology/Description	Stage of Development	Economics
Chemical Transformation (continued)		
Electrodialysis: Separation based on differential rates of diffusion through membranes. Electrical current applied to enhance ionic movement	Commercial technology, not commercial for hazardous material recovery	Moderately expensive
Chlorinolysis: Pyrolysis in atmosphere of excess chlorine	Commercially used in West Germany	Insufficient U.S. market for carbon tetrachloride
Reduction: Oxidative state of chemical changed through chemical reaction	Commercially applied to chromium; may need additional treatment	Inexpensive
Chemical dechlorination: Reagents selectively attack carbon-chlorine bonds	Common	Moderately expensive
Thermal oxidation: Thermal conversion of components	Extensively practiced	Relatively high

Physical Treatment Technologies

Gravity Settling. Gravity settling (flocculation) is a process that removes solid particles that are suspended in a liquid. Special tanks or holding ponds provide sufficient time and adequate space for the solids to settle because of normal gravitational force. Many solid materials settle out of water naturally; others can be converted into settling solids through the addition of something else. Metal salts precipitate out of solutions of heavy metals that result from plating processes.

The amount of air emissions resulting from high concentrations of volatile compounds (usually organic solvents) is an environmental concern associated with gravity settling. Covering the surfaces of open tanks or holding ponds to prevent emissions is sometimes necessary; many locales require that tanks

Types of Waste Streams	Separation Efficiency[a]	Industrial Applications
Separation/concentration of ions from aqueous streams; application to chromium recovery	Fairly high	Separation of acids and metallic solutions
Chlorocarbon waste	Good	Carbon tetrachloride manufacturing
Metals, mercury in dilute streams	Good	Chrome-plating solutions and tanning operations
PCB-contaminated oils	High	Transformer oils
Chlorinated organic liquids; silver	Fairly high	Recovery of sulfur, HCl

SOURCE: Office of Technology Assessment, *Technologies and Management Strategies for Hazardous Waste Control*, OTA-M-196 (1983).

a. Good implies 50 to 80 percent efficiency, fairly high implies 80 percent, and very high implies 90 percent.

and ponds that hold volatile compounds be covered. Regulations also require that surface impoundments be protected against breaches by use of dikes, liners, and frequent maintenance. Ponds are particularly susceptible to flooding.

Flotation. Dissolved-air flotation separates solids and grease-like viscous materials from liquids by attaching tiny air bubbles to them. The particles then rise to the surface and can be skimmed off. The addition of a foaming agent can enhance bubbling. Dissolved-air flotation is especially useful for removing suspended solids, oils, and greases from wastes generated by petroleum refining, meat-packing, paint manufacturing, poultry processing, paper milling, and baking.

The drawback to flotation is that air pollution may originate from the large open flotation tanks. Placing the flotation tanks inside buildings or covering

TABLE 5-3. *Commercially Applied Recovery Technologies*

Generic Waste	Typical Source of Effluent	Recovery Technologies
Solids in aqueous suspension	Salt/soda ash	Filtration
Heavy metals	Metal hydroxides from metal-plating waste; sludge from steel-pickling operations	Electrolysis
Organic liquids	Petrochemicals/mixed alcohol	Distillation
Inorganic aqueous solutions	Concentration of inorganic salts/acids	Evaporation
Separate phase solids, grease/oils	Tannery waste/petroleum waste	Sedimentation skimming
Chrome salt solutions	Chromium-plating solutions; tanning solutions	Reduction, reverse osmosis
Metals, phosphates, sulfates	Steel-pickling operations	Precipitation

SOURCE: Office of Technology Assessment, *Technologies and Management Strategies for Hazardous Waste Control*, OTA-M-196 (1983).

them can almost eliminate the release of air pollutants. But flotation devices can never be entirely sealed because air is constantly forced into the system from underneath to create the bubbles.

Centrifugation. Centrifugation is the process of separating the solid and liquid components of a waste stream by rapidly rotating the mixture in a vessel called a centrifuge. The solids collect on the walls of the centrifuge. Removing water (dewatering) from sludges is the chief use for centrifugation in waste treatment. Because the centrifuge is an enclosed device, there are few air emissions. Like gravity settling, however, the resulting solids and liquids may need further treatment.

Several closely related methods rely on membranes to separate the components of a waste stream. The membrane's chemical and physical composition and the mechanical or electrical force driving the process can be tailored to the specific waste.

Filtration. When a mixture of solids suspended in liquids passes through a porous medium, many of the solids are intercepted. The porous medium, the filter, can be granular (such as sand or coal), or can be of woven cloth or fibrous mats.

Filtration usually is a step taken after gravity settling or flotation has removed most of the solids. Various filters are used to dewater sludges and to treat water for drinking purposes. The few environmental problems associated with filtration have arisen in cases where filters with captured solids were disposed of or recycled improperly.

TABLE 5-4. *The Major Chemical Waste Exchanges*

Exchange	Location	Contact
Waste Information Exchanges[a]		
California Waste Exchange Dept. of Health Services Toxic Substances Control Division	714/744 P St. Sacramento, CA 95814	Robert McCormick 916-324-1818
Colorado Waste Exchange Colorado Association of Commerce & Industry	1390 Logan St. Denver, CO 80203	Olie Webb 303-831-7411
World Association for Safe Transfer & Exchange	130 Freight St. Waterbury, CT 06702	Marcel Veroneau 203-574-2463
Southern Waste Information Exchange	Box 6437 Tallahassee, FL 32313	Roy C. Herndon 904-644-5516
Georgia Waste Exchange Business Council of Georgia	181 Washington St. SW Atlanta, GA 30303	Robert D. Sumner 404-223-2264
Industrial Material Exchange Service Illinois State Chamber of Commerce	IEPA-D LPC-24 2200 Churchill Rd. Springfield, IL 62706	Margot Ferguson 217-782-0450
Louisville Area Industrial Waste Exchange Louisville Chamber of Commerce	1 Riverfront Plaza, 4th Floor Louisville, KY 40202	Terry Feathers 502-566-5000
Great Lakes Regional Waste Exchange Waste Systems Institute of Michigan, Inc.	3250 Townsend NE Grand Rapids, MI 49505	William A. Stough 616-363-7367
Midwest Industrial Waste Exchange Regional Commerce & Growth Association	10 Broadway St. Louis, MO 63102	Clyde H. Wiseman 314-231-5555
New England Materials Exchange	34 North Main St. Farmington, NH 03835	David Green 603-755-4442 or 603-755-9962
New Jersey State Waste Exchange New Jersey Chamber of Commerce	5 Commerce St. Newark, NJ 07102	William E. Payne 201-623-7070

a. Waste information exchanges promote waste transfer through newsletters or bulletins.

(continued)

TABLE 5-4 *(continued)*

Exchange	Location	Contact
Waste Information Exchanges[a] (continued)		
Industrial Commodities Bulletin, Enkarn Corp.	Box 590 Albany, NY 12210	J. T. Engster 518-436-9684
Northeast Industrial Waste Exchange	90 Presidential Plaza, Suite 22 Syracuse, NY 13202	Walker Banning 315-422-6572 Stephen Hoefer 315-474-4201
Piedmont Waste Exchange Urban Institute	University of North Carolina Charlotte, NC 28223	Mary A. McDaniel 704-597-2307
Tennessee Waste Exchange Tennessee Manufacturers Association	501 Union St., Suite 601 Nashville, TN 28223	Sharon L. Bell 615-256-5141
Chemical Recycle Information Program Houston Chamber of Commerce	110 Milam Bldg., 25th Floor Houston, TX 77002	Jack Westney 713-658-2462 or 713-658-2459
Inter-Mountain Waste Exchange W. S. Hatch Co.	643 South 800 West Woods Cross, UT 84087	Joseph Parkinson 801-295-5511
Waste Materials Exchanges[b]		
Zero Waste Systems	2928 Poplar St. Oakland, CA 94608	Trevor Pitts 415-893-8257
ICM Chemical	20 Cordova St., Suite 3 St. Augustine, FL 32084	Anthony L. Tripi 904-824-7247
Alkem	25 Glendale Rd. Summit, NJ 07901	Alan W. Schneider 201-277-0060
Ore Corp.	2415 Woodmere Dr. Cleveland, OH 44106	Richard Immerman 216-371-4869
Techrad Industrial Waste Exchange	4619 North Santa Fe Oklahoma City, OK 73118	Ernest L. Koerner 405-528-7016

SOURCE: Laurie A. Rich, "Waste Exchanges: An Underutilized Way to Help the Bottom Line," in *Chemical Week* (New York, N.Y.: McGraw-Hill, May 16, 1984), p. 58.

a. Waste information exchanges promote waste transfer through newsletters or bulletins.

b. Waste materials exchanges employ brokers who arrange the sale of waste and earn a fee or commission.

Dialysis. The membrane separation process known as dialysis uses osmosis, the diffusion of one solution through a semipermeable membrane into a more concentrated solution, as the driving force. Components pass from one side of the membrane to the other, then they can be collected and removed.

The use of dialysis in the hazardous waste industry is generally limited to aqueous wastes containing high concentrations of dissolved compounds with low molecular weight (cyanides, acids, and caustics). Few environmental problems result from this process.

Reverse Osmosis. Reverse osmosis is a membrane separation process that employs a membrane and a mechanical pumping force to transport the solution from one side of the membrane to the other. The membrane, usually an acetate or nylon sheet, allows a solvent such as water to pass through but not the substances dissolved in the solvent. The process is called reverse osmosis because the force applied increases the concentration of the dissolved substances. In normal osmosis the concentration of the dissolved substances decreases.

This separation technique yields drinking water from brackish or saline surface water or ground water. This technique can also treat heavy metal solutions and organic solutions. The electroplating industry has developed membrane cartridges in compact units that use reverse osmosis to separate chemicals from plating rinse waters.

Environmental exposure from reverse osmosis is limited to the concentrated solution of dissolved chemicals. Unless it can be reused, this solution requires further treatment for ultimate disposal. Reuse is common in the plating industry.

Evaporation. *Evaporation* in the hazardous waste business means the vaporization of a liquid from a slurry, a thin watery mixture. Without the excess liquid what remains of the slurry is concentrated and thus has less volume. The vapor may or may not be recovered.

One common, inexpensive method of evaporation is the use of ponds or lagoons where liquids are directly exposed to the sun. The solar-evaporated vapor is emitted directly into the atmosphere. Another method uses a closed system in which heat is applied to a metal chamber where the contents are often under reduced pressure. (Reduced pressure lowers the boiling point and the energy costs.) The liquids evaporate in the order of ascending boiling points.

In hazardous waste management, evaporation is often used to dewater sludges and slurries and to concentrate rinse waters from electroplating operations. In the western United States, particularly in California, solar-evaporation ponds are widely used to reduce the amounts of water in hazardous wastes before they are placed into landfills. The high net evaporation in the western states removes large amounts of water. In the East, however, rainfall often matches evaporation, so solar evaporation is less effective.

Strong odors can arise when the evaporated substances are volatile organics, especially compounds that are toxic or photochemically reactive (smog-forming)—hydrocarbons and organic nitrogen compounds with a low molecular weight are notorious for their smell. The concentrated solutions, sludges, or slurries may have environmental effects when they are recovered, treated, or placed in the land. If they are still decomposing, they may emit a smell that is musty at best. At worst, area residents may be plagued by the rotten egg smell of hydrogen sulfides.

Evaporation ponds or lagoons that have been in operation for several years are of special concern because they are more likely to have associated ground water contamination. Better design and construction methods and the use of monitoring wells can greatly reduce these problems. New lagoons and ponds have to meet stricter requirements under RCRA, but leaks can still occur. Retrofitting older lagoons takes place as facility operators go through the RCRA permitting process.

Distillation and Steam Stripping. Distillation and steam stripping can separate components from a liquid mixture because the components have different boiling points. The techniques are often used to purify organic products or to separate byproducts. Figure 5-2 presents a simplified rendering of a typical distillation plant.

Steam stripping is especially appropriate for removing organic liquids that have low boiling points and that are in dilute concentrations in water. Distillation is similar, but it can separate a wider range of components. The techniques are important means of solvent recovery in the hazardous waste management industry. Stripping removes organic solvents prior to discharging waste waters under a federal or state permit.

The residue that remains in the equipment after distillation or steam stripping may present a hazard. The remaining heavy organics, tars, and sludges may be toxic; they can usually be treated by high-temperature incineration.

Adsorption. Adsorption (adhesion) is used frequently to remove low concentrations of organics from aqueous waste streams. The most common adsorbent in use is activated carbon—carbon particles rendered extra porous by heat treatment. Waste streams pass through a tank or column filled with carbon granules or powder. The carbon traps many contaminants in the pores by the natural forces of surface attraction. Adsorption is a common method of removing organics, chlorinated pesticides, PCBs, and phenols from waste water. More recently, activated carbon and several synthetic resins have been used to remove solvents from contaminated ground water supplies.

Environmental impacts resulting from adsorption most often occur when toxic compounds trapped in the carbon are cleaned or burned. Temperatures

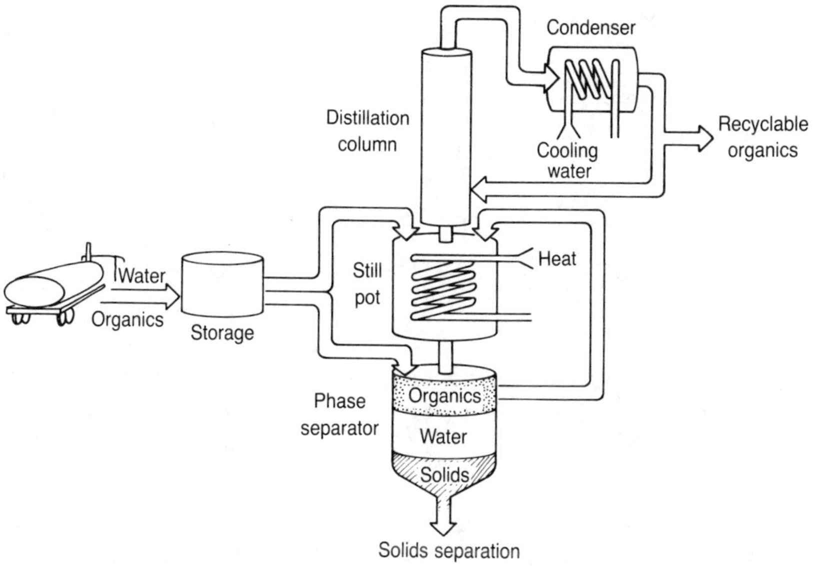

FIGURE 5-2. *Physical Treatment: Distillation Process for Recovering Recyclable Organics*

SOURCE: Toxics Assessment Group, Governor's Office of Appropriate Technology, *Alternatives to Land Disposal of Hazardous Waste, An Assessment for California* (1981).

need to be high enough to incinerate the contaminants totally but not so high that the carbon itself burns, a process that could result in the formation of dioxins. Burning off the contaminants is called regeneration; it regenerates the carbon for reuse.

Biological Treatment Technologies

Biological treatment processes have been used to treat conventional wastes for a century. Several biological treatment methods for treating toxic wastes now exist as well. Figure 5-3 shows the steps in the biological treatment of industrial waste. Living microorganisms decompose wastes by "eating" them and transforming them into water; carbon dioxide; and simpler, less dangerous molecules. Inorganic wastes are not destroyed by these processes; therefore, managers must use great care in applying biological treatments to inorganic compounds. As the bacteria die they form a sludge. The sludges have a reduced volume and associated hazard, but they may themselves be toxic. Sludges may contain some inorganic constituents (especially metals), so in many cases they must be sent for additional treatment or for disposal as toxic wastes.

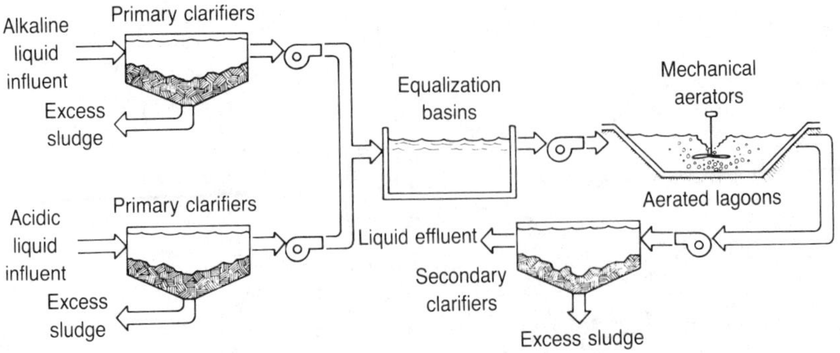

FIGURE 5-3. *Biological Treatment of Industrial Waste*

SOURCE: Toxics Assessment Group, Governor's Office of Appropriate Technology, *Alternatives to Land Disposal of Hazardous Waste, An Assessment for California* (1981).

Activated Sludge. High concentrations of microorganisms and ample oxygen can treat organic contaminants as long as the solids content is low. Any metals in low concentrations that are not toxic to the bacteria can also be removed from the organic wastes. Special bacterial cultures are even capable of treating mercury-containing wastes. The bacteria end up in the sludge, so some of the sludge is recycled back into the incoming waste as a source of bacteria.

Trickling Filters. In a system with a trickling filter, organic waste waters trickle through a bed or rock or synthetic material coated with a slime of microorganisms. The microorganisms destroy organic wastes such as cyanides, formaldehyde, ketones, resins, acetone, and acetic and formic acids. Systems with trickling filters can be quite odorous if operated improperly. They destroy more types of waste than activated sludge systems, they take up less room, and they can treat a greater volume of waste in a given time.

Aerated Lagoons. Aerated lagoons are usually big earthen ponds that contain a large number of microorganisms. Mechanical agitation with mixers or water sprays introduce enough oxygen into the waste water to allow decomposition of wastes. Aerated lagoons treat the same types of organics and metals as activated sludge plants. Waste in aerated lagoons must contain a low solids concentration, however. Residue from lagoons is not recycled and the lagoons take much more space than activated sludge systems; therefore, lagoons are more common in areas where land is cheap.

Waste Stabilization Ponds. Organic waste waters containing solids in low concentrations of less than 1 percent are placed in large, shallow ponds where decomposition takes place over a long time—often months. Waste stabilization is often a finishing process after most of the organics and all of the inorganics have been removed. Like lagoons, waste stabilization ponds take up much land.

Landfarming. Landfarming is a biological treatment technique in which microorganisms decompose aqueous organic wastes and sludges in soil rather than in water. EPA considers landfarming a disposal technique because the wastes are either decomposed or mixed with the soil. The liquid components may infiltrate to the ground water. The soil should be relatively porous and permeable. The land should be almost level in order to prevent runoff and erosion. Application rates vary with the nature of the soil, overall climate, depth to ground water, and composition of the wastes.

Landfarming is uncommon. It is suitable for only a few types of organic wastes, including biological sludges with a low concentration of metals; cooling waters; and dilute, oily wastes from petroleum refineries. If heavy metals are present in the waste, soil particles can adsorb significant amounts, and the heavy metals accumulate in the organic layer of the soil. Evaporating volatile organics can cause air-quality problems.

Chemical Treatment Technologies

Neutralization. Neutralization is a common technique in the treatment of wastes that are hazardous because they are too acidic or too caustic. By mixing the proper proportions of wastes of low pH (acids) or high pH (bases), the resulting wastes approach a neutral pH of 7.0 and are not hazardous. Figure 5-4 shows the sequence of mixing in which neutralization occurs. Acidic wastes are usually neutralized with sodium or calcium hydroxide or limestone. Caustic wastes can be neutralized by hydrochloric or sulfuric acid. Substances that precipitate during neutralization (metal salts, typically) need further treatment.

Hazardous waste streams from steelmaking, tanning, and petrochemical industries can be treated by neutralization. The process is effective in treating the incineration of certain hazardous substances. Burning plastics, for example, release hydrogen chloride, which forms hydrochloric acid when mixed with water. A lime slurry, a caustic solution, can "scrub" or neutralize the gases.

Precipitation. Precipitation refers to the removal of dissolved substances from a solution by changing the solubility relationships in the solution. Equilibrium is usually altered by changing temperature, pH, or by adding other chemicals to initiate the formation of insoluble compounds. In the hazardous waste industry, precipitation is usually combined with physical processes such

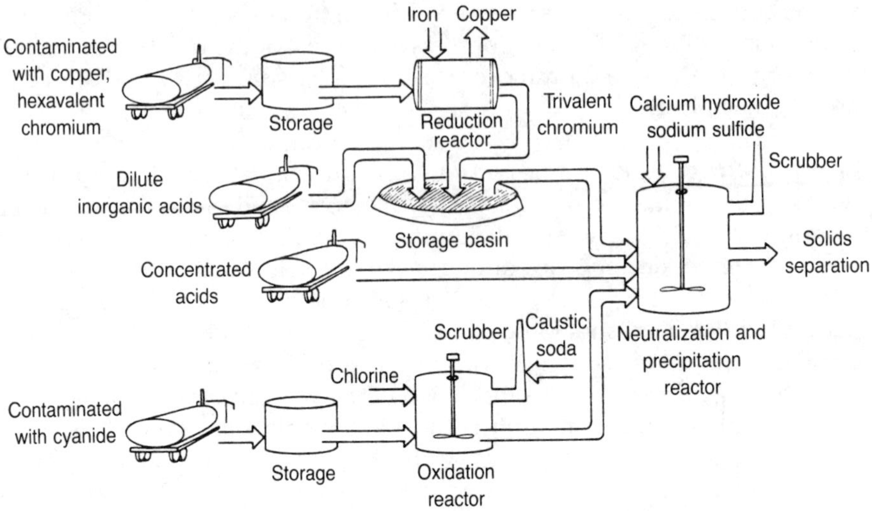

FIGURE 5-4. *Chemical Treatment: Neutralization, Precipitation, and Chemical Oxidation/Reduction*

SOURCE: Toxics Assessment Group, Governor's Office of Appropriate Technology, *Alternatives to Land Disposal of Hazardous Waste, An Assessment for California* (1981).

as flotation or settling, which remove the settled solids from the waste stream. Precipitation is frequently used in electronics and metal finishing, especially for electroplating processes and wastes. Precipitation removes toxic metals such as chromium, cadmium, or nickel from concentrated waste waters. The inorganic chemical industry also uses precipitation to remove metals from waste streams.

The solids resulting from precipitation usually require further treatment or disposal. The metal-rich sludges often undergo chemical reaction or stabilization so that little toxic leachate will result if they are deposited in landfills. Sometimes the sludges can be recycled.

Oxidation-Reduction. Oxidation and reduction processes use common chemical reactions to change materials. Some of the most widely used oxidizing agents are air, ozone, chlorine, hydrogen peroxide, sodium hypochlorite, and nitrous acid. Oxidation treats organic and inorganic wastes including cyanides, phenols, lead, pesticides, and aromatic hydrocarbons.

Oxidation-reduction processes are most effective on aqueous waste streams that contain low concentrations of toxics. The rinse water used in the electroplating process contains a low concentration of cyanide, for example, and is a good candidate for oxidation-reduction.

Toxic sludges or air emissions that contain gases such as chlorine may result from oxidation-reduction. These products and the aqueous effluent from the process may need further treatment to remove solids or metal ions that remain.

Ion Exchange. In ion exchange dissolved inorganic substances separate from an aqueous liquid. The liquid passes through layers of natural or synthetic resins, allowing ions in the resin to be exchanged with inorganics in the liquid. The electroplating industry uses ion exchange to extract chromium and cyanide ions. The process is also useful for treating certain acid solutions and dissolved salts.

The inorganics removed from aqueous waste streams may cause environmental damage. If the inorganics are hazardous, they must be deposited in a RCRA facility.

Thermal Treatment Technologies

Incineration. Incineration is hardly a new technology. Many municipalities used it to dispose of solid waste until the 1960s. During the 1970s, however, more stringent air-quality regulations forced incinerators to close, and new highway systems allowed easier access to landfill sites. Nevertheless, incineration is one of the better ways to manage hazardous wastes because it can destroy most wastes completely. Incineration also eliminates most future environmental liabilities. Landfills, in contrast, shift risks to future generations. Individual incinerators must be designed and operated to meet stringent operating standards and emissions limitations.

Present incineration costs are generally much higher than the costs of placing wastes in landfills or deep wells. But as various wastes are banned from deep wells or landfills in the years ahead, incineration will become more competitive in spite of more stringent emissions standards.

Most organic hazardous waste can be incinerated at temperatures between 1000°F and 3000°F. Both the temperature and the time the waste stays in the incinerator (the residence time) depend on the waste mixture and the design of the incinerator. (Many conventional incinerators were adapted to treat hazardous wastes.) Some wastes have a flash point that is so low that no additional fuel is needed; others need large amounts of fuel before the wastes burn. If some of the wastes can be used as fuel, operating costs are lower, and the price of incineration becomes more competitive with the price of landfills and deep wells. U.S. EPA rules now require that more than 99.99 percent of many wastes be destroyed prior to emission. The requirement for PCB-contaminated wastes is destruction of more than 99.9999 percent of the material.

Liquid Injection. In this, the most common method of hazardous waste incineration, wastes are broken into fine droplets (atomized) as they are injected into a combustion chamber. The technique is similar to how oil enters a home-heating furnace. Liquid-injection systems can burn almost any combustible waste that can be pumped. At temperatures of 1200° F to 3000° F, the atomized wastes remain in the combustion chamber for seconds at longest. If necessary, additional fuel is injected into the chamber to promote combustion.

Liquid injection has been used to burn PCBs, solvents, phenols, pesticides, herbicides, and distillation residues. The burn rates must be carefully adjusted for specific wastes; otherwise, explosions or hazardous emissions may result.

Rotary Kilns. Rotary kilns were originally used for fire-curing and glazing clay in the cement industry. In the hazardous waste industry, solid, liquid, or gaseous wastes are injected into one end of a rotating chamber. Because the central chamber of the kiln is at a slight slope, gravity moves the wastes into the combustion zones where the temperature can be as high as 3000° F. Depending on the type of waste, the material remains in the kiln from a few seconds to several hours. Residues are continuously removed at the lower end of the kiln. Combustion gases usually pass through an afterburner or some type of scrubber before they are emitted to the atmosphere. Figure 5-5 shows a typical setup that employs a rotary kiln and an afterburner. A wide range of hazardous wastes—solvents; PCBs in capacitors; waste gases from chemical weapons; paints; and others—have been burned in rotary kilns. Kiln burning is usually unsuitable for heavy metal sludges and wastes high in organics, however.

Some companies are purchasing older, nonoperating cement kilns and retrofitting them to handle hazardous wastes. The use of existing kilns eliminates some of the high capital costs of building new kilns. Dow Chemical claims to have had notable success with kiln disposal of toxic materials.

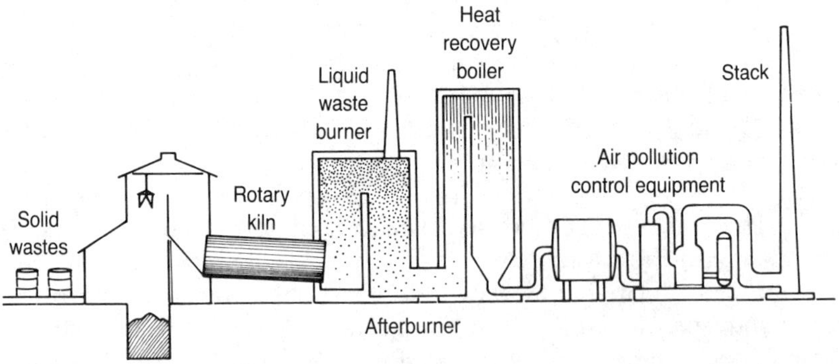

FIGURE 5-5. *Rotary Kiln Incinerator with Heat Recovery and Air-Pollution Control*

SOURCE: Toxics Assessment Group, Governor's Office of Appropriate Technology, *Alternatives to Land Disposal of Hazardous Waste, An Assessment for California* (1981).

Fluidized-Bed Incinerators. Fluidized-bed incinerators have a cylinder that contains a bed of inert granular material on a perforated plate. The bed of inert material is heated and a gas (usually air) passes through the plate to "fluidize" the bed to the consistency of quicksand. At temperatures of 800° F to 1300° F, liquid or solid wastes are usually incinerated within a few seconds of coming into contact with the hot grains.

Fluidized-bed incinerators were developed in the 1960s and 1970s to handle wastes from pulp and paper mills and oil refineries. Their use as hazardous waste incinerators has been limited because, like the rotary kiln, they are expensive. They are safe for disposal of wastes that do not require very high temperatures for destruction. Fluidized-bed incinerators are not capable of destroying PCBs, for example, although an incinerator design that employs a derivative of molten salt can destroy them and may point to a new trend in technology.

Multiple Hearth Incinerators. Multiple hearth incinerators consist of tiered hearths. Wastes enter at the top and cascade downward into ever-hotter hearths. The cylindrical shells are lined with refractory material that contributes to fuel efficiency. Mining operations first used multiple hearth incinerators to roast ores. In the 1930s altered designs began to burn waste-water treatment sludges (sewage) and nonhazardous industrial sludges. Designs introduced in recent years can process hazardous wastes as well.

Incinerator technologies continue to develop; the discussion of emerging technologies cites the trends in innovation.

DISPOSAL TECHNOLOGIES

Landfills. The commercial hazardous waste industry has used landfills to dispose of an extremely wide range of hazardous wastes. From a long-term perspective, however, landfills should be considered as a means of storage, not disposal. The bulk of the buried wastes now being placed in landfills could be removed in the future and treated to reduce or eliminate hazardous properties.

The state-of-the-art "secure" chemical waste landfill is an engineered burial site with a nearly impermeable set of liners between the waste and the surrounding environment. The liners are usually composed of several feet of low-permeability clays, one or more synthetic plastic liners, or both. Above and between the liners a leachate collection system collects any liquids that move downward through the landfill. Most of the leachate comes from rain that infiltrates the cover material at the landfill. Some of the leachate is produced by the liquids disposed of in the landfill itself. This leachate is pumped out and treated.

The typical chemical waste landfill is comprised of individual cells. These cells are filled in sequence to keep the number of active uncovered portions to a minimum. Cells may be designated for specific waste types. Accurate maps must be kept in case leaks develop to facilitate possible waste removal.

A completely filled landfill is covered by a cap of clay, which is sometimes used in tandem with a synthetic liner. The top of the clay is sculpted and graded to prevent any accumulation of rain or snow. Shallow-rooted grasses or shrubs are planted to prevent erosion—a significant problem during the years after closure of any landfill.

Landfills offer the potential for environmental problems in two areas.

Siting

- poor soil and geological conditions
- proximity to ground water and surface water
- a high-density population nearby
- inadequate, accident-prone transportation routes (traffic congestion and narrow, winding highways).

Construction and operation

- low-quality construction materials
- improper or inadequate design
- inadequate operating and maintenance procedures.

Figure 5-6 illustrates the construction of a landfill that meets minimum technology standards. Landfills have the potential for producing adverse environmental impacts along all of the pathways described in the U.S. EPA's HRS. Exposure along any one of these pathways could endanger human health and lower environmental quality. Ground water quality has been the focus for much of the public's concern. Rarely have landfills been able to completely contain all wastes or to prevent some loss of leachate. A new type of landfill—the above-ground vault—may be able to completely keep its contents away from the surrounding environment in some circumstances, however.

Surface Impoundments. Closely associated with disposal in landfills is the practice of disposal of liquids, sludges, and slurries in surface impoundments. Surface impoundments, known as pits, ponds, and lagoons, store and dispose of wastes as well as treat them. Surface impoundments have been used by industry and municipalities for many years, but usage certainly increased during the 1960s and 1970s. U.S. EPA conducted the most comprehensive survey of the quantity and quality of surface impoundments from 1978 to 1980, before much of the strong concern over ground water contamination

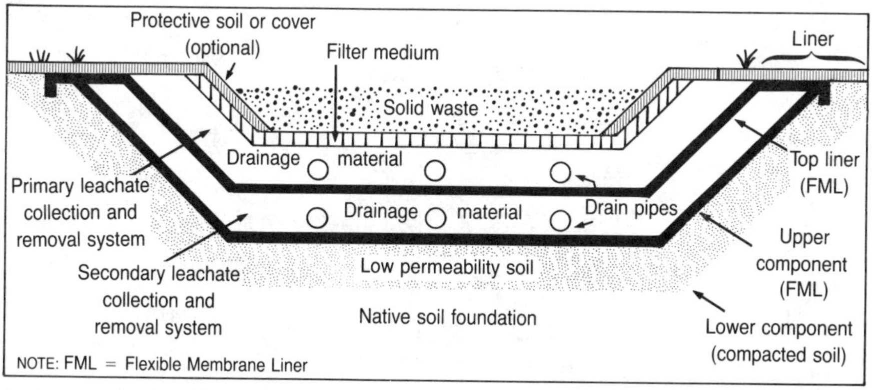

FIGURE 5-6. *Schematic of an FML/Composite Double Liner System for a Landfill*

SOURCE: U.S. EPA, "Draft Minimum Technology Guidance on Double Liner Systems for Landfills and Surface Impoundments—Design, Construction, and Operation" (May 24, 1985).

arose.[8] The data were collected before specific hazardous wastes (as defined by RCRA) were listed by the information sources that the agency used. The following survey findings are relevant.

- The survey identified more than 180,000 active surface impoundments (65,000 were oil and gas brine pits).
- Over 98 percent of the sites were in areas with drinking water within one mile or where the water contains less than 19,000 parts per million total dissolved solids and is therefore a potential source for drinking water in the future.
- The survey revealed that 70 percent of the surface impoundments were unlined.
- From 50 percent to 70 percent were in areas where contaminants will travel quickly through the ground.
- Approximately 15 percent of all sites (30 percent in the industrial category) contained waste considered hazardous under the Surface Impoundment Assessment (SIA) hazard rating system.

Deep-Well Injection. High volumes of liquid wastes are injected into deep disposal wells in several sections of the country. They are most common in Texas, Louisiana, and the Midwest. But New Jersey, California, and other states have their share. Since the 1930s, thousands of wells have received petroleum brines, sewage, stormwater runoff, and aqueous chemical wastes. Brines comprise the largest category of wastes in deep wells.

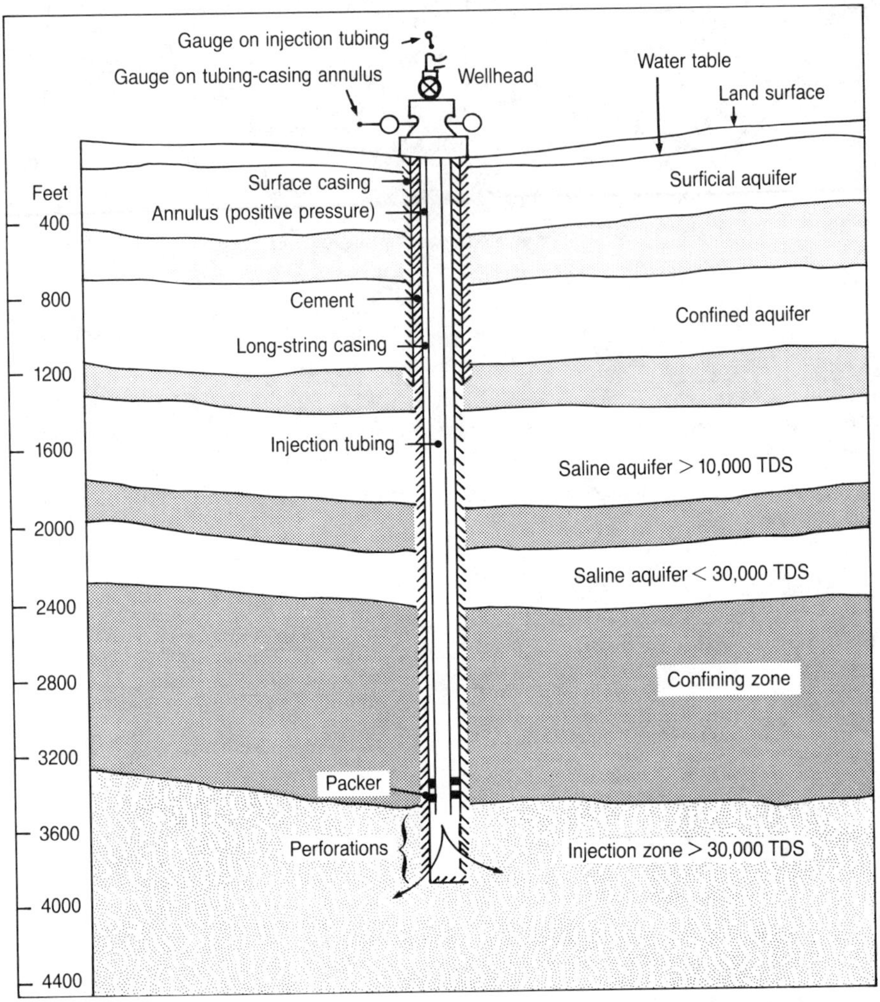

FIGURE 5-7. *Ideal Injection Well and Site*

SOURCE: U.S. EPA, Office of Drinking Water, "Report to Congress on Injection of Hazardous Waste," EPA 570/9-85-003 (July 1985).

Deep wells put wastes at depths from 1000 to 6000 feet. The wastes are injected into porous rocks such as sandstone, limestone, or dolomites. Figure 5-7 illustrates U.S. EPA's notion of an ideal deep-well site. The formation into which a particular well injects wastes should meet the following criteria.

- Ground water should be of very low quality in terms of potability or at a great depth in relation to useable shallower aquifers.

- One or more impermeable layers should separate injected wastes from shallower ground water.
- The surrounding formation should have sufficient porosity to accept large volumes of waste.
- The deep wells should be in a stable geologic area in which seismic and tectonic activities and any resulting faults, fractures, or discontinuous layers are at a minimum. (Examples of such areas are along the Gulf Coast and in the upper Midwestern states.)

The Safe Drinking Water Act stipulates that the Underground Injection Control (UIC) program regulate underground injection. Under the hazardous waste management permitting system, RCRA also regulates injection into deep wells. Attempts to coordinate these two federal programs are ongoing.

The major causes of environmental problems associated with deep well injection are poor construction methods and improper operational techniques. Operational problems usually occur during transportation of wastes to the well or during unloading and transfer of the wastes. Careful handling and constant monitoring of all operations can help prevent almost all losses. If no wastes are spilled, then the uppermost aquifers (where drinking water might be drawn from) are not usually contaminated.

Ocean Disposal. Until the early 1970s near-shore areas were commonly used as dumping sites for several types of hazardous wastes. With the passage of the Marine Protection, Research, and Sanctuaries Act of 1972 and the regulatory system for waste discharges within territorial seas established by CWA, the amount of disposal at sea has reduced drastically.

At present, the wastes dumped into the ocean fall into one of the following categories:

- dredge spoils containing small amounts of hazardous materials
- sewage sludge from several northeastern waste-water treatment plants (Sewage usually contains varying amounts of toxic or pathogenic materials.)
- treated municipal and industrial waste waters, which travel through pipelines to ocean outfalls
- acid wastes that are usually neutralized quickly by seawater.

Although ocean disposal has been reduced, it is likely that municipal sewage sludge and dredge spoils will be dumped for several more years. New plants are not allowed to use ocean disposal.

Above-Ground Vaults. One variant of the "secure" landfill is the above-ground vault. Vaults freestanding on a plain or burrowed into a hillside should be well above local water tables. Because the roof remains intact, the vaults should be relatively immune from leachate problems. The concrete roof and walls of such structures may be more impervious to the elements than the liner and clay cap of a landfill cell. In addition, leaks from above-ground vaults would probably be more obvious and damages more easily repairable than leaks from a landfill. Finally, the design of the vaults could allow easier retrieval of wastes in the event of a leak or some other unforeseen occurrence.

EMERGING TECHNOLOGIES

As the trend of banning certain wastes from landfills or requiring treatment prior to land disposal continues, industry must develop new technologies to handle wastes. The remainder of this chapter describes some of the emerging technologies that will shape the future of waste management. Many of these technologies are still in the development stage; others are mature technologies that now offer better waste "destruction" but may not yet be competitive in terms of cost.

Pyrolysis. Pyrolysis has been used for several years for treating nonhazardous waste. Pyrolysis does not actually burn wastes. The wastes are "roasted," subjected to high temperatures in an environment with insufficient oxygen to support combustion. Pyrolysis has been used to handle process-specific wastes rather than the broad range of mixed wastes destroyed by more common incineration technologies.

Plasma Arcs. Solid or liquid wastes are pyrolyzed into combustible gases by exposure to a plasma gas. The super-hot plasma gas results from exposure to high-energy radiation and has localized temperatures ranging from 50,000° F to 90,000° F. The plasma arc can break up the molecular structure of the waste without any traditional chemical reactions at all. This process is compact, so the development of mobile units is likely. The ability to move the unit to where the waste is generated could eliminate transportation problems. Because of the extremely high temperatures involved and the low volume of waste input, the formation of hazardous interim products is unlikely. Research is also exploring the use of lasers in association with plasma-arc technology.

High-Temperature Fluid Walls. Liquid or solid wastes pass down a cylinder of materials tolerant to high temperatures, and are subjected to electrically generated heat of up to 4000° F. Heat transfers to the waste through an inert gas, which protects the cylinder wall. Fluid-wall technology may enhance pyrolysis processes in regard to the disposal of organics and wastes containing some metals.

Molten Salts. The molten-salt "incineration" process is similar to the plasma arc furnace and is sometimes likened to fluidized-bed systems, but it operates at much lower temperatures—from 1300° F to 1800° F. Molten salts such as sodium carbonate act as catalysts and serve as a heat-transfer medium. The molten salt traps ash and acidic gases, so levels of particulate emissions are low. Demonstrations of molten-salt technology show that it is capable of handling a variety of liquids and slurries, especially PCBs and other chlorinated hydrocarbon wastes. Wastes with high content of ash or wastes that are high in chlorine (such as some plastics and salts) are not as suitable for this process.

Supercritical Water. Supercritical water is water at temperatures above 374° F and pressures above 218 atmospheres. The process is based on the fact that many organic compounds and many gases dissolve in supercritical water. The process is effective in treating aqueous streams of dilute chlorinated organic wastes. Organics can be rapidly oxidized and converted to carbon dioxide and water. Inorganics will separate out as solid salts that can be reused or disposed. The supercritical-water process also has destroyed PCBs and some pesticides during pilot demonstrations of the technology.

Wet Oxidation. Wet oxidation reactions are similar to those that occur in combustion, but they occur in the liquid state. They have been able to destroy cyanides, phenols, and nonhalogenated pesticides. Reactions take place at temperatures ranging from 430° F to 660° F at pressures from 1000 to 2000 pounds per square inch (70 to 140 atmospheres). The process can treat very dilute aqueous wastes such as those produced by the petrochemical industries. The process is also applicable to highly toxic wastes prior to treatment in biological plants and to nonchlorinated organics. Versions of this process are used at many sewage plants for sludge destruction; it has been commercialized since the mid-1960s.

Ocean Incineration. Ocean incineration has been used commercially off the European coast for 15 years. Several U.S. companies have also built seagoing incinerator ships to burn a range of solid and liquid organic wastes. In most cases a liquid-injection incinerator or rotary kiln on the ship operates at temperatures ranging from 1200° F to 3000° F. Residence time varies from several seconds to several hours depending upon the waste type. Experimenters performed test burns on PCBs and several herbicides since 1974, but full-scale commercial operation now awaits U.S. EPA regulation and permitting.

Seagoing incinerators are controversial because they contain less pollution-control equipment than would be required on land. Their stacks are also very short. Proponents claim that removal of acidic gaseous emissions is unnecessary since gases disperse over the ocean where seawater neutralizes them. Proposed incineration would take place more than a hundred miles from the shoreline and populated areas. Still, concern regarding potential adverse impacts on the ocean ecosystem remains. In addition, on- and off-shore handling—transportation, storage, loading, and unloading—pose significant health and safety concerns.

Chemical Dechlorination. An emerging group of chemical technologies strips chlorine ions from highly chlorinated compounds such as PCBs and pesticides to produce a less hazardous residue.

Some of these technologies have portable treatment units and the processes produce no air emission. Chemical dechlorination has been effective in removing most PCBs from transformer oils. The oils can be reused.

CHAPTER SIX
How to Choose:
Tools for Evaluating Facilities

This chapter reviews various approaches for evaluating hazardous waste facilities. The circumstances and the nature of the user largely determine which approach is best. A facility selection method should:

- apply to real-world, site-specific evaluations
- establish a consistent protocol for multiple applications
- outline areas critical to the inquiry
- differentiate the relative significance of the critical areas
- synthesize data
- assess overall relative benefits and risks
- compare and rank options.

Regulatory efforts under RCRA and its amendments perform many of these functions, but government regulation is not intended to mimic the consumer. Rather, the government merely establishes minimum standards for the marketplace. Comparative analysis is left to the customer—the waste generator or the public official.

RELEASE RANKING SYSTEMS

Release ranking systems can provide an adequate mechanism for the evaluation of a facility's site, management, and technology. These systems are based on standard probability theory. They evaluate indicators of probability rather than calculating actual probabilities, however. Release ranking systems are fundamentally comparative in that they do not estimate absolute risks; rather, they provide an estimation of the relative potential for damage to human health or the environment.

Release ranking systems would be insufficient for estimating the adequacy of a waste facility's insurance coverage because they do not estimate actual probabilities. But so many variables figure into a full-scale hazard analysis that associated costs are often prohibitive. Release ranking systems split the difference. They provide a relatively cost-effective method for screening, ranking, and selecting a hazardous waste management contractor. In short, they can be a practical site-specific tool.

If, for example, the wastes buried in a landfill are within the highest seasonal level of the saturated zone (water table), then there is a higher probability of ground water contamination than in a situation where more than 150 feet lie between the lowest level of waste disposal and the highest seasonal level of the saturated zone. Thus, a landfill with zero distance between ground water and the hazardous waste would be assigned a value, 3, which represents the greatest hazard level. Where greater depths separate the wastes from the ground water, the assigned value would decrease proportionally to 0.

Unfortunately, abstraction introduces errors. The purpose for which a system is used should determine how much one is willing to pay for accuracy. In a world with imperfect data sources and imperfect analytical tools, one should consider human health and environment as well as reasonable costs.

Release ranking systems perform all seven tasks outlined at the beginning of this section. Therefore, CEP chose to use a release ranking system to evaluate ten facilities owned by the eight largest commercial hazardous waste management firms. CEP modified the system to compensate for certain weaknesses, however.

CEP chose U.S. EPA's HRS from among several release ranking systems currently in use. CEP judged that HRS represented the national standard. Certain states, such as Michigan, have systems that are tailored to special state reporting requirements and use slightly different analytical models. All are fundamentally similar to the HRS in assigning numerical values to technical information.[1] Furthermore, the HRS is used to make decisions of national importance in selecting sites for the Superfund NPL.

U.S. EPA'S HAZARD RANKING SYSTEM

To accomplish the urgent task of deciding which uncontrolled hazardous waste sites to clean up, U.S. EPA funded the design of an "analytical" methodology for comparing environmental impact evaluations of uncontrolled hazardous waste sites. The goal was to develop criteria that EPA could apply to all sites. The result was the Hazard Ranking System (HRS), a mathematical model that assigns risk values to:

- the hazard of the material at a site
- the adequacy of containment
- the characteristics of routes through which the material can escape
- the magnitude and sensitivity of potential targets that could be adversely affected by exposure to the waste material.

By associating numerical values with a broad range of variables that affect the hazard posed by a site, U.S. EPA derives a final score to compare one site with any other. The HRS does not quantify the absolute probability of harm from a facility or the potential absolute magnitude of impact; rather, it estimates relative risks. The HRS is an effective tool for making decisions with limited information. Most important for its legislated purpose, the system provides a consistent method for ranking sites.

The final HRS score is based upon three potential routes by which hazardous wastes can migrate from a site and cause contamination or exposure:

- ground water
- surface water
- air.

The possibilities of fire and explosion or injury by direct contact at the site are considered separately because these hazards usually do not require Superfund remedial measures. Simpler removal actions suffice. Table 6-1 provides a comprehensive list of the HRS rating factors. Because HRS is designed for screening sites about which little may be known, the model has limitations in evaluating active facilities. The next sections of this chapter address these concerns. Chapter 8 outlines CEP's modifications to the HRS for use in evaluating active facilities.

How U.S. EPA Applies the HRS

U.S. EPA uses HRS to designate a site for cleanup. U.S. EPA awards contracts to remedial action consulting firms and active hazardous waste management facilities to dispose of the wastes, usually in landfills and sometimes by incineration or other forms of treatment. Landfills often offer the most practical means of disposal because the wastes tend to be mixed with soil making treatment more difficult.

U.S. EPA does not use a predefined model to determine consistently the type of remediation a Superfund site requires. In one case EPA may decide to remove the wastes; in another case EPA may improve containment at the same site. U.S. EPA must also determine the extent of remediation needed to minimize the potential for environmental harm (In other words, how clean is clean?) or the adequacy of the facility to which the materials may be moved. To make these determinations U.S. EPA usually hires remedial action consulting firms to conduct a case-specific Remedial Investigation/Feasibility Study (RI/FS). These studies are diagnostic and cost-benefit comparisons of alternative strategies for remedying an uncontrolled site.

Finally, the decision maker performs a detailed analysis of alternatives and selects the remedy. These evaluations are intensive and in most cases are far

TABLE 6-1. *Hazard Ranking System Rating Factors*

Hazard Mode	**Factor Category**	**Factors**		
		Ground Water Route	*Surface Water Route*	*Air Route*
Migration	Route characteristics	Depth to aquifer of concern Net precipitation Permeability of unsaturated zone Physical states	Facility slope and intervening terrain One-year 24-hour rainfall Distance to nearest surface water Physical state	
	Containment	Containment	Containment	
	Waste characteristics	Toxicity/persistence Hazardous waste quantity	Toxicity/persistence Hazardous waste quantity	Reactivity/incompatibility Toxicity Hazardous waste quantity
	Targets	Ground water use Distance to nearest well/population served	Surface water use Distance to sensitive environment Population served/distance to water intake downstream	Land use Population within 4-mile radius Distance to sensitive environment

Hazard Mode	Factor Category	Factors
Fire and explosion	Containment	Containment
	Waste characteristics	Direct evidence Ignitability Reactivity Incompatability Hazardous waste quantity
	Targets	Distance to nearest population Distance to nearest building Distance to nearest sensitive environment Land use Population within 2-mile radius Number of buildings within 2-mile radius
Direct contact	Observed incident	Observed incident
	Accessibility	Accessibility of hazardous substances
	Containment	Containment
	Toxicity	Toxicity
	Targets	Population within 1-mile radius Distance to critical habitat

SOURCE: *United States Code of Federal Regulations*, Title 40, Part 300, Appendix A.

beyond the means of the average industrial user of an active RCRA facility.[2] Cost is an overriding criterion in the language of CERCLA. In the past this has meant, by Lee M. Thomas's own admission, that "some removals have been scaled down below the level needed to achieve a cost-effective response."[3] By that he meant that despite the money spent, not enough was committed to ensure a long-term cure.

Applying the HRS to Controlled and Uncontrolled Sites

Superfund has come to be identified with the cleanup or control of wastes in abandoned sites. If a private party is prepared to pay for remedial action, there is no need for the federal government to do so. In cases in which a responsible private party cannot be found at the time, the president of the United States is authorized to use Superfund to pay for remedial action at an uncontrolled site if "such assistance will not otherwise be provided on a timely basis."[4] For this reason the term *uncontrolled site* has come to be equated with abandoned dumps for which responsible parties are difficult to identify. But closed and abandoned sites can be controlled sites if they are well contained. Sites of this kind do not need a Superfund cleanup action. Active facilities, on the other hand, not only can become uncontrolled after they are closed, but they can be uncontrolled while operating.

If an active facility is not managed properly, it can cause the same amount of damage as an abandoned dump. U.S. EPA's finding in 1984 that 82 percent of active facilities undergoing assessment of ground water contamination may not have been in compliance with federal monitoring requirements illuminates the degree to which active facilities can be uncontrolled.

U.S. EPA has occasionally evaluated active facilities using the HRS, although after RCRA was in place, most of these operating sites accepted nontoxic solid wastes only—not hazardous wastes.

U.S. EPA Region 9 applied HRS to the BKK facility in West Covina, California, for example. On June 28, 1984 the administrator of Region 9 summarized the evaluation in a memo to Lee M. Thomas. He wrote that "documented ground water and air releases [of vinyl chloride] beyond the 'facility boundaries' " led to a score of 48.24, a rating well above the minimum needed to qualify for Superfund cleanup. Mitre Corporation (which played a major role in designing the HRS in the first place) performed a quality-assurance check and found the reported score to be valid. EPA decided not to place the site on the NPL. The reasoning had nothing to do with HRS limitations, however. Instead, the June 28 memo argued against placing the site on the NPL for the following reasons:

> Because similar air releases may be occurring at other RCRA facilities, listing the BKK facility based on an air release may set a precedent involving the listing

> of many other RCRA facilities. . . . We also suggest that the policy encourage the exhaustion of RCRA and CERCLA enforcement remedies before listing of RCRA facilities.[5]

One month after the dispatch of the memo, 19 families had to be evacuated from their homes because of releases of explosive methane gas from the 540-acre landfill.[6]

WEAKNESSES IN THE APPLICATION OF HRS TO ACTIVE FACILITIES

The Time Factor in HRS Ratings

Thus, HRS can and has been used to evaluate active sites. CEP found the HRS to be a useful tool for comparing active facilities in terms of three general variables: management, technology, and site. The major difference between evaluating active facilities and assessing closed or abandoned sites is that the operating sites change in ways that unattended closed sites do not. The owners and operators of a facility can change its course at any time by applying technology and management skills. The HRS, in contrast, takes a snapshot of a site at one moment in time and quantifies hazard relative to snapshots of other sites. The HRS is vague in reference to the length of time it portrays. Conditions that affect the hazard level of both active and inactive sites do change, however.

Management Performance and the HRS

Management performance at active facilities is perhaps the most difficult variable to quantify. Part B of the RCRA permit application requires pages of material outlining management guidelines for personnel training, contingency planning, quality assurance, preparedness and prevention, security provisions, inspection procedures, laboratory analysis, and so forth. The information the applicant presents in Part B is useful to a degree. It certifies in writing that facility management understands federal requirements. But a well-written permit application in itself does not reveal the quality of an operation. Actions and events—not words—tell whether a site is managed competently.

The HRS accounts for management practices in its hazard score as causes of releases or as factors affecting containment. If a site scores poorly using HRS, it is an indication that remedial action must be taken. The higher the score, the more serious the need for corrective measures. One weakness of the HRS for ranking operating sites is that companies improve their operations at different rates. Nevertheless, a poor ranking raises the question: How did the facility get into this condition in the first place? The rankings of the sites can reveal the relative, if not absolute, historical management performance at a hazardous waste facility. If the site has not changed hands, the rankings may be valid for current management as well.

How the HRS Rates Technology

HRS ranks the hazard levels associated with less sophisticated disposal methods, such as whether a landfill or surface impoundment has a relatively impermeable liner and if there is a system to collect leachate. Because there is no consideration of containment in the air route section, however, the HRS does not distinguish between an incinerator with a scrubber and one without or between a covered and open mixing pit. All of these examples of different technologies represent variations within waste management processes and not distinctions among processes. CEP had to correct for these as well as other shortcomings of HRS when applying it to active facilities (see chapter 8).

In the long run, does a landfill that meets current requirements by having a double liner and a leachate collection system pose a hazard similar to a state-of-the-art incinerator? During their operating lives—which could vary by a few dozen years—the hazards they pose as ranked by the HRS would be indistinguishable if management prevents releases, if the quantity and hazardous characteristics of the wastes are the same, and if their siting parameters are identical. But once the sites' operations cease, there are clear differences in hazard potential that HRS scores ignore. A closed landfill, no matter how well constructed, probably contains ignitable, corrosive, reactive, acutely hazardous, or toxic wastes. And recall that, if untreated, wastes containing heavy metals can remain toxic permanently. But the owner of the facility is responsible for postclosure care for only 30 years after operations cease.

If, prior to disposal, all wastes are treated to render them less hazardous, then both the present and long-term hazards posed by the disposal site are reduced. HRS accounts for this variance with the waste characteristics score. But unless a site reduces the toxicity and persistence of all highly toxic and persistent wastes prior to disposal, the HRS does not reflect any amount of treatment that the facility conducts; it scores only the most hazardous materials. The waste characteristics score for most operating commercial hazardous waste landfills is the highest possible because these facilities store a great quantity and wide range of wastes. The significance of the inability of the HRS to detect the degree of treatment at a site is muted by three observations. First, because most current hazardous waste disposal facilities began operations before any significant treatment technologies were implemented, their landfills are packed with precisely those wastes that the HRS rightfully reflects as posing significant hazard. Second, most facilities are currently attempting to develop and implement treatment processes—business will depend on it as U.S. EPA activates landfill restrictions through 1989. Emphasizing current treatment capacity in a site's score could reduce the durability of the ranking to one year or less. After that time the inventory of available treatment options could be outdated. A site's record of introducing new technologies is, nevertheless, an important component of a site's overall attractiveness for which the HRS does not account. Third, most variables affecting the hazard level of an active facility that only treats hazardous waste are reflected in the HRS score. After closure the variance in the hazard level posed by incinerators would hold true for other treatment facilities—assuming no wastes are left on site after closure.[7]

During operation the HRS does not include the hazard characteristics of additive materials stored for use in chemical and physical treatment processes because these materials are not wastes. Some of this hazard is reflected by the containment score, however. For example, the HRS scores the integrity of treatment vessels, the capacity of dikes, and so on. The ignitability and reactivity scores of the waste characteristics section reflect the hazards that result from physical and chemical reactions promoted by the treatment processes.

The technology used at a site has impact on one factor that is crucial to the hazard level of an active facility but nonexistent at a closed site—transportation. Vehicles that bring wastes to an active land disposal facility usually leave empty. If a site incinerates, treats, or recycles wastes, however, vehicles may leave with a load. If a flammable waste leaves a treatment facility as a cleaner but equally flammable product, then the facility could pose greater transportation hazards than a landfill where the material is rendered nonflammable by solidification and buried. If a treatment facility renders a waste legally nonhazardous but creates an inert residue that must be disposed of off site, the hazard is the same as if the truck left empty or is even less of a hazard. Even though foodstuffs should not be loaded into the same trailer as hazardous materials, residents observing trucks leaving landfills in farmland areas often report witnessing such an event. Therefore, an important factor in assessing transport hazards is whether a facility has a truck wash, another technological variable for which the HRS does not account.

HRS Ratings and the Site

The characteristics of the site also affect the hazard of a facility. Site parameters affect the hazard level of both active and closed sites similarly; therefore, the HRS evaluates sites more effectively than it evaluates management or technology. Whereas most management and technology variables are dependent on time (and thus are unique to active facilities), many prominent features of the site do not alter radically. Over time, however, marginal environmental and demographic changes do occur, and these changes affect the hazard level of both active and closed facilities. A school and a sewage line were built over the Love Canal site, for instance. These changes affected the site's imperviousness to water. The HRS distinguishes ranges of population density and distance. If the facility is located near a possible future water source, the system assigns greater potential hazard.

On the other hand, the hazard posed by transportation is affected even more by the site than by technology. The HRS does not account for this completely. Although the system does register the increased transportation hazard associated with proximity to population and surface water, the HRS does not account for other site-specific variables that significantly affect hazards: traffic density, speed limits, weather, and road conditions.

INHERENT DEFICIENCIES OF THE HRS

The HRS also has a number of inherent shortcomings that apply to evaluations of active and inactive sites. U.S. EPA did not rectify many of these deficiencies because the added expense undermines its legislated purpose, that is, to select sites that need remedial action swiftly and consistently. The HRS embodies a trade-off between data availability and accuracy. More complex models generally require more information, but a greater need for information is accompanied by a greater need for time and money to collect and prepare the information.

One should be aware of these shortcomings when considering a particular site. Theoretical shortcomings were presented in the general discussion of release ranking systems. A description of the most significant technical problems follows. Many of the problems outlined were first raised during public comment on the HRS; others came to light through CEP's independent use of the methodology in scoring sites.[8]

Waste Characteristics

The HRS does not associate specific waste streams with the quantity of that waste estimated at a site. This problem, first voiced by the coal industry, reflects the fact that coal-mining operations produce a large volume of waste that contains only a small percentage that is hazardous. U.S. EPA maintains, however, that the additional expense of determining the quantities of individual wastes are inordinate. The HRS continues to represent the hazards at a facility on the basis of the most hazardous compound.

Ground Water

Many believe that the HRS does not examine site-specific characteristics that affect ground water in sufficient detail. Critics list the failure of the HRS to address:

- the sorption capacity of soil and waste
- the hydraulic gradient of ground water
- the rate of recharge (how long it takes an aquifer to be replenished).

Hydrogeological conditions greatly affect both the direction and rate of ground water migration. Relatively distant persons "downstream" could be more seriously affected by contamination than closer users of the ground water who are "upstream." Instead of examining specific hydrogeological conditions, U.S. EPA established an arbitrary boundary. U.S. EPA defines the potential target of ground water contamination as the entire population served by the aquifer within a 3-mile radius of the hazardous waste.

Michigan's site assessment system, on the other hand, uses a boundary of 3 miles, but only for the ground water users within 90 degrees either side of the down gradient of the wastes. For HRS this distinction would cut an HRS target estimate in half if the population were evenly distributed. The distinction would have an erratic effect in the final score because the HRS groups population size by varied levels of magnitude (1 to 100 persons, 101 to 1000 persons, 1001 to 3000, 3001 to 10,000, and more than 10,000). A change from 100 to 50 persons or 200 to 101 would not affect the target score, but a change from 150 to 75 would affect it.

Many believe the HRS also fails to address

- the severity of ground water contamination.

HRS criteria establish that if an observed release has occurred, then there is a 100 percent probability that any material that is not adequately contained will leak. Thus, the waste with the highest hazard level is used to factor toxicity and persistence, even if it has not been released. This approach may overstate existing contamination; it maximizes calculated risk.

- If there are two aquifers beneath a site, and one is contaminated, the HRS can significantly understate the hazard.

If an unused or little-used shallow aquifer has been contaminated while a major source of water from a deeper aquifer shows no contaminants, the HRS user can score the ground water migration route in one of two ways. First, one can focus on the shallow water-bearing zone and identify it as the aquifer of concern. The HRS score would reflect an observed release. Because the aquifer is not presently a major water source, however, the total route score would show a relatively low hazard. The problem with this approach is that the deeper aquifer is not considered, even though contaminants could migrate from the upper to the lower permeable zones. The second approach identifies the major ground water source as the aquifer of concern, which in fact it usually is. In this case, however, there has been no observed release because contaminants were detected only in the shallow aquifer. Again the total route score would underestimate hazard.

To correct distortions resulting from the contamination of the less-used aquifer would require adjustment of three scoring factors: the depth from the lowest point of waste disposal to the uppermost level of the aquifer of concern, the permeability of the unsaturated zone (the approximate rate at which the waste will migrate to the aquifer of concern), and containment (the factors preventing a release). The HRS users' manual only accounts for the first of these adjustments (see figure 6-1) and for the permeability from ground surface to the top of the shallower aquifer.

HRS should alter the containment value if wastes have seeped into an upper aquifer, but the event is not considered an observed release into an aquifer of concern. If contaminants have already leaked from a site, it is absurd to score containment as posing low hazard, even if the facility has liners and a

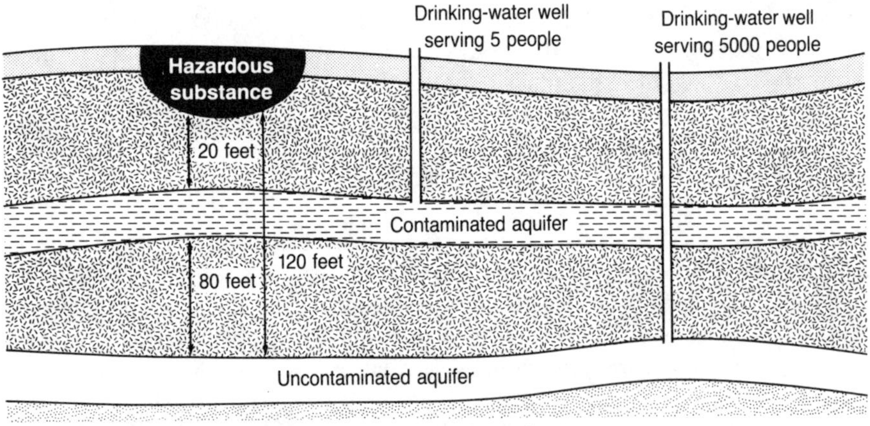

NOTE: If the upper aquifer is the aquifer of concern, then the depth to aquifer of concern is 20 feet and the population served is five persons. If the lower aquifer is of concern, the depth is 120 feet (assuming no known contamination below the indicated hazardous substance) and the population is 5000 persons. If the upper aquifer is contaminated and the lower aquifer is of concern, the depth would be 80 feet (vertical distance between hazardous substance and aquifer of concern, and the population would be 5000 persons.

FIGURE 6-1. *How HRS Deals with Depth to Aquifer of Concern*

SOURCE: *United States Code of Federal Regulations*, Title 40, Part 300, Appendix A.

leachate collection system. Once the contaminants have escaped, the geological formation between the two aquifers becomes the principal determinant of the degree of containment.

Of course, other geological factors, such as artesian pressure (which can cause ground water to flow from a lower to an upper aquifer) and sorption capacity might be included if the complexity of the model is increased. Practical considerations limit the complexity of the existing model, but CEP maintains that adjusting the scoring of containment, depth to aquifer of concern, and permeability would better reflect the hazard level without significantly adding complexity.

- Critics of HRS maintain that the system underestimates the hazard level posed by contamination of unused and unuseable aquifers.

Contamination of ground water that is unused and unuseable can cause serious hazards that HRS overlooks completely. The ground water migration route scores zero hazard if a facility contributes to the contamination of an unuseable water-bearing zone; the HRS scores no risk of exposure because the water is not used. Liquids are not the only forms of matter that travel through more permeable geologic formations, however. Gases travel too and often at much faster rates than liquids. Direct contact can result from solvents and gases that seep out of the ground, not just from drinking contaminated ground water. The evacuation of Love Canal occurred because of the threat of direct contact with migrating toxic gases; ground water under the site was both unused and unuseable. During the summer of 1984, another threat of migrating volatile organic chemicals caused the evacuation of residents near the BKK Landfill in West Covina, California.

- The HRS does not differentiate impact of contamination on various qualities of ground water.

Severe ground water contamination can permanently damage a high-yielding useable aquifer, yet the HRS can assign a low score to the hazard. In response to public comment, U.S. EPA differentiated aquifers that are in use from unused and unuseable aquifers. Still, if a low-yield, low-quality aquifer is currently in use, then the HRS rates the hazard level as greater than if an unused high-quality aquifer is damaged.

Surface Water

Critics have also registered complaints about HRS scoring in regard to surface water.

- HRS ranking treats pristine surface water that is currently unused identically to unuseable surface water.

Although the ground water pathway distinguishes between used and unused aquifers, the HRS does not make similar distinctions regarding surface water. The HRS only differentiates how a body of surface water is used—for drinking, irrigation, recreation, or industry, for example. If wetlands surround unused surface water, they affect the HRS score indirectly by affecting the factor that reflects a sensitive environment. Without surrounding wetlands unused surface water receives the lowest target score.

- The HRS does not consider the size and rate of flow of endangered surface water.

The size and rate of flow greatly affect the ability of a body of surface water to assimilate and dilute contaminants. These problems have parallels to those mentioned for ground water; however, data on surface water is easier to gather. The HRS, for example, only considers down-gradient users of surface water.

Air

In regard to air emissions, the HRS does not consider the containment of volatile materials. As a result

- The HRS does not reflect the probability of an air release unless a release has been observed in the past.

If officials have not observed an air release in the past, then the HRS considers the probability of a release in the future to be zero. The inverse also holds true: there is a 100 percent probability of additional air releases if one release has occurred at a site. (Note that the HRS does not observe emissions of methane. Methane is not photochemically reactive, it is not toxic in low concentrations, and it is common in nature.) U.S. EPA did not provide route characteristics for the air pathway because existing databases were inadequate to define the characteristics of air migration. But containment factors, such as the existence and permeability of covers, could be included in the HRS.

- Severity and frequency of air releases do not affect the HRS hazard score.

One very toxic emission can result in much greater damage than a series of smaller releases that add up to the same contaminant total. On the other hand, repeated gas releases are a nuisance to local residents and may cause long-term health problems. The severity and frequency of air releases are significant, but both are difficult to determine and require standardized, long-term monitoring.

For all migration routes, the HRS users' manual states, "the hazardous substance that was observed for scoring the release category may be different from the substance used to score waste characteristics," providing the most conservative risk assessment.[9] The HRS manual also tells the user to estimate the quantity of only those wastes that are incompletely contained. But there is no containment score for the air route! Thus, the value for "quantity" and "hazard level" of wastes does not reflect what could be released to the air.

- The HRS does not provide guidance for consistent air sampling or analytical protocol.

Source tests and monitoring of ambient air are prescribed in other U.S. EPA regulations, but most of the tests are not directly applicable to monitoring at landfills. Few hazardous waste management facilities in the United States have air-monitoring devices. When responsible local and state government agencies receive complaints about odor, the usual response is to send out an investigator who is experienced in distinguishing the smells of industrial chemicals to confirm or deny an air release. The notion remains that the human sense of smell is a better indicator of a release than technologies designed to detect air contaminants.

The HRS users' manual states that, "the only acceptable evidence of release for the air route is data that show levels of contaminants at or in the vicinity of the facility that significantly exceed background levels."[10] Strong identifiable odors could certainly be considered data indicating significant increase in background levels, but odors do not necessarily indicate a specific chemical. Furthermore, the guidance in the HRS users' manual for observing releases to ground water—the most detailed for any of the three migration routes—states that "qualitative evidence of release (an oily or otherwise objectionable taste or smell in well water, for example) constitutes direct evidence only if it can be confirmed that it results from a release at the facility in question."[11] This guidance suggests that odors that are officially confirmed constitute an observed release to the air. This logical interpretation of acceptable sampling and analytical protocol does not distinguish between a nuisance and an emission that threatens health or the environment. U.S. EPA also maintains that odors are not likely to provide reproducible evidence that would be considered conclusive in court, although this may not be the case in practice.

Scoring Biases and Omissions

There are general biases in overall scores of the HRS.

- The HRS total migration score is more sensitive to air releases than to releases by other routes.

Because of the "all-or-nothing" approach to air, the variance caused by an air release is greater than that caused by other releases. This bias may not reflect the relative hazard to the environment or public health posed by contaminants released to the air in comparison with releases to ground or surface waters. The philosophy is, however, that the existence of an air release at an abandoned facility often does represent a more serious threat than releases via other pathways. U.S. EPA reasons that if obvious air problems exist at a site long after activities have ceased, it is likely that there has been long-term human exposure.[12]

- The total migration score is more sensitive to hazards posed by sites in rural areas than in urban locations, despite the extra target populations in urban area.

A net effect of the shortcomings in all three migration routes is to assign hazards posed by rural sites a worse score than those posed by facilities in urban environments. Municipal wells generally tap much deeper aquifers than domestic rural wells. The use of aquifers for irrigation almost always occurs in rural areas. Thus, ground water contamination in an urban area tends not to affect any targets as defined in the HRS. Similarly, even though surface water bodies travel through urban environments, they are often used only for transportation.

Finally, air releases, fire and explosion, and direct contact pose the greatest threats to human health in urban areas because of dense populations. The model's inability to estimate the potential of an air release results in an estimation of zero hazard if a release has not occurred. Because fire and explosion and direct contact are not factored into the total migration route, they do not mathematically affect a site's hazard ranking either. U.S. EPA's reasoning for this exclusion is that fire, explosion, and direct contact hazards can usually be mitigated with removal or other clean-up actions that are easier and less costly than full-scale remediation.

- The HRS does not account for geological and climatic instability at a site.

Just as the HRS estimates average annual rainfall and 24-hour rainfall, so too could rough estimates of the distance to active faults, and the frequency of earthquakes, hurricanes, tornadoes, and volcanic activity contribute to evaluation of the hazard level of a site. U.S. EPA maintains that the likelihood of these events is not as significant as the other factors considered by the HRS.

CHECKLISTS

Companies that generate substantial volumes of hazardous waste often use checklists similar to the checklist that U.S. EPA uses to review permit applications and to ascertain compliance. The checklists usually cite the following areas:

- type and scale of operation
- character of area surrounding the plant
- existing permits
- waste analysis program
- security
- inspection requirements
- personnel training
- health and safety preparedness and accident prevention
- contingency plans and emergency procedures
- closure and postclosure plans
- recordkeeping arrangements
- compliance status.

Appendix A provides an example of a comprehensive checklist for use in evaluating active facilities or potential sites.

Although checklists are a relatively inexpensive means to establish a consistent review, they are only as good as the people who fill them out. Checklists limit questioning to a fixed set of variables. Consequently, less experienced users may overlook a strength or weakness that is significant at a site but absent from the checklist. In addition, checklists usually do not weigh the relative importance of each question or provide a quantifiable scoring feature. Checklists often allow only for yes or no answers. The characteristics of two sites may result in 20 positive and 20 negative answers for each. How does one choose in such a situation? Some large generators call for a consensus of company technical experts. Smaller companies may not have the technical expertise to evaluate the results.

Large generators are usually unwilling to tell less experienced generating companies which waste contractors to use. Many consider their checklists or other selection methods proprietary business information. Management may also fear libel suits from poorly rated facilities, and liability suits from users of recommended facilities that later proved to be problematic. According to one source in the Chemical Manufacturers Association, in 1984 the association delayed publication of a guide to solid waste management practices and abridged the section containing a detailed contractor selection checklist. Liability was the reason the source gave for the changes; companies might sue the association if the checklist led them to use a future Superfund site. Remember, even though a checklist may appear to cover all angles, the user, to a large degree, determines the quality of the application of the checklist.

In the wake of the disastrous leak of toxic fumes in Bhopal, India, the Chemical Manufacturers Association was having difficulty renegotiating a liability policy. The difficulty in getting environmental impairment liability insurance is symptomatic of the enormous difficulty in assessing the risks associated with the management of hazardous waste and of the enormity of potential damages. Interested insurers often require companies seeking pollution liability insurance to hire a team of consultants that includes experts in toxicology, chemistry, hydrogeology and engineering. The on-site risk analyses that such a team conducts can cost more than $20,000 per analysis.[13] These analyses include sophisticated analytical models based on data collected at the site.

FAULT TREES

Constructing a fault tree is a means that uses logic to organize complex realities to estimate the risk of personal injury resulting from an industrial operation. What the process calls initiating events can result in a number of different outcomes. The consequences are determined at chance nodes. Each potential outcome, or branch, is associated with a probability and a payoff. Using a fault tree a waste generator or manager can map out a pattern of potential alternative events that could lead to a liability claim. Faults are those events that may cause adverse effects. Unlike the checklist, a fault tree can offer a rough estimate of the chances and costs of a potential accident.

Figure 6-2 gives as an example a fault tree that depicts the possibility of human exposure via ground water contamination. The dissociation of a liquid hazardous waste from a solidifying agent constitutes an initiating event. There is a certain probability of whether or not the containment system will prevent the liquid waste from migrating. Relatively small costs (a payoff of -5) are accrued if the waste is contained. If the wastes leak, however, and the chain of events continues, the payoffs needed to correct the situation become more and more expensive. The insurer must make the largest payoff if the party is found liable.

Fault trees present the same limitation as a checklist: the user may overlook important factors. The fault tree provides no implicit method for identifying omissions. Furthermore, the probabilities associated with faults are difficult to assign because of the rarity of catastrophic events, the resultant lack of data, and the insufficiencies of hydrology, hydrogeology, and toxicology. From an insurer's perspective the problem of estimating potential losses is compounded by the uncertainty of court decisions regarding liability.[14]

HAZOP AND HAZAN

Fault trees can be incorporated into a Hazard and Operability Study (HAZOP) or a Hazard Analysis (HAZAN), two analytical techniques that minimize the possibility of omission by providing open-ended systems of inquiry into the constituent parts of hazards. HAZOP and HAZAN techniques go beyond check-

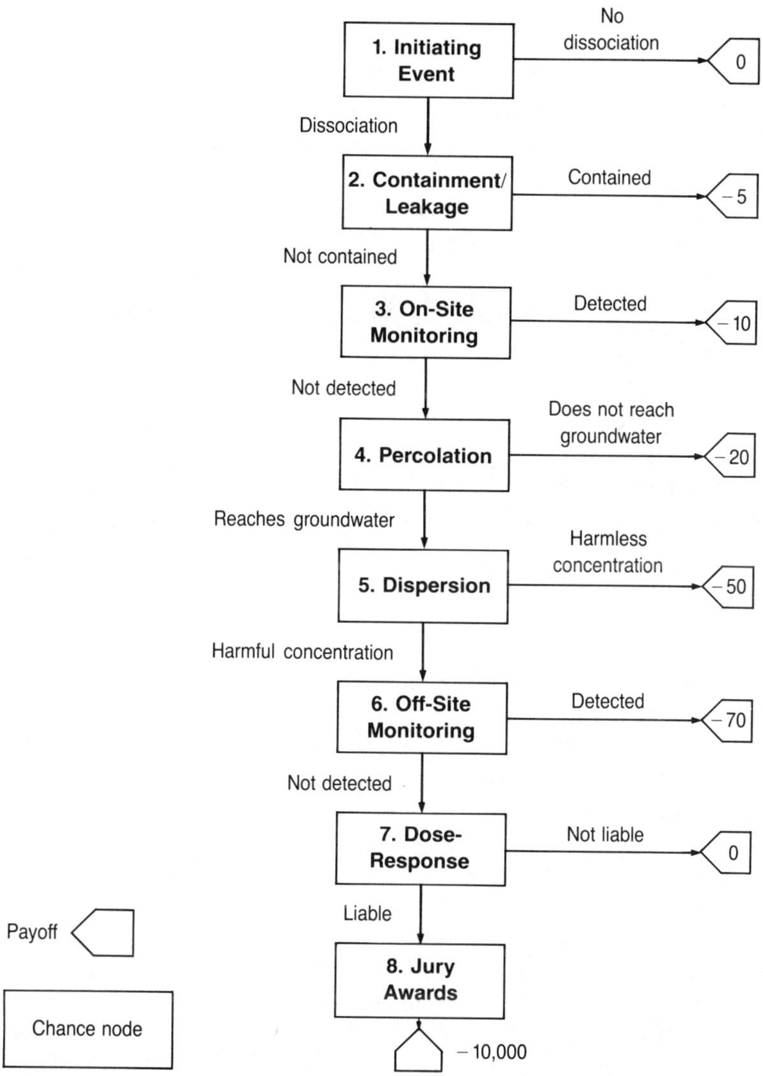

FIGURE 6-2. *Fault Tree That Traces the Potential for and Cost of Human Exposure to Ground Water Contamination*

ADAPTED FROM: Martin T. Katzman, *Chemical Catastrophes: Regulating Environmental Risk through Pollution Liability Insurance* (Homewood, Ill.: Richard D. Irwin, 1985), p. 117.

lists and economic risk assessments by providing an auditing process that aims at eliminating plant hazards. The frameworks for hazard assessment provided by HAZOP and HAZAN are possibly the most exhaustive that are currently available. These types of analyses tend to be more appropriate for facility management or public officials than outside auditors because in most

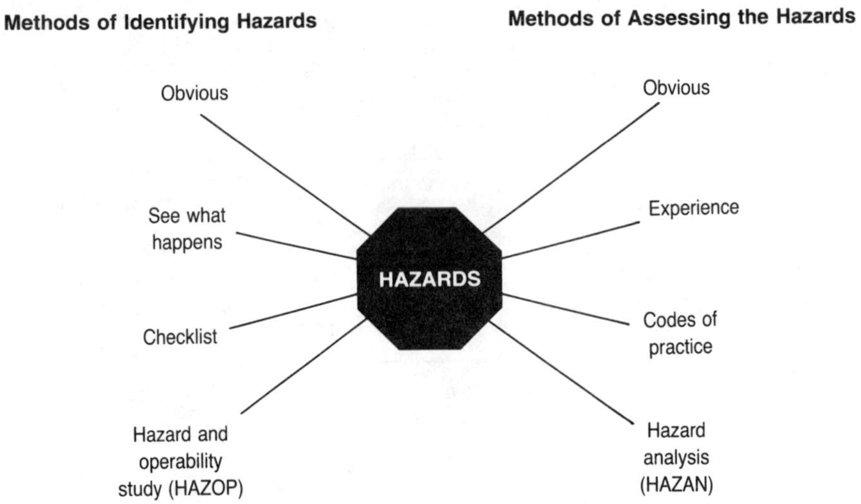

FIGURE 6-3. *Methods for Identifying and Assessing Plant Hazards*

ADAPTED FROM: Trevor A. Kletz, "Eliminating Potential Process Hazards," *Chemical Engineering* (New York, N.Y.: McGraw-Hill, April 1, 1985), p. 48.

cases only management and regulators have the authority to invest in or order improvements.

HAZOPs can be applied during the development of a plant design or to an existing facility. A HAZOP identifies hazards and obstacles to efficient operability. After a HAZOP identifies a problem, a HAZAN determines appropriate corrective measures (see figure 6-3).[15] Managers usually apply HAZOP and HAZAN techniques to chemical plants, but they can be equally useful for evaluating the processes used at chemical waste treatment and destruction facilities.

A HAZOP at an existing facility calls for a team. The team consists of the plant supervisor, the foreman, an engineer, an instrument manager, the process investigation manager, and an independent team leader who is a HAZOP expert. The team spends up to 3 hours asking questions about each plant unit—the still, the reactor, and the heater, for example. The team also looks at components that join these units—the pipelines, the pumps, and the valves. The method increases the likelihood that the participants will think of all possible ways in which hazards or operating problems might occur.

A series of guide words that indicate potential problems are applied to each plant item in turn (see table 6-2). Using each guide word, the team asks: Could this problem happen? If so, how? With what consequences? Are the consequences hazardous or do they prevent efficient operation? How can the deviation be prevented?

Following the HAZOP the team conducts a HAZAN, focusing on the ques-

TABLE 6-2. *Deviations Generated by Various Guide Words*

Guide Word	Possible Deviations
None	No forward flow when there should be, i.e., no flow or reverse flow.
More of	More of any relevant physical property than there should be, e.g., higher flow (rate or total quantity), higher temperature, higher pressure, higher viscosity, etc.
Less of	Less of any relevant physical property than there should be, e.g., lower flow (rate or total quantity), lower temperature, lower pressure, etc.
Part of	Composition of system different from what it should be, e.g., change in ratio of components, missing component, etc.
More than	More components present in the system than there should be, e.g., extra phase present (vapor, solid), impurities (air, water, acids, corrosion products), etc.
Other than	What else can happen apart from normal operation, e.g., start up, shutdown, uprating, low running, alternative operation mode, failure of plant services, maintenance, catalyst change, etc.

SOURCE: Trevor A. Kletz, "Eliminating Potential Process Hazards," *Chemical Engineering* (New York, N.Y.: McGraw-Hill, April 1, 1985), p. 49.

tion: Does the scale and probability of the consequence justify the expense of correction? To answer this the HAZAN comprises three steps:

1. How often?
   estimating how often the incident will occur
2. How big?
   estimating the consequences to employees, the public, plants, and profits
3. So what?
   comparing the results of step 1 and step 2 with a target or criterion to determine if management can ignore the hazard or if action to reduce the probability of occurrence or to minimize consequences is desirable.

Conducting a HAZOP or HAZAN can be time consuming and expensive. As Trevor A. Kletz, senior research fellow and professor at University of Technology, Loughborough, England, points out:

> These techniques assume that management is competent, and that the plant will be operated and maintained in the manner assumed by the design team and in accordance with good management and engineering practice. In particular, it is assumed that protective systems will be tested regularly and, when necessary, repaired promptly. If these assumptions are not true, then HAZOP and/or HAZAN are a waste of time. It is no use identifying hazards or estimating their probability if no one wants to do anything about them; it is no use installing trips and alarms if they are going to be ignored or not maintained.[16]

A potential customer of a facility cannot merely assume that the contractor's management is competent. The fact that the potential waste contractor has performed such analyses, however, may be an indication of management quality. On the other hand, a review of resulting documents may reveal potential risks. At the end of a HAZAN the decision maker is still left in a murky area of policy. The manager must set a target that defines what the institution considers an acceptable risk.

RCRA RISK-COST ANALYSIS

The RCRA Risk-Cost Analysis Model is an attempt by U.S. EPA's Office of Solid Waste (OSW) to design a tool that can help policy makers estimate the risk associated with different hazardous waste management practices. The model's fundamental strength is that it is a device for comparative analysis.

The RCRA Risk-Cost Analysis Model was developed under U.S. EPA contract by ICF, Inc., a consulting firm. The model is also known as the WET model, an acronym derived from the three factors that it comprises.

- W stands for 154 industrial wastes and includes the physical and chemical characteristics and quantities of each waste stream;
- E stands for the environments in which the wastes are generated, treated, and disposed. The WET model incorporates characterization of environment according to three-digit ZIP codes, 12 possible settings for evaluating human health risks, and seven settings for evaluating risks to ecological systems.
- T stands for the types of technologies used to treat, transport, store, and dispose of the waste. The model defines 15 treatment technologies, two types of transportation vehicles, five disposal methods, incineration, and four general sources of environmental release and corrective action.

Each possible combination of a specific waste, environment, and technology define a cell within a three-dimensional matrix. Each cell represents a specific waste management practice for which the model calculates the associated risks and costs. Figure 6-4 outlines the logic that the model applies to evaluate each cell.[17] The model can be updated with scientific data to determine optimal waste management practices with respect to risk and/or cost.[18]

The assignment of risks and costs requires extensive data on national waste generation, chemical toxicity, the costs and effectiveness of various technologies, characteristics of geographical locations, and so forth. Therefore, the model is also an extensive database.

During 1982 the Science Advisory Board of U.S. EPA, OTA of Congress, industry associations, environmental organizations, and the interested public

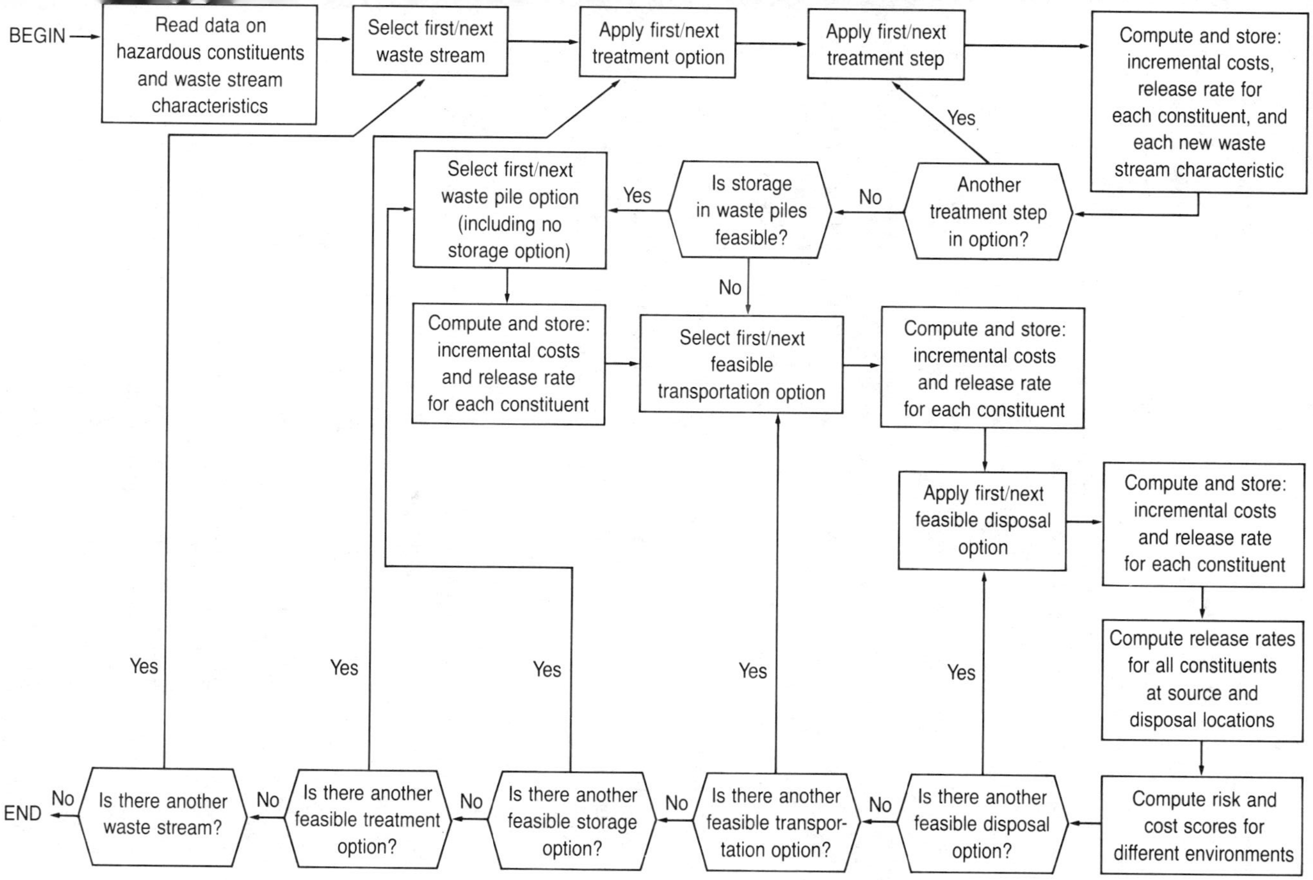

FIGURE 6-4. *WET Computing Diagram* SOURCE: ICF, Inc., *The RCRA Risk-Cost Analysis Model Phase III Report* (Washington, DC: U.S. EPA, March 1, 1984).

strongly criticized the draft model. The Chemical Manufacturers Association faulted the model's method of establishing guidelines for choosing technology. The association maintained that the model was particularly weak in defining technology for the plastics and synthetic fibers industries that would conform to standards of CWA.[19] OTA charged that the data used by the model was "too simplistic" for establishing regulations.[20] Environmental organizations charged that the model's definition of risk was not consistent with that used by the scientific community and that the model failed to adequately consider damage to the natural environment.[21] The model was improved in response to many of the objections.[22]

The Phase III version of the model (March 1984) did not meet with the same degree of criticism as the earlier draft. One fundamental problem appears to continue to plague WET, however, even after 4 years of effort and millions of dollars. The problem of applicability remained; what questions can the model answer? When U.S. EPA had to choose a model to evaluate land disposal bans, the agency chose a fate-and-transport model rather than WET. EPA rejected WET because of the model's insufficient data on the 450 waste streams that EPA had to consider banning from land disposal. The agency selected a simpler approach.[23]

U.S. EPA is still developing WET. In the future, for example, the model may be used to compare incineration to land disposal in order to ascertain whether land disposal restrictions reduce hazard or simply transfer it. Meanwhile WET has influenced other evaluation systems. OSW's Liner Location Risk Analysis Model, for example, is a tailored evaluation of landfills that is based on WET and uses waste stream characteristics and dose-response data from the WET model.

Eric Malés, U.S. EPA's supervisor of Phase III of the WET project, points out that the data currently available and the equations that WET currently employs are sufficiently imprecise to render the model unsuitable for use on a site-specific basis.[24]

SITE EVALUATION SYSTEMS

As of September 1984 approximately 24 states and a number of regional commissions had established programs and systematic criteria for selecting sites for hazardous waste management facilities.[25] These siting systems are often intended to apply to either proposed facilities or expansions to existing operations.

Many states have multistage review processes. The process begins with preliminary screening of an application for completeness, then for technical soundness, and finally for feasibility. Other states consider a set of general criteria or guidelines that give the reviewers substantial leeway in making their decision. Others synthesize a proposal's desirability with numerical values.

Environmental Resources Management, Inc., a consulting firm based in West Chester, Pennsylvania, designed a system of site evaluation for the Delaware River Basin Commission and for the New Jersey Department of Environmental Protection. The firm chose a screening and ranking system.[26] A proposed facility is evaluated against three levels of criteria that become more specific as a proposal passes to the higher levels. Table 6-3 outlines this siting system and Figure 6-5 provides a hypothetical application.

TABLE 6-3. *Criteria for Screening, Ranking, and Selecting Locations for Hazardous Waste Management Facilities*

Level 1 Criteria. Technically defensible criteria that identify and eliminate clearly unacceptable areas and in some instances that identify areas with higher potential for containing suitable sites including:

Coastal flood hazard areas; Coastal wetlands; Diabase or basalt bedrock areas; Anthracite mining areas; Public water supply watersheds; Aquifer/well yield; Critical recharge areas; Aquicludes; seismic risk zones; Natural areas of designated county, state, regional, or national significance.

Level 2 Criteria. Criteria such as land use, zoning, and infrastructure that identify candidate sites suitable for detailed screening, including:

Lands designated for industrial use; Sites of existing facilities; Dedicated lands in public trust; Arterial highways.

Level 3 Criteria. Technically based and detailed site-specific criteria that (a) exclude outright a site from further consideration, and (b) indicate general suitability in terms of "negative," "neutral," and "positive" ratings including:

Riverine flood hazard areas; Stream proximity and use; Stream flow and quality; Aquifer use; Ground water flow systems; Ground water quality; Geologic faults; Unconsolidated deposits; Bedrock; Areas of mineral development; Slope; Prime agricultural land; Soil permeability; Soil pH; Soil cation exchange capacity; Freshwater wetlands; Critical habitat for rare and endangered species; Nearshore commercial shellfish resources; Historic places; Air quality designation; Structures along transportation corridor; Transportation restrictions; Population density in vicinity of facility; Proximity to incompatible facilities/structures; Site ownership.

Ranking System: (a) Level 3 criteria ratings are determined for each factor at each site; (b) The number of "negative" and "positive" ratings are totaled for each site; (c) The sites are ranked in two lists, one list for the number of "negative" ratings and the other for the number of "positive" ratings; (d) The quartile (up to a maximum of 15 sites) having the least "negative" ratings and the most "positive" ratings are identified; (e) The identified sites from each list are combined, in no particular order, in a new list.

Final Site Selection: For the final list of suitable sites, field and laboratory investigations are conducted to verify data and to get more detail; public hearings are held to consider political and social concerns.

DERIVED FROM: Environmental Resource Management, Inc., "Technical Criteria for Identification and Screening of Sites for Hazardous Waste Facilities" (West Chester, PA: Delaware River Basin Commission and New Jersey Department of Environmental Protection, August 1981).

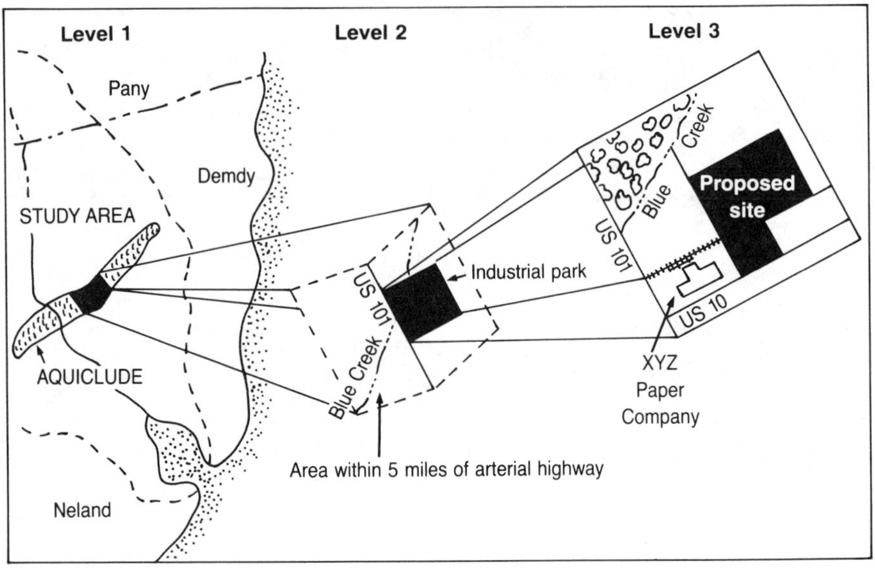

FIGURE 6-5. *Hypothetical Application of Siting Criteria*

ADAPTED FROM: Environmental Resource Management, Inc., "Technical Criteria for Identification and Screening of Sites for Hazardous Waste Facilities" (West Chester, Pa: Delaware River Basin Commission and New Jersey Department of Environmental Protection, August 1981).

Like checklists, screening and ranking systems limit the inquiry to outlined criteria. Public review of the findings, however, can help to identify missing elements. Another drawback arises because the systems are often oriented to assessing proposed facilities. As a result they often do not evaluate past management performance.

WHICH SYSTEM DID CEP CHOOSE?

All of the methods discussed in this chapter have strengths and weaknesses that define their usefulness. CEP chose a release ranking system because these systems can achieve the seven goals required of a comparative facility selection within a limited budget. Chapter 8 outlines the significant modifications that CEP implemented to conduct facility evaluations.

CASE STUDIES

CHAPTER SEVEN
The Big Eight: Corporate Comparisons

CEP chose eight commercial hazardous waste management companies for close review. These companies provide waste generators with off-site waste management services. They also provide transportation, consulting and laboratory analysis, field services for emergency response, and remedial cleanup. Some of them provide the most important element of all in the long term: research and development of new techniques for dealing with hazardous wastes.

The Big Eight firms are all major corporate entities; they are publicly owned with shares traded on the major stock exchanges. Their names are not household words, but they are known to the industries they serve. Their corporate structures continue to change—acquisitions and spin-offs abound. The eight companies dominate the industry and are visible to the media and to the public.

THE COMMERCIAL MARKET

The eight major commercial firms and their smaller competitors took in revenues of $1.2 billion in 1983 according to the Congressional Budget Office (CBO). Those revenues are nearly 20 percent of the total spent by generators for hazardous waste management for on-site and off-site management costs.[1] Estimates for cleanup under the Superfund program range from $10 billion to $100 billion, a substantial additional source of revenue for the commercial firms.[2]

In 1981 62 percent of all generators sent their wastes off site for treatment and disposal, and an additional 22 percent sent some wastes off site. U.S. EPA estimated the total number of generators at that time to be over 14,000. The 1984 RCRA amendments substantially expand the number of generators that must comply with its provisions, and U.S. EPA estimates an additional 175,000 companies will require hazardous waste management services as a result.

These newly regulated companies are potential customers for commercial hazardous waste managers (see chapter 4).

THE BIG EIGHT

CEP chose the eight companies from hundreds of commercial hazardous waste managers as the only ones that met the following criteria.

1. Each is publicly owned and has its headquarters in the United States. All must report to the United States Securities and Exchange Commission (U.S. SEC).
2. All provide commercial treatment treatment, storage, and disposal services that include incineration or land emplacement. These companies handle wastes generated on site, but more than 50 percent of the wastes they handle come from off-site generators.

In 1985 a survey compiled by Environmental Information, Ltd. listed 447 commercial TSDFs, 9 percent of the 5056 on-site and off-site TSDFs listed by U.S. EPA in May 1985.[3] Directly or through subsidiaries the Big Eight own and operate 84 active facilities—nearly 20 percent of those available to generators who choose to have their wastes shipped off site.

EVALUATION OF THE BIG EIGHT

CEP gathered comparative information on several aspects of corporate performance: financial viability, regulatory compliance, exposure to hazardous waste liability, corporate and public relations, and technological capability. CEP also evaluated at least one active company-owned or operated facility (see chapter 9).

From the wide range of information gathered, CEP chose certain parameters for its overall company evaluation. The parameters were based on:

1. relevance of information to company financial, technological, and overall performance
2. reliability of available information
3. availability of information for all companies.

CEP's evaluation, which led to company grades, is based on 21 parameters in seven categories. The information evaluated was the latest available to CEP. Chapter 10 discusses deficiencies and inconsistencies in the data. CEP submitted the data used in the evaluation to the companies for review, and the changes they made were incorporated. The companies did not receive CEP's methodology, opinions, or conclusions; the fact that the Big Eight reviewed CEP data in no way implies company acceptance of methodology or conclusions. In fact, some of the firms strongly objected to our approach. The reader can update the evaluation by substituting newer information as it becomes available.

Tables 7-1 through 7-7 present the data that CEP used to evaluate the Big Eight and specific facilities that the companies own. The discussion in the remainder of this chapter gives CEP's conclusions and shows how CEP used the data to arrive at its results.

Average Rankings and Grades

For each parameter CEP ranked the companies from 1 through 8 to represent poorest to best. CEP added the ranks for the parameters and divided the sum by the applicable number of parameters to obtain an average ranking.[4] Of 168 rankings only four were not applicable. USPCI, Inc., the highest-ranking company, had no documented leaking sites and thus no rank for ground water corrective action; the total of USPCI rankings was divided by 20 rather than 21. AMEC, the company scoring seventh, had three unranked items; CEP divided AMEC's total by 18 rather than 21.

Tables 7-1 and 7-2 show the ranking totals for each of the seven categories. The average of all rankings for each company was placed on a scale of 1 to 8. The scale was divided into 13 grades, F to A+. USPCI, Inc., the company with the highest average rank, achieves a B. It ranks at the top (rank 7 or 8) in 12 of the 21 parameters and lowest (rank 1 or 2) in only two. Low ranks in some parameters kept USPCI's average grade below an A. IU, the company with the lowest average rank, achieves a D. IU ranks lowest (rank 1 or 2) in 12 of the 21 parameters and highest (rank 7 or 8) in only two. Most of the eight companies ranked near the top of some categories but near the bottom of others; thus, their average rank fell in the C to C+ range.

The following table shows the grades of the Big Eight.

Group	*Company Name*	*CEP Grade*
1.	USPCI, Inc.	B
2.	International Technology Corp. (IT)	C+
	Environmental Systems Co. (ENSCO)	C+
	Rollins Environmental Services, Inc. (RES)	C+
3.	Browning–Ferris Industries, Inc. (BFI)	C
	Waste Management, Inc. (WMI)	C
4.	American Ecology, Inc. (AMEC)	D+
5.	IU International Corp.	D

TABLE 7-1. *The Big Eight: Summary of Company Rankings*[a]

Areas of Comparison	AMEC USEC	BFI CECOS	ENSCO same	IT same	IU ESI	RES[b]	USPCI same	WMI CWM
Corporate financial viability	20.0	25.0	25.0	24.0	10.0	25.0	26.0	26.0
Potential hazardous waste liabilities	4.5	9.5	21.5	16.0	15.5	18.0	11.0	12.0
Corporate and public relations	3.5	9.0	14.0	11.0	3.5	8.5	14.5	8.0
Compliance with government regulations	14.5	25.5	18.5	34.5	21.5	17.5	30.5	15.5
Ground-water monitoring	10.5	17.5	9.5	10.5	14.0	14.5	14.5	16.0
Technology	3.0	7.5	10.0	6.7	4.0	11.5	11.5	12.0
CEP facility evaluations	4.0	3.0	6.0	2.0	1.0	5.0	8.0	7.0
Total of ranks	60.0	97.0	104.5	104.5	69.5	100.0	116.0	96.5
Average rank[c]	3.33	4.62	4.98	4.98	3.31	4.76	5.80	4.60
Grade	D+	C	C+	C+	D	C+	B	C

Grade scale for average ranks

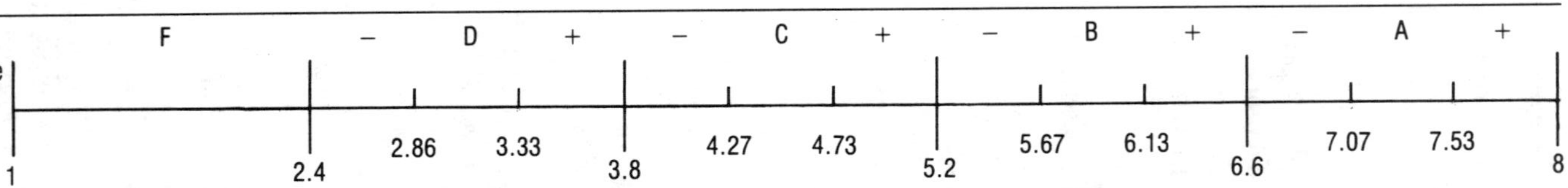

NOTE: Abbreviations of parent company and principal hazardous waste entity are listed in alphabetical order by parent.

a. Under each company are the sum of ranks for all parameters within each category. Larger numbers indicate higher overall ranks.

b. Rollins Environmental Services' subsidiaries operate its hazardous waste facilities, including: RES (LA), RES of LA, RES (NJ), and RES (TX).

c. Average ranks equal sum of ranks divided by number of parameters (21 total) less ne (not evaluated) and na (not applicable).

Current Revenues and Financial Viability

Table 7-3 presents the average annual growth rates for total revenues, pretax income, and total assets for 1980 to 1984. The growth of total revenues is used as an evaluation parameter. The other two growth rates are provided as information. Basic information about the parent corporations' financial condition is from the annual 10-K reports submitted to U.S. SEC as of August 1985.

The other values that CEP used as evaluation parameters are annual hazardous waste revenues, net income divided by average equity and total debt divided by one year's cash flow. Table 7-3 shows other amounts as information only. Comparative figures for total revenues, pretax income, provision for income taxes, and net income per common share are from the annual consolidated statement of income. Figures for hazardous waste revenues and hazardous waste pretax income are from the annual report or 10-K breakdown by industrial segment in notes from consolidated financial statements. For the four companies whose only business is hazardous waste management—ENSCO, IT, RES, and USPCI, Inc.—the amounts are the same.

CEP evaluated the financial viability of the eight parent companies and their hazardous waste subsidiaries. Financial viability is an indication of a company's ability to repair damage caused by its operations, its ability to finance facility repair or maintenance, and its ability to invest in new technology. In the CEP evaluation USPCI, Inc. and Waste Management, Inc. tie for highest rank. IU ranks lowest.

CEP used five standard measures of corporate health—standards that are applicable to service industries as a whole. CEP obtained information from the annual 10-K reports that the parent companies filed with the U.S. SEC as of August 1985. (The 10-K report is a summary of a public company's annual report in a format standardized by the SEC.) Table 7-3 shows the data that served as the basis for these ratios and numbers. The following list outlines the measures of corporate health that CEP used.

1. Average annual growth of total revenues—CEP found the percentage increase in total revenues from year to year over the 5-year period from 1980 to 1984 and averaged the four percentages.
2. The evaluation included consideration of annual hazardous waste revenues for the latest available fiscal year.
3. Return on average equity—CEP divided the ratio of net income for the most recent fiscal year by the average of total common stockholders' equity at the end of the most recent fiscal year and the end of the previous fiscal year.
4. Debt to cash flow—CEP divided total debt for the latest fiscal year by that year's cash flow, defined as the sum of net income, depreciation and amortization, and provision for deferred income taxes.

TABLE 7-2. *The Big Eight: Company Evaluations*[a]

Areas of Comparison	Type of Parameter[b]	AMEC USEC		BFI CECOS	
Corporate financial viability	t		20.0		25.0
average annual growth total revenues	%+	13%	2	14%	3
$ hazardous waste annual revenues	$+	8	1	92	6
$ net income/average equity	r+	.66	8	.21	5
$ total debt/one year's cash flow	r−	.00	8	.61	6
unqualified auditor's opinion	n+	0	1	1	5
Potential hazardous waste liabilities	t		4.5		9.5
# financial requirements inadeq/# adeqcy eval	r−	ne	ne	.3	3
$ HW revenues/HW land emplacement facilities	r+	2800	2	3417	4
# CERCLA sites owned or operated	n−	2	3	2	3
Corporate and public relations	t		3.5		9.0
# questions answered on CEP questionnaire	r+	.0	2	.1	3
# sites opposed/# sites evaluated	r−	1.0	2	.7	6
Compliance with government regulations	t		14.5		25.5
U.S. EPA Regulations					
# enforcement actions/# inspected facilities	r−	1.8	5	1.6	6
$ penalties/# inspected facilities	r−	.0	8	1.1	6
U.S. OSHA Regulations					
# citations/# inspected facilities	r−	ne	ne	3.1	5
$ penalties/# inspected facilities	r−	ne	ne	.6	3
U.S. DOT Regulations					
# HW spills/# active HW facilities	r−	7.5	2	.3	6
Ground-water monitoring	t		10.5		17.5
# monitoring inadequacies/# adequacies eval	r−	.6	3	.5	5
# docmntd leaking sites/# subject facilities	r−	.7	3	.2	7
# corrective actions/# leaking sites	r+	.5	5	.8	6
Technology	t		3.0		7.5
# treatment options/# active facilities	r+	.0	1	1.3	6
# advanced treatment options	n+	.0	2	.0	2
CEP facility evaluations	t	2.7	4.0	2.8	3.0
Average rank[d]			3.33		4.62

NOTE: Abbreviations of parent company and principal hazardous waste entity are listed in alphabetical order by parent.

a. Under each company, left column shows value used to determine ranks found in right column. Rounded mean is shown for equal ranks. ne—not evaluated; na—not applicable

b. n—number; r—ratio; t—total (sum of ranks in each category); %—percentage; $—millions of dollars; +—positive value (higher rank for larger value); −—negative values (higher rank for smaller value).

ENSCO same	IT same	IU ESI	RES[c]	USPCI same	WMI CWM
25.0	24.0	10.0	25.0	26.0	26.0
100% 8	29% 6	4% 1	22% 5	42% 7	19% 4
33 4	144 7	10 2	69 5	17 3	199 8
.23 6	.16 3	.02 1	.16 3	.60 7	.17 4
3.02 2	3.01 3	3.30 1	.35 7	2.14 4	1.07 5
1 5	1 5	1 5	1 5	1 5	1 5
21.5	16.0	15.5	18.0	11.0	12.0
.0 8	.2 5	.0 8	.2 5	1.0 2	.2 5
33483 8	6249 5	1000 2	11563 7	2881 3	8311 6
0 6	0 6	0 6	0 6	0 6	5 1
14.0	11.0	3.5	8.5	14.5	8.0
1.0 8	.8 5	.0 2	.9 7	.9 7	.7 4
.7 6	.7 6	1.0 2	1.0 2	.0 8	.8 4
18.5	34.5	21.5	17.5	30.5	15.5
1.0 8	1.0 8	2.5 4	3.3 2	5.0 1	3.2 3
19.2 3	8.0 5	42.5 2	12.1 4	.0 8	225.4 1
38.0 2	.0 8	7.5 3	3.0 6	.0 8	5.0 4
2.2 2	.0 7	.0 7	.2 5	.0 7	.2 5
1.3 4	.0 8	.3 6	8.0 1	.0 8	2.3 3
9.5	10.5	14	14.5	14.5	16.0
1.0 1	.7 2	.0 8	.5 5	.4 7	.4 7
1.0 1	.3 6	.5 4	.7 3	.0 8	.3 6
1.0 8	.3 3	.0 2	1.0 8	na na	.4 4
10.0	6.5	4.0	11.5	11.5	12.0
1.3 6	.7 3	.3 2	2.0 8	1.7 7	1.0 4
2.0 5	1.0 4	.0 2	1.0 4	2.0 5	3.0 8
2.4 6.0	3.0 2.0	3.1 1.0	2.5 5.0	0.8 8.0	2.0 7.0
4.98	4.98	3.31	4.76	5.80	4.60

c. Rollins Environmental Services' subsidiaries operate its hazardous waste facilities, including: RES (LA), RES of LA, RES (NJ), and RES (TX).

d. Average ranks equal sum of ranks divided by number of parameters (21 total) less ne (not evaluated) and na (not applicable).

TABLE 7-3. *The Big Eight: Financial Overview*

Areas of Comparison	Source	AMEC USEC	BFI CECOS
Growth rates			
% av annual growth rate tot revenues 80–84[b]	k,c	12.9%	13.6%
% av annual growth rate pretax income 80–84	k,c	39.9%[c]	18.6%
% av annual growth rate total assets 80–84	k,c	38.5%	14.6%
Annual data			
$ thous total revenues—fiscal 1984	k	45383	1000814
$ thous hazardous waste revenues[b]	k	8400	92258
% hw/tot revenue	k,c	1.4%	9.2%
$ thous net income[b]	k	5024	89168
$ thous average equity[b]	k	7598	431385
$ thous total debt[b]	k	0	119517
$ thous 1 year's cash flow[b]	k	4442	195936
$ thous pretax income—fiscal 1984	k	9922	161536
$ thous hazardous waste pretax income	k	d	4650
% hw/total pretax income	k,c	d	2.9%
% (provision for) income taxes/pretax income	k,c	49.4%	44.8%
Fiscal year ends	k	12/31	9/30
Stock information			
Exchange on which stock traded—symbol		OTC ECOL	NYSE-BFI
# common & thous common equivalent shares	k	2899	34249
$ stockholders equity/# common equiv. shares	k,c	$.39	$13.60
$ common stock highest price during 1984	k	$8.13	$44.50
$ common stock lowest price during 1984	k	$3.50	$26.50
$ common stock price 8/1/85	wsj	$27.13	$52.88
$ earnings per common equivalent share—1984	k	$1.73	$2.60
$ dividends per common equivalent share—1984	k	$.00	$.96

c: Calculated
k: Fiscal year 1984 10-K report (except for IT, whose fiscal year 1985 ended 3/31/85)
wsj: *Wall Street Journal* (8/1/85).

NOTE: Abbreviations of parent company and principal hazardous waste entity listed in alphabetical order by parent.

5. Auditor's opinion—To an unqualified auditor's opinion of a company's latest annual financial report, CEP assigned the value of one. A qualified opinion received the value of zero.

CEP derived items 1, 3, and 4 from the consolidated balance sheet, item 2 may be found in notes to the 10-K report, and item 5 is from the auditor's statement in the annual report.

ENSCO same	IT same	IU ESI	RES[a]	USPCI same	WMI CWM
99.8%	29.0%	4.3%	22.4%	41.8%	19.2%
156.7%	128.1%[c]	(19.2%)	15.3%	(20255%)	24.0%
99.7%	51.4%	.2%	26.0%	57.9%	31.9%
33483	143728	2550074	69375	17287	1314761
same	same	>10000[d]	same	same	199462
100%	100%	0.3%[d]	100%	100%	15.2%
1327	6316	7900	6477	1557	142509
5767	39668	369120	39516	2593	826768
17738	39674	370937	4605	5941	211563
5879	13171	112548	13028	2772	197952
1526	11483	25244	11229	1654	247977
same	same		same	same	25654
100%	100%	<5%[d]	100%	100%	10.3%
13.1%	45.0%	56.3%	42.3%	5.8%	42.5%
10/31	3/31	12/31	9/30	12/31	12/31
OTC-ESCO	NYSE-ITX	NYSE-IU	NYSE-REN	OTC-UPCI	NYSE-WMX
2510	5227	968370	101554	3859	48696
$.26	$7.03	$.37	$4.15	$.88	$18.19
$18.00	$12.13	$25.63	$11.38	$7.00	$47.88
$7.25	$8.13	$15.13	$6.13	$5.17	$31.38
$22.50	$24.50	$13.50	$12.13	$20.75	$59.75
$.50	$.67	$.27	$.64	$.42	$2.93
$.00	$.00	$1.18	$.00	$.00	$.77

a. Rollins Environmental Services' subsidiaries operate its hazardous waste facilities, including: RES (LA), RES of LA, RES (NJ), and RES (TX).

b. Parameter used for ranking companies (see table 7-2).

c. Growth rate 82–84 only.

d. CEP estimates. AMEC 1984 10-K report states a chemical waste pretax loss; CEP estimates AMEC's loss to be greater than $3 thousand.

Potential Hazardous Waste Liabilities

CEP used three measures to evaluate potential exposure to environmental liabilities. ENSCO ranks best in this category by showing the least potential exposure to hazardous waste liabilities. Although AMEC ranks lowest for this category, the number is not comparable to the others because U.S. EPA had not evaluated financial requirements for any of AMEC's sites. However, the final overall ranking adjusts for this lack of information.

Tables 7-5 and 7-6 show the data from which CEP calculated the ratios. The following list outlines the parameters of the data.

1. CEP divided the facility insurance and assurance coverage found not to meet U.S. EPA requirements by the number of evaluations made by U.S. EPA. CEP used the most recent data available from U.S. EPA's database for this ratio.[5] The U.S. EPA made these determinations from 1983 to 1985. Table 7-5 shows the data that served as the basis for these ratios. This ratio is also a measure of regulatory compliance.
2. CEP divided hazardous waste revenues by the number of land emplacement options available at company facilities. The number of land emplacements is derived from Part A of RCRA applications and is the sum of options for waste piles (SO3), surface impoundments used for storage (SO4), injection wells (D79), landfills (D80), and surface impoundments used for disposal (D83) at active and closed company-owned hazardous waste facilities.[6] CEP used this ratio as a measure of the ability of a company to pay for correcting potential damages arising from land-emplaced wastes.
3. The parameters also include the number of company-owned or company-operated sites listed or proposed as Superfund sites (see table 7-7 for site names).[7]

Corporate and Public Relations

CEP used two measures to evaluate corporate and public relations.

1. the number of questions the company answered on CEP's corporate questionnaire divided by 28 (the number of questions asked)
2. the number of sites for which CEP had identified local opposition divided by the number of sites for which information was available (see table 7-4). (In July 1985 CEP made a telephone survey of local citizen groups that could be contacted from lists provided by national public interest groups.[8] The sites surveyed included all of the Big Eight's 81 treatment and disposal facilities of which CEP was aware at the time. The survey did not include facilities that only store hazardous wastes. CEP was able to get information about 90 percent of the sites.)

In regard to corporate and public relations, USPCI, Inc. ranks best and IU and AMEC tie for the lowest rank.

Compliance with Government Regulations

By performing computations shown in table 7-2 of the data in table 7-5, CEP ranked the companies according to compliance with government regulations. IT ranks highest in this category, AMEC ranks lowest. The measures CEP used to derive the rankings are:

1. compliance with U.S. EPA regulations—the number of enforcement actions in U.S. EPA's database from 1980 to 1984 divided by the number of inspected facilities.[9]
2. penalties assessed for noncompliance with U.S. EPA regulations—the dollar amount of U.S. EPA penalties assessed from 1980 to 1984 divided by the number of inspected facilities.[10]
3. compliance with U.S. Occupational Safety and Health Administration (U.S. OSHA) regulations—the number of OSHA citations listed in U.S. OSHA's database from 1972 to 1984 divided by the number of inspected facilities.[11]
4. penalties assessed for noncompliance with U.S. OSHA regulations—the dollar amount of penalties assessed from 1972 to 1984 divided by the number of inspected facilities.[12]
5. damages assessed for transportation spills—the dollar amount of damages assessed for hazardous waste spills listed in U.S. Department of Transportation (U.S. DOT) databases from 1978 to 1984 divided by the number of active facilities.[13]

Ground Water Monitoring

Because ground water contamination is one of the most serious and costly risks posed by improper hazardous waste management, CEP considered it separately from the other parameters of compliance with government regulations. CEP focused on protection of ground water in relation to compliance with U.S. EPA requirements intended (1) to prevent ground water contamination, (2) to provide adequate monitoring for identification of contamination when it does occur, and (3) to clean up contamination before it can spread.

The CEP measures used data:

1. from U.S. EPA documents—the total number of ground water monitoring inadequacies for each company's facilities that are subject to ground water monitoring requirements divided by the number of requirements for which adequacy had been evaluated. Requirements cover the following areas: well system, sampling, analysis and evaluation program, assessment report, waiver demonstration, and records.

CEP used 1984 and 1985 U.S. EPA determinations, the most recent available.[14]

2. from U.S. EPA documents and a congressional survey—the number of sites at which contamination had been documented as of 1985 divided by the number of facilities subject to ground water-monitoring requirements.[15]
3. from U.S. EPA documents—the sum of values assigned to sites having corrective actions in progress (these sites had received 1 point each) or having corrective actions planned (these sites had received 0.5 point each) as of 1985 divided by the number of leaking sites.[16]

Browning–Ferris ranks highest in regard to ground water monitoring. ENSCO ranks lowest.

Technology

WMI ranks highest in the technology category; AMEC ranks lowest. Table 7-5 shows the data CEP used in evaluations. CEP derived the data:

1. from Part A of U.S. EPA RCRA applications—the number of treatment options as identified by U.S. EPA process codes (including treatment in tanks (T01); surface impoundments (T02); incinerators (T03); other physical, chemical or biological treatment processes (T04); and land application (D81)) divided by the number of active facilities.[17] Where possible, CEP distinguished treatment options presently in use from those planned and counted only those options in use.
2. from CEP point assignments—each facility that owned a U.S. EPA-approved PCB incinerator and each solvent recovery facility listed in the 1985 Environmental Information survey received one point.[18]

Management Overview

Table 7-4 gives some basic information on the companies and their management's response to the public and to requests for information. The table also presents an overview of the total number of employees, the number of hazardous waste management employees, and the number of employees with college degrees in science or engineering. Comparable information was not available for all companies. These data from the annual 10-K reports and from the company questionnaire are provided for information only.

CEP sent a corporate questionnaire that contained 28 questions relating to available services, personnel, environmental management, unionization, pricing and marketing, acquisitions and divestitures, and facilities. Four companies answered 20 or more questions. Three answered two or fewer. CEP used the number of questions answered as an evaluation parameter. ENSCO answered all the questions; AMEC and IU answered none of the questions.

CEP made a telephone survey in July 1985 to identify the number of treatment and disposal facilities for which there was organized local citizen opposition. The number of sites opposed is the numerator over the total number surveyed. This ratio served as one parameter of evaluation. Table 7-4 presents the number of facilities for which no information was available in parentheses. USPCI, Inc. ranked highest with no identifiable citizen opposition; IU and AMEC tied for lowest.

To compile the data in table 7-4, CEP also retrieved legal data through LEXIS, an on-line database owned by Mead Data Central Corp. LEXIS enables the user to search the full text of old and recent court cases and statutes from the federal and state level. The user can search for references to cases, statutes, and regulatory material by matching relevant words and phrases to occurrences in the primary sources. LEXIS also includes specialized "libraries" of materials, including *Environment Reporter* and *Chemical Regulation Reporter,* both publications of Bureau of National Affairs, Inc., a private service. During March and April 1985, CEP used LEXIS to search for the names of the eight companies and names of subsidiaries. Although LEXIS provides one of the most comprehensive listings of litigation available, some cases may be missing from the system. The number of cases is given for information only and is not used in the evaluation.[19]

In the public's perception, hazardous waste management is sometimes tied to organized crime. CEP has searched legal cases, literature, and newspaper articles but did not find substantive allegations directly relating any of the Big Eight to organized crime.

FACILITY EVALUATIONS

The best of the ten facilities evaluated was USPCI's Grassy Mountain facility in Clive, Utah. The worst was BFI's CECOS facility in Williamsburg, Ohio. Table 7-2 presents the scores that support these conclusions; these scores are the result of CEP's site evaluations, which chapter 9 discusses in detail. If CEP evaluated more than one site owned by the same company, the grade is based on an average of their scores. BFI's higher average score leaves IU with the poorest facility ranking.

Table 7-5 provides the data used for making the facility-specific ratios in table 7-2. Table 7-6 provides statistics on the total number and types of facilities owned by the Big Eight. Table 7-7 shows the names, hazard scores and hazard groups of the Superfund sites for which the Big Eight have potential responsibility. Their responsibility can range from being the owners, partial owners, or operators to being one of many companies identified as a generator of

TABLE 7-4. *The Big Eight: Management Overview*

Areas of Comparison	Source	AMEC USEC	BFI CECOS
Corporate management			
# total employees—parent & subsidiaries 1984	k	302	16,800
# employees-hazardous waste management 1984	k,r	ns	ns
# empl—scientific/engineering college degr	k,r	ns	ns
y/n public disclosure-questionnaire retrnd	r	no	yes
% questions answered on CEP questionnaire	r	0%	4%
# months until response received by CEP	r	na	9
# sites local citizen grps oppose/# sites (no information)	p	2/3(1)	14/21(2)
Involvement in litigation			
# cases antitrust or securities fraud	l	0	4
# cases workers, victims compensation	l	0	4
# cases trade contract, labor relations	l	0	8
# cases Racketeer Infl & Crimnl Orgnz	l	0	2
# cases permits and siting	l	1	10
# cases as plaintiff	l	1	8

k: fiscal year 1984 10-K report
l: Lexis search
p: phone research
na: not applicable
ns: not specified
r: CEP questionnaire

NOTE: Abbreviations of parent company and principal hazardous waste entity listed in alphabetical order by parent.

waste found at the site. Table 7-7 distinguishes between an owner or operator which table 7-2 considered as an evaluation parameter and a potentially responsible party (PRP). Unless otherwise indicated, the data in tables 7-5, 7-6, and 7-7 are from U.S. EPA databases.[20]

PROFILE: AMEC AND ITS SUBSIDIARY, USEC

The revenue of U.S. Ecology's parent company, AMEC, comes entirely from waste management and principally from radioactive waste management. U.S. Ecology (USEC) is AMEC's radioactive and chemical waste management subsidiary.

USEC's involvement in chemical waste management is one of the two smallest of the eight companies in this study. The annual revenue of USEC's chemical operations is estimated at $8.4 million, an amount smaller than CEP's estimated revenue for IU (see table 7-3).[21] Over 56 percent of AMEC's 1984 revenue of $45.4 million comes from managing the nuclear waste repository on 100 acres leased from the U.S. Department of Energy (U.S. DOE) at the Hanford Reservation in Washington.[22]

ENSCO same	IT same	IU ESI	RES[a]	USPCI same	WMI CWM[b]
400	1,650	37,000	400	146	12,900
all	all	ns	all	all	1,970
60	354	ns	95	7	206
yes	yes	no	yes	yes	yes
100%	80%	0%	86%	89%	77%
8	4	na	4	8	6
2/3(0)	5/9(2)	6/7(1)	4/4(0)	0/2(0)	24/32(1)
0	0	0	0	0	4
0	0	1	3	0	4
1	0	1	3	0	1
0	0	0	0	0	0
0	0	0	2	0	6
1	0	0	2	0	4

a. Rollins Environmental Services' subsidiaries operate its hazardous waste facilities, including: RES (LA), RES of LA, RES (NJ), and RES (TX).
b. SCA Services, Inc. involvement in litigation is not included under Waste Management, Inc.

Corporate Evaluation. AMEC received a D+, the second lowest grade of the eight companies in CEP's corporate performance evaluation. Overall it ranked just above IU but well below Waste Management and Browning–Ferris. AMEC was among the lowest three companies in five categories placed fifth in CEP's site evaluation. AMEC received its best rank and tied for sixth in ground water monitoring (see table 7-2).

USEC's chemical waste operations have been limited to landfill and storage options, although its facility in Robstown, Texas—Texas Ecologists (TECO TX)—uses carbon adsorption to treat waste water and remedy ground water contamination.

AMEC is one of the three Big Eight companies that owns or operates sites proposed for inclusion under Superfund. AMEC owned the USEC chemical site in Sheffield, Illinois, and operated the Maxey Flats radioactive waste site in Hillsboro, Kentucky (see table 7-7).[23] At the Sheffield site corrective action will proceed under RCRA according to a consent agreement finalized in the fall of 1985 between USEC and U.S. EPA. If this corrective action is successful, the Sheffield site will not be listed on the NPL.[24]

CEP evaluated one of AMEC's two chemical hazardous waste sites, TECO TX, in April 1985. The site ranked lowest of those visited in the South by earning a D+. Chapter 9 presents the evaluation.

TABLE 7-5. *The Big Eight: Facility-Specific Parameters*

Areas of Comparison	Source	AMEC USEC	BFI CECOS
Compliance with government regulations			
U.S. EPA regulations			
# enforcement actions/# inspected facilities	e	7/4	45/29
$ thous EPA penalties	e	0	31
U.S. OSHA regulations			
# OSHA citations/# inspected facilities	o	0/0	96/31
$ thous OSHA penalties	o	0	18.6
U.S. DOT regulations			
$ thous spill damages/# spills	d	0.3/15	0/6
Ground water monitoring			
# monitoring inadequacies/# adequacies evaluated	e	4/7	11/23
# documented leaking sites/# subject facilities	e	2/3	2/11
# corrective actions/# leaking facilities	e	1/2	1.5/2
Financial instruments			
# financial inadequacies/# adequacies evaluated	e	0/0	10/33
Technology			
# treatment options/# active hazwaste facilities	e	0/2	10/8
# high temp incinrt-PCB liquids, U.S. EPA permit	k	0	0
# solvent recovery facilities, U.S. EPA permit	k	0	0

e: U.S. EPA databases
d: U.S. DOT databases
k: fiscal year 1984 10-K report
o: U.S. OSHA database

NOTE: Abbreviations of parent company and principal hazardous waste entity listed in alphabetical order by parent.

History. AMEC was formed by Teledyne, Inc. in October 1983 as a holding company for two subsidiaries, USEC and National Ecology. AMEC stock was distributed to Teledyne stockholders, and since March 31, 1984, AMEC has operated independently.

USEC, formerly the Nuclear Engineering Company (NECO), was purchased by Teledyne in 1975. Founded in 1952 NECO purchased a site at Maxey Flats, Kentucky, in 1962. The state of Kentucky purchased the site and leased it back to NECO to operate. A total of 4.75 million cubic feet of radioactive materials are buried there. The state stopped NECO's operations at Maxey Flats in 1977 when radioactive materials were found to be migrating from the storage areas. The state's 1984 estimate for decommissioning costs at the site ranges from $52 to $121 million. The cleanup is now being paid for with federal and state funds.[25]

NECO was the only operator of the low-level radioactive waste facility.[26] Kentucky paid NECO $1.2 million for the remainder of the operating contract.

ENSCO same	IT same	IU ESI	RES[a]	USPCI same	WMI CWM
2/2	13/13	12/5	13/4	15/3	118/37
38.4	104.3	212.5	48.4	0	8339
38/1	0/2	30/4	12/4	0/0	95/19
2.2	0	1	0.7	0	3.2
0/4	0/0	0/1	1/24	0/0	537.3/68
2/2	14/19	0/2	4/8	2/5	22/52
1/1	2/8	2/4	2/3	0/2	7/22
1/1	0.5/2	0/1	2/2	0/0	2.5/6
0/4	8/38	0/5	2/12	2/2	7/46
4/3	14/21	1/3	8/4	5/3	29/28
1-AR[b]	0	0	1-TX[b]	0	1-IL[b]
1-AR[b]	1-CA[b]	0	0	2-OK[b]	2-OH,AL[b]

a. Rollins Environmental Services' subsidiaries operate its hazardous waste facilities, including: RES (LA), RES of LA, RES (NJ), and RES (TX).

b. State abbreviations identify locations of facilities.

The contract includes a clause that holds NECO harmless for its operation of the site.[27]

Hazardous Waste Management. USEC has a chemical waste operation at Beatty, Nevada, on 20 acres adjacent to its radioactive waste site. Both sites are leased from the state. This facility has been authorized by U.S. EPA to receive certain types of solid PCBs for burial. USEC has applied to expand the permitted area to 36 acres.

Business Activities. USEC has begun to offer other hazardous waste services. It has established U.S. Ecology Consultants and a Special Services Division to plan and conduct cleanups of chemical waste sites.[28] USEC has a laboratory in Louisville, Kentucky, where its corporate headquarters are located.

USEC's major business is the operation of repositories for low-level radioactive waste. It operates two of the three commercial facilities in the country

TABLE 7-6. *The Big Eight: Facility Overview*

Areas of Comparison	Source	AMEC USEC	BFI CECOS
Total facilities			
# solid waste facilities	e	1	92
# hazardous waste facilities	e	4	36
# active facilities	e	4	117
# inactive facilities	e	1	13
# active hazardous waste facilities	e	2	8
# major facilities	e	4	21
# proposed hazardous waste facilities	e	0	7
# facilities that submitted Part A	e	3	42
# active and inactive hazardous waste land emplacement facilities (S03 + S04 + D79 + D80 + D83)	e	3	27
Active hazardous waste transportation/storage			
# vehicles used for transportation	k	ns	190
# facilities offering transportation services	e	3	29
# storage terminals:			
# container storage facilities (S01)	e	1	8
# tank storage facilities (S02)	e	1	7
# waste pile storage facilities (S03)	e	0	1
# surface impoundment storage facilities (S04)	e	0	8
Active hazardous waste treatment			
# tank treatment facilities (T01)	e	0	3
# surface impoundment facilities (T02)	e	0	4
# incineration facilities (T03)	e	0	0
# other treatment facilities (T04)	e	0	2
# chemical treatment facilities	m	0	1
# biological treatment facilities	m	0	4
# physical treatment facilities	m	0	6
# thermal treatment facilities	e	0	0
# high temp incinrt-PCB solids, EPA permit	k	0	0
# other incineration capacity	m	0	0
y/n mobile high temp incinerator available	k	no	no
Active hazardous waste disposal			
# deep-injection wells Class I facilities (D79)	e	0	2-TX,LA
# landfill facilities (D80)	e	2	3
# land application facilities (D81)	e	0	1
# ocean disposal facilities (D82)	e	0	0
# surface impoundment disposal facilities (D83)	e	0	0

e: U.S. EPA databases
m: Environmental Information, Ltd.
k: fiscal year 1984 10-K report

NOTE: Abbreviations of parent company and principal hazardous waste entity listed in alphabetical order by parent.

ENSCO same	IT same	IU ESI	RES[a]	USPCI same	WMI CWM
0	0	0	0	0	87
5	22	7	4	7	43
2	22	4	4	7	111
0	0	3	0	0	23
3	21	3	4	3	28
2	17	4	4	4	28
0	2	2	0	0	4
2	22	8	4	3	38
1	23	10	6	6	44
25	234	ns	ns	60	ns
2	13	5	3	7	10
3	9	3	3	4	26
1	18	3	4	4	19
1	2	1	0	1	0
0	9	1	3	2	8
1	4	0	2	2	8
1	6	0	2	0	4
1	3	0	3	0	2
1	0	0	1	2	9
0	3	0	1	0	1
1	4	2	2	2	7
1	4	3	3	3	13
1	1	0	3	0	2
1-AR	0	0	1-TX	0	1-IL
0	1-CA	0	2-LA,NJ	0	1-IL
b	b	no	b	no	no
0	0	0	1-LA	0	3
0	5	2	2	2	9
0	1	1	0	1	0
0	0	0	0	0	1
0	7	0	0	1	4

a. Rollins Environmental Services' subsidiaries operate its hazardous waste facilities, including: RES (LA), RES of LA, RES (NJ), and RES (TX).

b. Under development and testing or operating.

TABLE 7-7. *The Big Eight: Potential Responsibility for CERCLA Sites*[a]

CERCLA Site Name	Location	Company is Owner and/or Operator	Company is Potentially Responsible Party	USEPA HRS Score	USEPA HRS Group
American Ecology, Inc.					
Seymour Recycling	Seymour, IN		x	58.15	2[b]
Chem-Dyne	Hamilton, OH		x	47.78	2[b]
U.S. Ecology	Sheffield, IL	x		34.22	9
Maxey Flats Nuclear Disposal	Hillsboro, KY	x		31.71	10
Browning–Ferris Industries, Inc.					
McAdoo Associates	McAdoo Borough, PA		x	65.32	1[b]
Western Processing	Kent, WA		x	36.30	1[c]
Marshall Landfill	Boulder County, CO		x	41.00	2[b]
Chem-Dyne	Hamilton, OH		x	47.78	2[b]
South Brunswick Landfill	South Brunswick, NJ	x		53.42	3
Summit National	Deerfield Township, OH		x	52.28	3
Monroe Township Landfill	Monroe Township, NJ	x		42.37	6
Laskin/Poplar Oil Co.	Jefferson, OH		x	35.95	8
Bayou Sorrell	Bayou Sorrell, LA		x	38.15	9
Bio-Ecology Systems	Grand Prairie, TX		x	35.10	9
International Technology Corporation					
Stringfellow	Glen Avon Heights, CA		x	61.40	1[b]

Rollins Environmental Services, Inc.					
Price Landfill	Pleasantville, NJ		x	71.60	1[b]
Helen Kramer Landfill	Mantua Township, NJ		x	70.06	1
Kin-Buc Landfill	Edison Township, NJ		x	50.64	4
Bayou Sorrell	Bayou Sorrell, LA		x	38.15	7
Waste Management, Inc.[d]					
McAdoo Associates	McAdoo Borough, PA		x	65.32	1[b]
Omega Hills North	Menomonee Falls, WI	x		58.54	1
Seymour Recycling	Seymour, IN		x	58.15	2[b]
A. L. Taylor (Valley of Drums)	Brooks, KY		x	17.68	2[b]
Kin-Buc Landfill	Edison Township, NJ		x	50.64	4
Lowry Landfill	Arapahoe County, CO		x	48.36	4
Re-solve, Inc.	Dartmouth, MA		x	42.72	4
Silresium Chemical Corp.	Lowell, MA		x	47.71	6
Envirochem Corporation	Zionsville, IN		x	46.44	6
Lauer 1 Sanitary Landfill	Menomonee Fall, WI	x		42.69	6
Nat. Ind. Env. Services	Furley, KS	x		39.59	7
Lee's Lane Landfill	Louisville, KY		x	39.52	7
Bayou Sorrell	Bayou Sorrell, LA		x	38.15	7
Holland Lagoons	Holland, MI	x		37.20	8
Modern Sanitation Landfill	Lower Windsor, PA	x		33.93	9
Sand Springs Petrochemical	Sand Springs, OK		x	28.86	11

DERIVED FROM: U.S. EPA "Listing of Potentially Responsible Parties at CERCLA Sites" (9/13/84) and U.S. EPA, "National Priorities List" (10/84).

NOTE: Companies are listed in alphabetical order; sites are listed from highest to lowest hazard.

a. USPCI, Inc., ENSCO, Inc., and IU International, Corp. have not been identified as potentially responsible parties for any CERCLA sites. CERCLA sites include all sites proposed for and listed on the National Priorities List.

b. Site designated by state as top priority.

c. Site designated in a higher hazard group in consideration of factors other than migration score.

d. Because Genstar, Inc. is responsible for SCA Independent Landfill in Muskegon Heights, MI, the site is excluded.

licensed to receive low-level radioactive wastes—the site at Beatty, Nevada, and the Hanford Reservation in Washington. The low-level disposal facility at Hanford operates under a license from the state of Washington as well as other licenses.[29]

Operating income from Hanford makes up for losses in other operations (see table 7-3). The Nuclear Regulatory Commission (NRC) license under which the facility operates was subject to renegotiation and renewal in November 1985. The renewal request is still being reviewed by NRS; meanwhile, operations may continue under the old license. The company's positive operating income depends on the NRC license and on the continuation of its lease, which extends until 1990. In 1982 the Hanford disposal facility represented 209 percent of USEC's operating income.[30]

If implemented the federal Low Level Radioactive Waste Policy Act of 1980, which requires that each region of the country provide its own waste disposal facilities, would divert 95 percent of the waste now being sent to Hanford by January 1990. As a result USEC has submitted an application to North Carolina to build a low-level radioactive waste incinerator to serve the Southeast Regional Compact.[31]

The solid waste subsidiary, National Ecology, operates plants and provides engineering and solid waste management and resource recovery for local governments. National Ecology operates one municipal plant in Maryland and one in Utah. These activities provide 20 percent of AMEC's revenues but only 7 percent of its operating income.[32]

Financial Viability and Liability. USEC has incurred operating losses from its chemical waste disposal operations for a number of years. The losses were due to significant costs associated with license compliance at TECO TX and closure and monitoring activities at Sheffield (which currently produces no revenue).[33]

AMEC summarized the company's potential liabilities in its 10-K report.

> The State of Illinois has instituted two legal actions against U.S. Ecology, Inc. One action, relating to the closure of the Sheffield, IL low-level radioactive waste disposal site, seeks injunctive relief, including exhumation of all or part of the low-level radioactive waste, and damages of $97 million. The other suit relates to chemical landfill operations at the closed Sheffield, IL chemical waste disposal site and seeks injunctive relief, including exhumation of all the waste buried at the site. U.S. Ecology, Inc. believes that the damages alleged are grossly overstated and that it has substantial defenses to the claims made in both the low level and chemical actions.[34]

Tritium, a radioactive isotope of hydrogen, was detected in ground water off site at Sheffield in 1978. By 1982 tritium had migrated 665 feet from a closed burial trench. Recent tests found a narrow tritium plume, now alleged to be moving east at a rate of 2000 to 3000 feet per year.[35]

USEC contests the report of a moving stream and states:

> Also we believe that the joint tritium study committee comprised of representatives of the Nuclear Regulatory Commission, US Ecology, the Illinois Department of Nuclear Safety and Illinois Geological Survey have concluded that this plume poses no significant health or safety threat and the parameters of the contamination are well known and ultimately contained on US Ecology-owned property.[36]

USEC notes that the levels of tritium are well below release standards for nuclear facilities such as power plants.

USEC also points out that the suit alleging multimillion dollar damages has been pending since 1978, that in the opinion of counsel the suit is without merit, and that significant defenses are available.[37] AMEC states the effect of an adverse judgment on U.S. Ecology in its 10-K statement.

> While U.S. Ecology believes it has substantial defenses to these actions, an adverse judgment in either case could materially decrease U.S. Ecology's liquidity or render it insolvent. The Company has established a closure reserve of approximately $2 million in 1984 covering the Sheffield low-level facility.[38]

Potential costs are significant should Illinois win either of these legal proceedings against the company. In the action concerning radioactive waste, Illinois has asked $97 million in damages. In the action concerning the chemical waste at the site, the state has asked for removal of the wastes. The costs for exhumation of 450,000 cubic feet of waste, transportation to another site, and treatment or burial would have a materially adverse financial effect upon AMEC.[39] Chemical wastes have been detected in the ground water-monitoring wells. USEC and U.S. EPA recently signed a consent agreement concerning the site.[40] An approved RCRA closure plan will be necessary.

Major Litigation. The legal action for damages allegedly caused by chemical waste disposal at the Sheffield, Illinois, site was brought against U.S. Ecology under its previous name, Nuclear Engineering Company. AMEC points out in its 10-K report that there is precedent in Illinois courts to require removal of buried waste.[41]

Corporate and Public Relations. Through a telephone survey, CEP found that there is organized citizen opposition to the Sheffield and Robstown facilities (see table 7-4). CEP found no organized opposition to the Beatty facility. USEC points out that there is considerable support for the Robstown facility from local government leaders and educators.[42]

PROFILE: BFI AND ITS SUBSIDIARY, CECOS

BFI is one of the two largest publicly owned firms engaged primarily in managing solid and hazardous waste. Second only to WMI in solid waste revenue, BFI's total revenue for 1984 was over $1 billion. Of that revenue 88 percent was derived from solid waste management and only 9 percent from hazardous waste management. The hazardous waste revenue, $92.3 million, makes BFI's hazardous waste segment third in size, behind WMI's Chemical Waste Management Division and IT (see table 7-2).

BFI is ranked 60th in the list of diversified service companies. In 1984 it was ranked 68th. BFI is 49th in assets, 22nd in net income, and eighth in net income as a percent of sales.[43] In 1985 A+ Standard & Poor's gave BFI an A+,[44] a stronger rating than the A it received in 1984. In 1985 Moody's gave BFI an "upper medium grade."[45] The positive trends in BFI's financial history show the strength of its solid waste operations, 91 percent of company operations.

Corporate Evaluation. CEP's grade for BFI is average, C. BFI's rating is better than that of AMEC and IU, just above WMI but below ENSCO, IT, and RES (see table 7-1). BFI ranked first in ground water monitoring, fourth in corporate financial viability and in CEP's facility evaluation, and sixth in general compliance with government regulations. In potential liabilities, public disclosure, and technology, BFI fell below average (see table 7-1). CEP visited and evaluated two BFI hazardous waste facilities—CECOS International in Williamsburg, Ohio (CEP gave the facility an F), and CECOS in Niagara Falls, New York (CEP gave the facility a C−).

The size of BFI's hazardous waste operations doubled in 1983 with the acquisition of CECOS. But CECOS's long-term growth prospects remain to be seen, in view of the 1984 RCRA amendments. BFI points out in its annual report that the 1984 RCRA amendments will be of material significance to its operations. CECOS will incur considerable expense to bring its facilities up to the new RCRA standards, though the company has not issued specific dollar amounts.[46]

CECOS offers landfills and deep-well injection for wastes, but it offers fewer treatment options per facility than most of the other seven companies (see table 7-6). BFI faces a number of potential liabilities in regard to sites it owns. BFI owns two Superfund sites and approximately ten closed sites. It is one of the three Big Eight companies that owns or operated Superfund sites (see table 7-7).

BFI appears to have a significant role to play in handling the huge influx of small loads that will be regulated under the amended RCRA. The company has significant experience and capability in providing cleanup services.

History. BFI is a Houston-based company that has been incorporated in Delaware since 1970. It is a successor to Browning–Ferris Machinery, which was incorporated in Texas in 1949. BFI's operations have expanded rapidly over the last 20 years through the acquisition of numerous companies. Some businesses gained through acquisition were sold or spun off.[47]

Hazardous Waste Management. In January 1983 BFI acquired CECOS and NewCo Waste Systems of Buffalo, New York, both firms that manage hazardous wastes. The purchase increased its hazardous waste operations to eight facilities at seven sites: five landfill and treatment facilities and three waste-water treatment plants.[48]

CECOS has concentrated its development efforts on improving the means of packaging and storing hazardous wastes. It has promoted a ground-mounted air structure, the CECODOME. Six to seven stories high, a CECODOME covers storage operations. Another development, the CECOPAK, is a packaging system that encases hazardous materials in protective packaging.[49] The packaging system may be useful in transporting the large number of small shipments that will result from the 1984 RCRA amendments.

BFI's efforts to destroy waste are developing slowly. CECOS has developed an activated sludge system that uses stress-activated bacteria (bacteria bred to destroy waste). The system is installed on a pilot scale at the Niagara Falls facility. BFI's RECRA Research developed the system in Buffalo, New York, under a grant from the New York State Energy Research and Development Authority.[50] BFI does not have high-temperature incineration capacity.

CECOS, with 190 trailer tankers, is a major transporter of hazardous waste. It is developing its capacity to bring together smaller quantities of waste which can then be sorted and reshipped to treatment storage or disposal facilities (see table 7-6).[51]

In 1984 CECOS acquired Connecticut Treatment Corporation's facilities in Bristol, Connecticut, which include a complete analytical laboratory, a containment area, a truck-loading area, vehicle maintenance facilities and waste-water treatment. An additional site in Bristol has in-ground concrete tanks and also a loading area. It is convenient to major highways in New England.[52]

A similar operation, the Cincinnati Processing Facility, offers waste-water treatment using activated carbon adsorption. The facility serves mainly as temporary storage for wastes from small and intermediate generators. The wastes are placed in landfills or gathered, sorted, and shipped to incinerators. This type of facility will play an important role under the 1984 RCRA amendments, which regulate for the first time the generators of smaller quantities of hazardous waste.[53]

Remedial cleanup services are provided by a CECOS subsidiary, CECOS Environmental Services (CEI), which has been the contractor at several Superfund sites.[54]

CEP's review of U.S. EPA's database indicates that Browning–Ferris notified U.S. EPA of 37 hazardous waste sites (see table 7-6). Only eight facilities at seven sites actively manage hazardous waste. Seven of the 37 sites are proposed hazardous waste sites. Several of them now handle only solid waste. Eight are major sites as identified by their state governments (see chapter 10).[55]

Business Activities. BFI's major business is the collection, the processing and recovery, and the depositing of solid waste in landfills. Those operations encompass 180 North American locations and 10 other sites outside North America. BFI as a whole owns about 5100 collection trucks and employs 16,800 workers. BFI has waste collection operations for municipal, commercial, and industrial customers in more than 170 locations in 39 states, Canada, Kuwait, Puerto Rico, Saudi Arabia, Spain and Venezuela. BFI's solid waste landfill operations include some 80 sanitary landfills, 35 of which it owns outright. The rest are partially owned or operated or leased. BFI's solid waste processing operations consist of 26 transfer stations of which it owns six and operates the others under contract. One of these stations, in Toronto, recovers the marketable components of the waste stream.

BFI has an exclusive license for a grate system which has been used in installations in Europe. A joint venture has been formed to build solid waste incinerators, which would be placed in several facilities across the country. These incinerators would provide an alternative to land emplacement for the company's solid waste collection operations.[56]

Financial Viability and Liability. BFI's commercial and industrial solid waste collection services bring in about 60 percent of total revenues. Residential collection brings in 17 percent; solid waste processing and disposal brings in 11 percent. Hazardous waste management brings in 9 percent or $92.3 million. BFI is the third largest of the hazardous waste managers.[57]

BFI believes RCRA, as amended in 1984, and the new regulations will be of material significance to their solid waste and hazardous waste operations.

BFI has maintained substantial liability coverage. In 1984 its Environmental Impairment Liability (EIL) insurance policy for nonsudden, gradual events had limits of $30 million per event and $30 million annual aggregate for bodily injury and property damage caused by long-term gradual pollution of surface or subsurface properties adjacent to company facilities.[58]

In November 1984 Ohio EPA closed the CECOS RCRA landfill in Williamsburg, Ohio, for two weeks. EPA suspected discharge of contaminated rainwater from a capped cell into a nearby tributary. (The tributary flows into a river that supplied drinking water to a town 9 miles away.) Ohio EPA closed the RCRA facility again in 1985 because of inadequate ground water monitoring. A landfill at the Williamsburg site continued to receive TSCA-regulated wastes such as PCBs during these events. The CECOS landfill at Niagara Falls has some cells that use the latest liner and leachate collection system and older cells that have only clay liners (see chapter 9 for a discussion of both facilities).[59]

BFI owns and operated two Superfund sites: the South Brunswick Landfill in South Brunswick, New Jersey, and the Monroe Township Landfill in Monroe, New Jersey (see table 7-7).[60]

Major Litigation. BFI has been involved in more litigation both as plaintiff and defendant than any other of the companies according to CEP's LEXIS search (see table 7-4). It has been plaintiff in eight cases and defendant in 19.[61]

Corporate and Public Relations. In 1982 BFI advised the Louisiana Department of Natural Resources (LADNR) that an on-site ground water monitoring well at its Willow Springs facility showed evidence of contamination in the well. In February 1983 a consultant's report required by LADNR showed that a part of the Chicot Aquifer System, a major drinking-water source, was contaminated. The level of concentration of many of the contaminants in monitoring wells exceeded U.S. EPA guidelines for drinking water. The Louisiana Environmental Control Commission issued a compliance order and a schedule to BFI Chemical Services.[62]

The situation prompted organized citizen action. Neighboring residents in Lake Charles filed a suit alleging that several different chemicals had leaked into surrounding ground water. In 1983 citizens marched on the state capitol to protest inaction by the state.[63] In April 1984 the Louisiana Supreme Court handed down a decision.

> There is no proof in the record that there is a danger to the health and safety of citizens or to the integrity of the environment from the continued operation of the landfill. There is no proof in the record that the current operation of BFI's landfill cells cause the health problems of which some witnesses complained. In fact, there is no proof that the current operations either caused or contributed to those contaminants which were found. . . . The Department of Health and Human Resources also concluded, after testing in response to water quality complaints, that the local drinking water was safe. . . .[64]

CEP was able to identify local community opposition to ten of the 21 BFI sites investigated (see table 7-4).

PROFILE: ENSCO

For the most part ENSCO has been a development company for the technology of high-temperature incineration. Melvyn Bell joined the company in 1972 and acquired a refinery site in El Dorado, Arkansas, in 1974. There, he and

Dr. George Combs have developed a high-temperature incinerator.[65] It is ENSCO's major asset and is equipped with several proprietary design modifications. It is licensed by U.S. EPA to incinerate PCBs and is capable of destroying other problem organics as well. This capability gives ENSCO excellent growth potential.[66]

Corporate Evaluation. With a C+ ENSCO is tied with IT for second in the CEP evaluation. The two companies rank below USPCI, Inc. and above RES. ENSCO ranked at the top in one category and above average in three more. Its low ranking for ground water monitoring and financial parameters pulled down the overall ranking.

The company has excellent potential for growth because of its incinerator technology and its ability to destroy PCBs and other halogenated wastes. Its integrated facility cleans tank cars and drums, a service important to waste haulers. ENSCO's limitation is its reliance on a single facility for the bulk of its revenue. CEP visited and evaluated the El Dorado, Arkansas, facility and gave it a C−.

History. The company was originally incorporated in Minnesota in 1959 as Commercial Chemical Co., renamed Pollution Controls, Inc. in 1967 and ENSCO, Inc. in 1982. On reincorporation in Delaware in August 1983, the name was changed to Environmental Systems Co.[67] The company filed with the Arkansas secretary of state as Energy Systems Co., and it is now called ENSCO, Inc.[68]

ENSCO's original business was reclamation of commercial chemicals. In 1975 the company (then Pollution Controls, Inc. of Shakopee, Minnesota) was taken to court by the Scott county attorney for operating an incinerator without the required permits and for storing approximately 25,000 drums of wastes. The site was finally cleared and the matter closed by a consent decree in the Minnesota District Court in 1982. The contents of the drums were shipped to the ENSCO Arkansas incinerator as well as to other licensed chemical waste facilities.[69]

Hazardous Waste Management. The company's primary asset is ENSCO's proprietary incinerator which operates at 140 million BTUs per hour. The incinerator is composed of two shredders, a rotary kiln, two fixed combustion chambers, a waste-fired boiler, venturi jet scrubbing systems, and other ancillary components.

ENSCO's incinerator is capable of destroying PCBs to the degree required by U.S. EPA under TSCA regulations—a destruction efficiency of 99.9999 percent. The TSCA PCB authorization for this incinerator is at a flow rate of 3700 pounds per hour of pure PCBs.[70] As a result of a test burn in March 1985, authorization is expected for a flow rate of 5210 pounds per hour for both solid and liquid PCB-contaminated wastes.[71] This capability will prove valuable in the next few years when society will need to dispose of large amounts of PCBs. The incinerator can destroy other organics, too, a capability that will serve ENSCO well for a longer term.

The destruction of chlorinated organics produces carbon dioxide, water, and hydrochloric acid. The acid is removed from incinerator stack gases by a scrubber system and captured as calcium chloride, a marketable substance. There is an above-ground clarifier used for storage of recirculating scrubbing media which is periodically withdrawn and processed into marketable calcium chloride products.[72]

A railroad siding for tank cars and ENERTRAN, the subsidiary facility that cleans railway cars and tank cars, facilitate hazardous wastes shipping.[73]

ENSCO has storage facilities in White Bluff, Tennessee (owned and operated by a subsidiary), that drain and flush PCB-filled transformers. The resulting PCB-contaminated fluids are transported to El Dorado and destroyed in the incinerator. White Bluff also has a storage area for PCB-contaminated solids and a modular incinerator assembly area.

In 1982 ENSCO purchased Pyrotech Systems to acquire the technology to develop modular incinerators. Two small units are complete and are undergoing testing and modification. ENSCO has manufactured two larger units with a 35-million-BTUs-per-hour operating capacity. One of these units is in use in Tampa, Florida, to burn nonhazardous waste. The second unit underwent U.S. EPA-supervised test burns in November 1985, and a third unit will be completed in early 1986.[74]

Modular incinerators represent intriguing technology because they can be moved directly to sites that require cleanup and sites that generate wastes. The number and type of modules can be tailored to the type of materials at the site. Modular incineration has the potential to lower cleanup costs and to eliminate the transportation and transfer of materials with the attendant possibilities of spills.[75]

ENTEK, an ENSCO subsidiary, operates a commercial analytical testing laboratory for remedial services and compliance monitoring and maintains a site remediation unit at White Bluff, Tennessee, and Baton Rouge, Louisiana. ENSCO has an analytical laboratory at Little Rock, Arkansas.[76]

Financial Viability and Liability. ENSCO is the fifth largest company in terms of hazardous waste management revenues; ENSCO's hazardous waste management revenues are greater than USPCI, Inc. but less than RES (table 7-3). The company's total revenues have climbed at the highest rate of the Big Eight companies and so has its pretax income. Net income fell back in 1984 after a large jump in 1983. Transportation, storage, and disposal of PCBs, toxic, and hazardous wastes have provided over 80 percent of the company's revenues.[77]

Financial analysts' reviews of ENSCO have been positive. Jeffrey Klein of Kidder, Peabody & Co. rated the company "attractive for risk oriented investors" with strong earnings gains likely in spite of the limited liquidity and volatility of the stock. Klein's analysis points out that the company's variable earnings record reflects the entrepreneurial nature of the management and the absence of consistent planning and operations discipline. He also notes, however, that a new president and chief operating officer brought in by management has improved organization of operating divisions and management responsibilities.[78] Kay Hahn of the Chicago Corporation recommended the stock for "aggressive investors" in the hazardous waste industry.[79]

In the past the company had to borrow substantial sums for incinerator modifications, a practice that gave it a high level of debt in relation to capital.[80] ENSCO will need to use its capacity to the fullest to benefit from that investment. The company's main source of revenue is its incinerator. Trouble-free operation of that facility is critical to the company's earnings.[81]

The company completed a stock offering of 1,069,500 common shares at a price to the public of $17.75 on October 10, 1985. The proceeds are intended for capital improvements at El Dorado ($6 million) and for two additional modular incinerators ($6 million). The balance will be added to the company's working capital to reduce past debt.[82]

Liabilities from the company's site in Minnesota have been limited to continued monitoring until 1989 at a cost of $10,000 per year under an agreement with the The Minnesota Pollution Control Agency.

Major Litigation. ENSCO's involvement in litigation concerns remedial action at its site in El Dorado. (See chapter 9 for more details.)

Corporate and Public Relations. Citizens are concerned about air pollution and ground water contamination. Citizen actions are reviewed in CEP's evaluation of the El Dorado facility in chapter 9. ENSCO's CEP rating for corporate and public disclosure is one of the best (see table 7-2).

PROFILE: IT

Over the last 4 years, IT has expanded from its headquarters in California, where its land-based facilities are, to provide services in 14 other states. IT's yearly revenue of nearly $144 million makes it the second largest commercial hazardous waste management company in the country, second only to CWM (see table 7-3). IT's sales have nearly tripled in 5 years, and its net income has doubled. IT owns or operates seven facilities in California and has a 30 percent interest in two others.[83]

Corporate Evaluation. IT is well positioned to prosper from the expansion of hazardous waste activities under the 1984 RCRA amendments and the CERCLA clean-up activities. It has the technical expertise among its own staff to bring to bear on the problems of remedial site work.

IT is tied with ENSCO for second rank of the eight companies and received a C+ in CEP's evaluation (see table 7-1). IT ranks highest of the eight companies in one category, compliance with government regulation, in the mid range in five categories but next to lowest in CEP's site evaluation. The company has avoided certain potential liabilities by not owning or operating any injection wells. (Only three of the eight companies own or operate injection wells: BFI, WMI, and RES.) CEP gave the Benicia site a D as the result of a number of deficiencies (see chapter 9).

IT's strengths lie in site analysis and remedial action. These activities provide only 30 percent of IT's revenues, but are the growth segment of its business.

History. The company began in 1926 as Industrial Tank, a company that responded to oil spills with industrial and marine cleaning services. Starting in the late 1970s, IT purchased some sites which had been used as landfills and began operations at others.[84] Since 1978 IT has expanded its capabilities and now specializes in handling hazardous spills and remedial cleanup of contaminated sites.

Hazardous Waste Management. IT's major capability is emergency response and remedial cleanup of hazardous waste sites. It has administered a U.S. EPA Environmental Emergency Response Unit Contract (EERU) since 1981. This EERU contract represented net sales to the company of $12.9 million in fiscal year 1985.[85] The contract places the company in charge of responses for U.S. EPA at Edison, New Jersey, and serves much of the East Coast. The company also provides training on proper procedures and safety precautions in dealing with hazardous wastes. IT was the contractor to clean up a serious PCB spill in downtown San Francisco and for a major chemical spill caused by a 43-car train derailment in Louisiana.[86]

IT offers one of the widest ranges of analytical and remedial services of any of the eight companies. The company concentrates on assessment, mitigation, and decontamination for emergencies and for remedial work. The company offers qualitative and quantitative analytical chemistry services at six laboratories across the country. Its instrumentation includes gas chromatographs with mass spectrophotometers, the most sophisticated equipment for identifying individual organic chemical compounds. The company also has seven mobile laboratories. Of its 210 employees, 60 have degrees in chemistry, biology, microbiology, or physics. The company maintains engineering offices in nine states. Most of the 308 engineering services employees have technical degrees in chemical, environmental, and civil engineering, industrial hygiene, geology, or hydrology.

IT contracted to operate U.S. EPA's mobile incineration equipment, which is now successfully destroying PCBs at a site in Missouri.[87] IT has also developed other portable treatment equipment.[88] IT owns six treatment and disposal facilities and leases a seventh, all in California. The company also has a 30 percent ownership interest in Environmental Protection Corporation (EPC), which in turn owns and operates two additional sites in California. The seven IT sites comprise nearly 3350 acres, of which a third is acreage permitted by the state for hazardous waste management. All sites have over 20 years of estimated remaining life.

Only two of IT's facilities offer chemical treatment options: Harbor Treatment Facility, Wilmington, California, and the Martinez Treatment Plant in Martinez, California. The other sites serve primarily as storage facilities. The Imperial Valley Site at Westmoreland, California, offers land treatment as well (see table 7-6).

In 1980 IT acquired Hydroscience, Inc., a former subsidiary of the Dow Chemical Company. In 1984 IT acquired D'Appolonia Waste Management Services, Inc., a geoscience consulting organization, and Stoner Laboratories of Santa Clara, California, specialists in pesticide analysis.

Proposed Expansion. In 1979 IT announced plans to construct a new major treatment and disposal facility in Ascension Parish, Louisiana, that would offer incineration. The company secured the required federal and state permits during 1983, but the issuance of permits by the state of Louisiana has been challenged. This proposed facility would allow IT to serve major generators of waste in the Texas Gulf region. IT also attempted to place a facility in Massachusetts, but abandoned its efforts in October 1984 after spending 3 years in and out of court. IT withdrew an application to construct site a hazardous waste management facility in the Santa Clarita Valley in California in January 1982. The application had been in the name of Uni-Tech Systems, Inc.

Financial Viability and Liability. IT has shown continuing financial strength in its growth. It has been highly rated by two financial analysts, Augenthaler of E. F. Hutton[89] and Patrick McVeigh of Franklin Research and Development.[90] Both made strong long-term predictions about IT's growth. IT has been able to generate sufficient revenues to purchase other companies, to spend over $1 million on research and development, and to continue investment in the proposed Louisiana facility.[91]

The major part (33 percent) of the company's revenues come from decontamination and mitigation and transportation, treatment, and disposal (33 percent). In 1984 engineering services provided 25 percent of revenues—nearly triple what engineering services yielded in 1983. Analytical services provided nearly 9 percent of revenues in 1984—almost double the yield of 1983.

IT, like other hazardous waste companies, has not been able to renew many of its existing liability insurance policies at prior coverage limits. It has not renewed its EIL insurance. The only new policy IT has been able to arrange specifies $10 million in coverage, but it indemnifies the carrier for all losses and expenses under the policy, and supports the indemnity commitment with a $10 million letter of credit. According to IT this arrangement satisfies current federal regulatory requirements but not California requirements. IT obtained an order from the state which will allow it to continue operations at its California treatment and disposal facilities through May 1986. IT has potential responsibility for one CERCLA site and possible leakage at its California landfill facilities. Questions have been raised about the quality and suitability of IT's surface impoundment and landfill facilities.[92] IT contests the validity of the report and states that the technical background of the report is inadequate.[93]

Major Litigation. CEP's LEXIS search did not reveal major litigation involving IT listed under its full name. LEXIS did not include the abbreviation *IT,* however, so CEP could not include *IT* in the search. Cases mentioned in the 10-K concern IT's attempt to construct its new facility in Louisiana.[94] The issuance of the permits was challenged in a case by Save Ourselves, Inc. et al. against the Louisiana Environmental Control Commission, which issued the permits. The Supreme Court has required that the court of appeals review the application with fuller definition of its costs and benefits and consideration of alternative sites.[95] Earlier in 1983 IT was charged with conflict of interest for buying a site it was contracted to evaluate for the state.[96] State and federal permits were not revoked.[97]

Corporate and Public Relations. CEP was able to identify organized community opposition to four facilities: IT's Westmoreland and Benicia landfills, the Wilmington storage terminal, and EPC's Bakersfield landfill. CEP found no organized opposition to IT's Fellows or Kelseyville facilities and found no information about the two Martinez facilities or the Collinsville facilities.

PROFILE: IU AND ITS SUBSIDIARIES

IU has annual revenues of $2.5 billion; it is the largest of the Big Eight (see table 7-3). IU's hazardous waste revenue, principally from Envirosafe Services, Inc. (ESI), is one of the smallest of the eight companies reviewed; only AMEC's hazardous waste income is smaller, according to CEP's estimate. IU was not willing to state the revenues or income of ESI. CEP estimates ESI's revenues to be over $10 million.[98]

IU is 20th of *Fortune*'s 100 largest diversified service companies. IU is an aggregate of more than 240 corporations, including several trucking subsidiaries, a distribution business in the southeastern United States, agribusiness in Hawaii, water and sewage services in 14 states, and industrial and utility waste services here and abroad. Although IU placed 20th in revenues, it was only 73rd in net income. The company has experienced a 24 percent decline in total return to investors over the last 10 years.[99]

Corporate Evaluation. In the overall evaluation CEP gave IU a D, the lowest of the eight companies. The company ranks among the lowest four companies for five of the seven categories. In the site evaluation CEP gave IU's facility in Grandview, Idaho, an F, the lowest ranking in the West. The hazardous waste services offered by IU are limited.[100] Expansion of these services has been proposed by IU at sites in Narvon, Pennsylvania, and Devers, Texas.

CEP estimates that hazardous waste management yields 5 percent of IU's total revenue.[101] IU's environmental services provided about 9 percent of IU's revenue but 16.5 percent of operating earnings. In spite of the good results produced by ESI, IU's financial strength must be judged on the merits of its major components, the most important of which is trucking.

History. IU was first incorporated in 1926 as a utilities holding company (International Utilities). The company's first holdings included gas, electric, and water companies. IU acquired trucking and shipping companies later. In the last few years, IU has acquired several trucking lines. In fact, it is the nation's largest trucking group. It has attempted to integrate these lines, but operations have not been profitable. IU now plans a major restructuring, which includes spinning off Ryder–P.I.E. Nationwide, Inc. (Ryder–P.I.E. was sold in December 1985) and IU's 50 percent stake in General Waterworks, a move that will halve the quarterly dividend to reduce debt.[102]

IU's involvement in hazardous waste stems from its experience in managing utility wastes. An IU subsidiary, Conversion Systems, developed a microencapsulation process using fly ash from coal-burning utilities. Fly ash, lime, clays, and dewatered sludges are mixed to form a hard material. In June 1983 IU's Conversion Systems acquired Fondessy Enterprises, Inc. (FEI). FEI owns a facility in Oregon, Ohio, which had been operated as a landfill since 1954. In 1964 FEI formed Associated Chemical and Environmental Services (ACES) to handle spills and remedial projects and to transport oil and wastes.[103]

Hazardous Waste Management. ESI has two active TSDFs (see table 7-5). FEI—a 100-acre facility in Oregon, Ohio—opened in 1956. Part B of FEI's RCRA application was submitted in October 1983. Its ground water monitoring has not been evaluated by U.S. EPA.[104]

The Ohio EPA, in response to a fire at FEI in July 1984, issued an enforcement order in October 1985 to improve management of the facility.[105] The other facility—Envirosafe Services of Idaho, Inc. (ESII) site B in Grandview, Idaho—is a 200-acre facility located at a former Titan missile base. The site's former owner filled the missile silos with hazardous wastes.[106] (CEP visited and evaluated the Grandview facility; see chapter 9.) ESII owns another former missile base nearby in Bruneau, Idaho, which is now closed. Both sites were formerly owned and operated by Wes-Con, Inc. ESI owns a 450-acre clay mine and processing plant in Narvon, Pennsylvania, and 144 acres in New

Jersey, and an inactive facility for the treatment of industrial liquid waste in Marcus Hook, Pennsylvania.[107] Part B of the RCRA application for Marcus Hook was submitted in April 1985. Part B applications have been filed for a new facility near Lancaster, Pennsylvania, at the Narvon clay mine and for a proposed facility at Devers, Texas.[108] The Texas Air Control Board turned down the application for Devers in a precedent-setting case—it was the first permit denial based on the potential hazard of air pollution. ESI is still pursuing a license for the Texas facility.[109]

Subsidiary Structure. As shown in the Directory of Corporate Affiliations, the name Envirosafe is used by a variety of subsidiaries: by Envirosafe, Inc. of Delaware; Envirosafe Services, Inc. of Delaware and its three subsidiaries (ESI of Iowa, of Pennsylvania, and of Texas); by Envirosafe Services of Idaho, Inc.; and by Envirosafe Services of Georgia, Inc., a subsidiary of Earth Management, Inc. (a Georgia corporation of which IU owns 63.4 percent).[110] IU indicates that since December 1984, all hazardous waste management activities are under ESI with headquarters at Horsham, Pennsylvania.[111] The legal relationships of the various subsidiaries remain unclear.

Two other companies form the main component of IU's environmental services. International Mill Service processes slag, reclaims scrap metal for steel mills, and processes metal from aluminum drosses. Conversion Systems designs, constructs, and operates waste management systems to convert flue gas, desulfurization sludge, and fly ash from coal-burning electric utilities into landfill material.[112]

Financial Viability and Liability. IU's environmental services as a whole (including those for steel mills and coal-burning utilities) were profitable; they provided 9 percent of 1984 revenues and 16.5 percent of operating income. Those revenues increased 20 percent over 1983. ESI expenses for cleaning up a disposal site and higher waste processing and site closure costs kept total income down, however.[113]

The revenue gains of the industrial and utility waste subsidiaries made up for losses incurred by ESI, which owns the industrial waste disposal facilities. IU expects to spend $4.7 million over the next two years to meet legal requirements at ESI facilities. These expected expenses are a tiny fraction of IU's more than $1 billion in assets, and less than 1 percent of its annual capital outlays.[114]

Outlays for cleanups could mount. Of particular concern is leakage from missile silos and the potential for leakage from unlined trenches at the Idaho sites. From 1973 to 1981, before ESI owned the facility, Wes-Con used 18 trenches on the site. ESI believes that Wes-Con directly buried liquids and incompatible wastes (including PCBs) in the same trenches. Wes-Con also used the abandoned missile silos for approximately 17,850 tons of wastes which include 7300 tons of PCB wastes.[115] These and other aspects of this site are described in more detail in CEP's facility description and evaluation in chapter 9.

The other active ESI site is FEI in Oregon, Ohio, near Lake Erie. An intake pipe for Toledo's water supply travels under the site between two of the trenches. Over 480,000 people are supplied by water from these intakes.[116]

If the potential problems at IU's hazardous waste sites do not cause substantial harm, it would appear that annual capital outlays and expenses for cleanup or closure at the present level could be sustained by the profitability of the other waste-handling segments.

Major Litigation. CEP's search of legal cases found only two involving IU, although with the large number of subsidiaries involved, CEP could have missed some cases. Neither of the cases involved handling of hazardous wastes. The search was limited because LEXIS did not index the name *Environmental Services.*

Corporate and Public Relations. Although only two sites remain active, IU has operated hazardous waste operations at three other sites known to CEP: Honeybrook, Pennsylvania; Marcus Hook, Pennsylvania; and Bruneau, Idaho. In addition, IU is proposing two other facilities, one at Devers, Texas (Envirosafe Services of Texas, Inc.), and one at Narvon, Pennsylvania (Envirosafe Services of Pennsylvania).

CEP was able to identify organized local citizen opposition at six of IU's eight facilities, at ESII Site B, FEI, Honeybrook, Marcus Hook, and at both proposed facilities (see table 7-4).

PROFILE: RES AND ITS SUBSIDIARIES

RES turned early to incineration as a solution to problems it encountered in transporting hazardous wastes. It is one of the four firms (ENSCO, ITC, RES, USPCI, Inc.) that are in the hazardous waste management business exclusively and make concentrated efforts to develop technological solutions.

RES's annual revenues of over $69 million make RES the fourth largest commercial hazardous waste company, smaller than BFI and larger than ENSCO. Its pretax income from hazardous waste management is higher than BFI's and places the company in third place in earnings (see table 7-3).

Corporate Evaluation. RES has potential strength in its more than a decade of experience in the field of high-temperature incineration. It has, in addition to its incinerator that is authorized to burn PCBs, two more that have undergone expensive development with the potential of PCB incineration in mind.

RES ranked among the top four companies in six of CEP's seven categories, and sixth in the other category. CEP's evaluation gives RES an average grade of C+, in the same group as IT and ENSCO.

History. RES evolved as part of the RLC Corporation, a major transporter of bulk chemicals calling itself a pipeline on wheels. RLC had to develop methods to clean out its tankers safely and, in addition, had experienced difficulty in finding acceptable ways to dispose of the wastes it transported. The company chose incineration as the most effective method of disposal for many wastes and made a major investment in high-temperature incineration. It built the first large-scale commercial hazardous waste incinerator in the United States in New Jersey in 1969.

RLC owned RES from 1968 until mid 1982 when RES spun off. Six of the officers and/or directors of RLC Corporation are also officers and/or directors of RES, including John W. Rollins, chairman of the executive committee. Rollins owns 15.9 percent of the outstanding shares of RLC. John W. Rollins, Jr. is also on the board of both companies.[117]

RES was originally incorporated in 1968 as Rollins–Purle, Inc. Its name was changed to Rollins Environmental Services, Inc. (RES) in 1972, at which time it operated facilities at Bridgeport, New Jersey, at Deer Park, Texas, and at Baton Rouge, Louisiana. In 1978 RES organized subsidiary corporations to own and operate each of these disposal facilities. In 1978 RLC Corporation organized a subsidiary, Rollins Environmental Services of Louisiana, Inc. (RES of LA) to acquire a deep-injection well and related facilities located in Plaquemine, Louisiana, owned by a firm named CLAW. This deep-well facility is operated by RES of LA.[118]

Hazardous Waste Management. RES subsidiaries operate three of the five major commercial incineration facilities in the country. The three high-temperature incinerators can offer liquid injection and the capabilities of a rotary kiln. These three facilities are RES (NJ) in Bridgeport, New Jersey; RES (TX) at Deer Park, Texas; and RES (LA) at Baton Rouge, Louisiana (see table 7-5).

The incinerator at Deer Park, Texas is authorized to destroy PCBs to the level specified by TSCA regulations. According to RES "an on-site regulatory presence and advanced control systems assure that the incinerator operates within its permitted parameters."[119] Other capabilities at this site consist of an activated sludge system, chemical treatment, and landfill. There are also surface impoundments, storage containers, and treatment tanks.[120] According to RES the possibility of ground water contamination is presently being assessed at this site.[121] A U.S. EPA damage incident summary documented ground water and suspected surface-water, air, and soil contamination.[122]

The RES (NJ) facility lies on the west side of Interstate Highway 295. A high-temperature incinerator is the only commercial treatment option available at the site.[123]

An explosion caused by contractor personnel who were rebuilding out-of-service tanks resulted in several explosions in the tank farm area in 1977. Six people died and several were injured. The tank farm was rebuilt and operations resumed in 1978.

RES (NJ) was fined in 1980 for violating a consent decree to limit the Bridgeport incinerator's air emissions. An explosion in the rotary kiln caused state officials to shut down the incinerator temporarily in 1981.[124] The facility has operated without significant incident since, and U.S. EPA advised RES (NJ) in the fall of 1985 that its incinerator was the first to be authorized to receive Superfund wastes.[125]

The township of Logan, in which the Bridgeport facility is located, has passed an ordinance prohibiting the storage of PCBs. RES has contested this ordinance, but the Superior Court of New Jersey found it to be valid.[126] According to RES, that decision is currently the subject of an appeal in which the attorney general of New Jersey has joined RES (NJ).[127]

CEP visited the RES (LA) facility in Baton Rouge and graded it C− (see chapter 9). On August 6, 1985 the Baton Rouge incinerator was temporarily shut down by Louisiana officials when it was reportedly "running out of control."[128] The company has strongly denied this assertion and the Court of Appeals for the First Circuit, State of Louisiana, authorized RES (LA) to resume incineration operations in October 1985. The incinerator is now back in operation.

RES of LA in Bayou Sorrell, Louisiana, is a deep-injection well site with an annual capacity of 250 million gallons. Three drinking water wells are in the vicinity. The site is next to the Sullivan Lake Oil and Gas Field and the Bayou Sorrel Oil and Gas Field; each field contains a number of wells drilled to an unstated depth.[129]

Modular Incineration. Rollins has constructed a pilot, plant-scale fluidized bed rotary incinerator.[130] According to RES its rotation is faster than that of conventional rotary kilns, and the fluidized bed provides improved combustion and sulfur dioxide removal for low-BTU wastes. The kiln can be designed as a large stationary unit or as a smaller portable unit which could incinerate low-BTU wastes (sludges and contaminated soils) on site.[131] This technology is best suited for small and reasonably uniform volumes of wastes.[132]

RES has developed an above-ground concrete vault which is placed on a reinforced concrete base. RES believes the cost and other and requirements of keeping wastes in such a vault compare favorably with those of conventional landfills. The technology has been mentioned favorably in the 1984 RCRA amendments.[133]

Financial Viability and Liability. RES has shown solid growth in its total revenues, pretax income, and total assets. Analyst Jeffrey Klein rated RES favorably.[134] The company ranked among the top companies in CEP's financial analysis. Since most of its revenue derive from incineration, any evaluation should consider the problems RES has had with its incinerators. Complete information about the August incident in Baton Rouge is not available at the time of this writing. It has one of the better current ratios of the eight companies and one of the highest debt ratios. Its total assets are just over $100 million, $25 million more than its liabilities (see table 7-3).

Corporate and Public Relations. CEP identified organized citizen opposition to all four facilities owned by the RES subsidiaries.

PROFILE: USPCI, INC. AND ITS SUBSIDIARIES

Beard Oil Co., a small independent oil and gas company, owns roughly 70 percent of USPCI, Inc., the holding company for USPCI. Beard Oil organized USPCI in 1968 to provide disposal services for itself and other oil and gas companies. USPCI expanded in 1974 to provide waste transportation and disposal services to industries. USPCI, Inc. has grown with the increased need for disposal services, and, in the last several years, the company has expanded far faster than Beard Oil.[135] Since 1980 the revenues of the holding company and the rest of the hazardous waste treatment segment have more than tripled, while the oil and gas revenues have barely increased.

Corporate Evaluation. USPCI, Inc. places in the upper half in six of the seven CEP ranking categories. The company is at the top of the eight companies in CEP's corporate evaluation with grade of B (see table 7-1).

USPCI's two landfill sites opened under RCRA and they are in unpopulated areas of relatively low rainfall. They are well removed from potable ground water.

USPCI's laboratory capabilities, solvent and oil recovery technology, transportation capability, its expanding investments in solvent recovery, and waste water-treatment technology make it a leader in the commercial industry.

USPCI's facility in Clive, Utah, received a B+, the highest grade of all the facilities CEP evaluated.

History. Beard Oil Company was founded in 1923 by a Beard family member. W. M. Beard is now chairman of the board and chief executive officer and owns 36.9 percent of Beard Oil's stock. Beard Oil has undeveloped oil and gas holdings over 1.1 million acres in 21 states.

Beard Oil's net earnings turned negative in 1982 and became a nearly $4.5 million loss in 1983. The net returned to a half-million dollar gain in 1984 with a small increase in the oil and gas segment and an 87 percent increase in the hazardous waste segment. In October 1983, USPCI, Inc. was organized as a Delaware holding company for USPCI and its subsidiary Hydrocarbon Recyclers, Inc. (HRI). A 1985 distribution of USPCI, Inc. stock to Beard Oil Stockholders leaves Beard Oil with 72 percent of USPCI, Inc.

USPCI, Inc.'s subsidiary, USPCI, was originally organized in 1968. In 1969 it became a wholly owned subsidiary of Beard Oil. In 1981 it acquired 50 percent of Hydrocarbon Recyclers, Inc. of Tulsa and acquired the remaining stock in 1983.[136]

Hazardous Waste Management. USPCI's primary business is treating oil and chemical wastes; it does not treat or store nuclear or biological waste. USPCI owns treatment sites in Oklahoma and Utah where there are few competitors. It also owns transportation terminals, vehicles, and solvent and oil recovery facilities (see table 7-6).

The company does not own deep injection wells, but it has operated two injection wells and may be partially responsible, along with 250 other parties, for a site in southern Oklahoma to which it transported waste. It has installed a waste water-treatment system at its Oklahoma facility that is intended to clean waste water sufficiently to allow reuse and avoid the need for deep-well injection.[137]

USPCI operates two treatment facilities. Lone Mountain, which is northwest of Oklahoma City, has surface impoundments, waste piles, landfill cells, and solvent and oil recovery equipment. It is on a 560-acre tract of land leased through August 1988 with provision for renewal through August 2008. USPCI has filed a RCRA application.

The Oklahoma Department of Health placed a moratorium on acceptance of new waste streams for 2 months in 1984 because of alleged leakage. A new landfill cell has been opened under a consent decree that required corrective actions costing over $900,000. The Oklahoma Department of Health states that all items of the current agreement have been corrected.[138] The other facility, Grassy Mountain, is 80 miles west of Salt Lake City, Utah. The facility has landfill cells, surface impoundment, and land treatment areas. The facility opened in March 1982. One landfill cell is nearly filled to capacity. A second cell opened in September 1985. USPCI estimates that the second cell has a capacity to last until 1987. The land treatment area should have a remaining useful life of 40 years.[139] CEP visited the facility in April 1985. CEP's evaluation is in chapter 9. These two operations treated 110,000 tons of waste in 1984.[140]

Other USPCI, Inc. facilities include four truck terminals in Tulsa, Oklahoma City, Salt Lake City, and Columbia, South Carolina.

HRI, the solvent recovery facility in Tulsa, provides fractional distillation of mixed hydrocarbon waste and sells the recovered product. The recovered product may be solvents or oil that can be blended to dilute contaminants to meet federal standards. Operations at Lone Mountain provide partial refinement. The solid residue of these operations is stored in the company's landfill cells.

A special services group handles site mitigation and emergency response. USPCI owns National Analytical Laboratories (NAL) and an engineering division. Both provide support to other company services.[141]

Financial Viability and Liability. USPCI, Inc. is one of the smallest firms CEP analyzed, with annual revenues of $17 million. Its hazardous waste revenues are the sixth largest, above CEP's estimate of IU and AMEC's hazardous waste revenues (see table 7-2).

USPCI, Inc., with sales over $17 million, accounted for 40 percent of Beard Oil's revenues in 1984. Its pretax income was $1.65 million, an amount that turned Beard's 1984 earnings positive. Earnings derived from treating wastes at the company's two treatment and land emplacement facilities provided $7.35 million in revenues—40 percent of USPCI's 1984 hazardous waste management revenues. Waste Transportation Services, USPCI, Inc.'s second largest revenue producer, brought in revenues of nearly $7 million in 1984. This subsidiary's primary business is transporting customers' wastes to sites operated by other firms. Solvent recovery produces $1.3 million in revenue and adds an important technical capability. Special services brought in 7 percent ($1.2 million) of revenues and generated $1.8 million in revenues for other company operations. Laboratories and engineering accounted for $500,000 in revenues.[142]

USPCI, Inc. has good prospects for continued growth of its hazardous waste management operations if all facilities receive final permits from U.S. EPA, Oklahoma, or Utah. In USPCI, Inc.'s favor are locations of its sites and the wide variety of services—from laboratory analysis to emergency response to treatment and landfill—that the company can offer.

The question remains whether it can generate sufficient revenue on its own to fund construction at its present facilities. More important is whether the company can fund the corrective measures that may be needed at its Oklahoma facility, cover any potential liabilities at a Superfund site, and provide the financial guarantees required under RCRA.

Until 1984 Beard Oil provided most of the capital for USPCI's expansion, and before 1984 the capital advances were interest free. From 1984 on the advances will draw 12 percent interest. In view of Beard Oil's diminished revenues from its drilling and drilling services, predictions of lower drilling production, and the falling prices for oil, Beard will probably not provide future capital funds. Beard Oil has stated its intention to discontinue providing funds for USPCI, Inc.[143]

The growth of revenue of USPCI, Inc. has been positive (see table 7-3). CEP's calculation of its growth of pretax income is a high negative number because of a relatively large loss in 1982. Pretax income turned around in 1983 and the average annual growth rate to 1984 was 173 percent. Beard Oil received a B− rating from Standard & Poor's in July 1984.[144]

USPCI, Inc. first became a publicly held company in January 1984 when 12 percent of its stock was distributed to Beard stockholders. May 1985 brought further distribution of stock, leaving Beard Oil with only 72 percent of the company.[145] With so short an independent history and relatively few shares available, financial evaluation of the company has been limited.

Corporate and Public Relations. USPCI, Inc.'s two landfill facilities are far removed from populated areas. CEP did not find any organized local opposition to either facility (see table 7-4).

PROFILE: WMI AND ITS OPERATING GROUP

WMI is one of the two largest waste-handling firms; BFI is the other. By several measures WMI is larger than BFI—the largest, in fact, in the United States. WMI's total 1984 revenue was $1315 million. Chemical Waste Management (CWM), one WMI operation, accounted for 15 percent of revenue or nearly $200 million, the highest hazardous waste revenue figure of the Big Eight (see table 7-2).

In 1984 WMI took over many of the assets of SCA Services, Inc., which had been a major competitor. WMI, with SCA residential contracts, serves over 430 communities and over 326,000 commercial and industrial customers. It operates 89 solid waste landfills in 36 states and in Canada, of which it owns 52. BFI, in comparison, operates 80 solid waste landfills in 39 states and Canada, of which it owns 35.[146]

WMI, like BFI and several others of the Big Eight, owns closed solid and hazardous waste sites that may have undiscovered problems. WMI owns five Superfund sites. Most of WMI's hazardous waste management revenue derives from land emplacement in a time when legislation and regulations are beginning to limit land disposition. WMI, however, has positioned itself better than BFI by investing in new technology. One major asset gained in the SCA transaction was an incinerator in Chicago that can burn PCBs and other hard-to-destroy organics.

Corporate Evaluation. CEP's evaluation placed WMI in the middle group with a C grade. WMI's ranking average is above AMEC and IU but just below BFI. It ranks among the top four companies in four of the seven categories, fifth in two categories but seventh in compliance with government regulations.

CWM is well positioned to take advantage of the work that must be done to clean up hazardous waste problems. CWM's technological capabilities—high-temperature incineration, site remedial work, and solvent recovery—should enable WMI to generate enough capital to correct most of the site deficiencies. WMI ranks at the top of CEP's technology category. If management can now put the resources of the firm to work using and developing advanced technology to remedy the problems of rapid growth, CWM can experience sure and steady growth.

History. WMI was founded in 1968 when Dean Buntrock, now WMI's chairman of the board brought together two companies belonging to his inlaws—

Atlas Disposal Company of Chicago and H. Wayne Huizenga's Southern Sanitation Service of Fort Lauderdale and added a third, Acme Disposal Company of Chicago.[147] In the early 1970s WMI grew through acquisition of nearly 350 smaller trash companies. In the mid 1970s the corporation expanded overseas to start municipal waste programs in Saudi Arabia, Argentina, and Venezuela. In the late 1970s WMI bought a number of existing hazardous waste landfills.

WMI moved into radioactive waste disposal in 1982 by buying Chem-Nuclear Systems in Barnwell, South Carolina. The site is one of three commercial radioactive waste disposal facilities now in operation.

WMI acquired a large part of SCA services, a major competitor.[148]

Hazardous Waste Management. WMI obtained from SCA the high-temperature incinerator in Chicago, treatment facilities in New Jersey, a treatment and landfill facility in New York, a landfill in Indiana and a proposed treatment facility near Memphis. The most important of these facilities, the incinerator in Chicago, is the largest commercial hazardous waste incinerator in the United States and a major asset. It can destroy PCBs, solvents, organic residues, certain hydrocarbons, herbicides, and laboratory waste.

WMI notified U.S. EPA of a total of 151 hazardous and solid waste sites. Of these, 59 have been identified as hazardous waste sites and the remainder as solid waste sites. WMI has filed Part A of the RCRA application for eight of the hazardous waste sites. Seventeen sites are known or believed to be inactive or closed (see table 7-6).[149]

Twenty-eight of the 59 hazardous waste sites that WMI owns are categorized by the states as major facilities. Five of these major facilities are closed and two are Superfund sites: National Industrial Environmental Services landfill in Furley, Kansas, and Omega Hills North landfill in Germantown, Wisconsin. Two closed major facilities are not Superfund sites: G.R.O.W.S. Inc. Landfill in Morrisville, Pennsylvania, and Lyncott Corp. Landfill in New Milford, Pennsylvania. WMI owns three other Superfund sites: Holland Lagoons in Holland, Michigan; Modern Sanitation Landfill in Lower Windsor Township, Pennsylvania; and Lauer 1 Sanitary Landfill in Menomonee Falls, Wisconsin. Modern Sanitation was part of the SCA acquisition.[150]

Active Facilities. CWM's major southeastern facility is in Emelle, Alabama. There is also a transfer station in Florida. WMI intends to close its facility at the Port of Chickasaw, Alabama.

Emelle became the center of a national controversy when over 2.8 million gallons of PCB-contaminated liquid was brought there for transport to an incinerator ship in the Port of Mobile. The PCBs were to have been removed by January 1984. They were too concentrated to be mixed with kiln dust and buried in the landfill cells at Emelle. The concentrated liquid is now being transported to the incinerator in Chicago to be destroyed because EPA has been slow in certifying WMI's ship for burning PCBs.

Emelle was authorized to handle for hazardous waste in 1977 under the name Resource Industries of Alabama. CWM acquired the facility about 9 months after it received a permit.[151]

In the Northeast the major facilities are in Model City, New York (a site WMI gained from SCA); a deep-well injection facility in Vickery, Ohio; and the West Carrolton, Ohio, solvent recovery facility. In the Midwest CWM's major facilities are in Illinois—the landfills at Calumet City and Joliet and the incinerators at Chicago and Sauget. CWM's major regional laboratory is in Riverdale, Illinois. CWM's major western landfill facilities are in California at Kettleman Hills, Bakersfield, and Coalinga. CWM's subsidiary Chem-Security operates a state-owned facility in Arlington, Oregon. The south central sites are landfill facilities at Port Arthur, Texas, and Lake Charles, Louisiana; laboratory facilities at Dallas and deep-well facilities in Corpus Christi, Texas. For an overview of company-owned facilities see table 7-6.[152]

Business Activities. WMI, incorporated in Delaware, consists of over 200 wholly owned American operating groups, 15 foreign operating groups, and equity positions in various joint ventures. The hazardous waste operating group, CWM, operates through at least five direct subsidiaries and indirectly through whichever WMI subsidiary owns a particular facility.[153]

In 1984 WMI acquired SCA Services of Boston, the third largest solid waste company in the country at the time. After its review of the antitrust implications of the purchase, the U.S. Department of Justice (U.S. DOJ) required that WMI divest certain SCA facilities to maintain competition in waste operations. WMI contracted with Genstar, Inc., a Canadian company, for the sale of 40 percent of SCA's assets (including all domestic waste operations). The sale satisfied U.S. DOJ requirements, and Genstar became the third largest solid waste company in the United States as a result.

WMI's SCA acquisition included 25 solid waste landfill units, 43 solid waste collection units, a solid waste transfer station, and eight existing or proposed hazardous waste facilities.[154]

Financial Viability and Liability. In 1984 WMI was 57th in revenues of the 100 top diversified service companies ranked by *Fortune*. By 1985 WMI had moved to the 46th position. In assets it ranked 22nd, in net income 12th, and in net income as a percent of sales, it ranked fourth. Its 1985 Standard & Poor's ranking was A and its Moody's rating was "upper medium grade".[155] Its overall 1984 net income rose 18 percent over 1983.

CWM increased its revenues by a third in 1984 to $199 million and accounted for 15 percent of total revenues. Chem-Nuclear Systems, another expanding segment, accounted for 6 percent, foreign operations accounted for 10 percent, and American solid waste operations provided 69 percent of total revenues.[156]

WMI has major potential liabilities. It owns five Superfund sites, one of which has been made the responsibility of its former owners. CWM is suing generators for help in cleaning up another of these sites. Altogether, WMI may have some responsibility for 15 Superfund sites (see table 7-7).

The National Industrial Environmental Services landfill presented a major liability problem when waste leakage was discovered. The Justice Department found the former owners liable for cleanup costs at the site they sold to CWM. The U.S. District Court for Kansas issued a decision in March 1985 that required the former owners to pay CWM a total of $10.6 million in damages because of misrepresentation of the site's condition. The damages included $6.9 million for cleanup costs incurred by CWM, $1.7 million for lost profits during the period that CWM could not operate the facility, and $2 million to cover possible cleanup costs through March 1986.[157]

In another case CWM is setting a precedent by suing the generators of waste dumped at Lyncott Corp. Landfill to help pay for cleanup there.[158] CWM purchased the site in 1980 from Lyncott Corp. Soon after the purchase the site was found to be leaking. CWM has won annulment of the ownership agreement, but the costs of remedial cleanup remain. CWM was held responsible for 100 percent of the cleanup costs, although it was responsible for less than 0.1 percent of all wastes disposed there.

A nationwide focus on the hazardous waste industry in 1983 brought to public attention a series of allegations against WMI.[159] The allegations brought about a corporate-wide investigation of SWM by U.S. EPA's NEIC. NEIC can cross state lines to review a company's facilities and practices.

The *New York Times* summarized active and threatened state enforcement actions against WMI in March 1983. The article alleged that waste-handling practices at WMI facilities in Vickery; Calumet City; Lyncott; Wichita; Germantown; and Lowry, Colorado, were improper.[160] CWM employees at Vickery and Calumet City alleged that hazardous wastes were dumped illegally. The Pennsylvania Department of Environmental Resources suspended existing and proposed operations at Lyncott, Elwin Farms, and Sunny Farms. The Kansas State Department of Environment and Health shut down WMI's National Industrial Environmental Services facility near Wichita in 1982. The company stopped taking hazardous wastes at Germantown, where it had reportedly operated in violation of state laws for several years. In Alabama the attorney general filed suit against CWM and U.S. EPA to block construction of two 800,000-gallon PCB storage tanks in an area susceptible to flooding.

WMI released a summary of responses to each item in the *New York Times* report on March 23, 1983.[161] The company also hired the law firm Karaganis, Gail, & White, Ltd. to investigate the charges of mismanagement. In September 1983 the resulting survey, the Karaganis Report, found many technical violations. Some consisted of less than complete compliance with written requirements; some involved serious noncompliance. The Karaganis Report denied certain charges and outlined environmental damage that previous owners of the landfills had caused.[162]

Since the report there have been major changes in CWM's management. CWM hired a number of former high-ranking U.S. EPA officials from the Carter and Reagan administrations and lawyers from environmental organizations.

The publicity surrounding the problems brought a rapid change in the investment community's perception of WMI. The company's stock dropped about one billion dollars, a third of its market value, in 2 weeks.[163] The stock had recovered by the end of 1985.

WMI carries a high level of EIL insurance, with a limit of $50 million per claim and an annual aggregate limitation of $80 million (subject to certain deductibles). This policy will only be available through 1985 under a one-year extension option. The insurer has notified the company of its intention not to renew the policy.[164]

Major Litigation. WMI has been a defendant in 15 major litigations. Table 7-4 does not include litigation concerning SCA Services. SCA has been the defendant in six cases and the plaintiff in eight cases.

Corporate and Public Relations. CWM owns and operates the largest hazardous waste landfill in the country, a facility that covers 350 of 2400 acres owned by WMI, at Emelle AL. The burial pits are 150 feet deep in a natural layer of chalk 450 to 550 feet deep. It is the only active hazardous waste landfill in Alabama.[165]

The Emelle site has been the subject of controversy for social as well as technical reasons. It is in a low-income, agricultural area where most of the residents are black. In 1983 over 288,000 tons of waste were brought there from 42 states and from U.S. military bases overseas. Only 25 percent of that waste was from Alabama.[166] The neighbors of the largest hazardous waste landfill in the country receive the benefits of employment and the taxes that the companies pay to local governments, but they receive no other economic benefits. An economic study by a local university estimates that operations at Emelle contribute $23 to $29 million to the state's economy.[167]

The controversy became a wider issue when the United Church of Christ Commission for Racial Justice visited Emelle in the fall of 1984. It joined with a county organization, the Minority People's Council, in taking issue with the disposal of toxic waste from other states and countries in a poor and minority community. The issues of environmental degradation and of worker health served to bring together black and white residents in an otherwise racially polarized area. Local residents organized in spite of the fact that some workers at the site opposed the demonstration because the landfill offered the only jobs available.[168]

CEP identified organized citizen opposition to 24 of the 32 WMI sites surveyed.

CHAPTER EIGHT
Picking a Winner: CEP Comparative Facility Evaluation Method

AVAILABILITY OF INFORMATION

The ten hazardous waste facilities CEP chose for detailed evaluation were all facilities for which information was available. They also provided a range of technologies and site geographies. The CEP evaluation method provides a logical framework within which to analyze differing management performances, siting parameters, and technologies.

For a nongovernment investigation like ours, the only sources of information about a facility's operations and problems are the facility operators; local, state, and federal officials; and informed neighbors or public interest groups. Insufficient government resources and the decentralization of information management often make it difficult to obtain enough data to perform a thorough evaluation of a site.

Visits to regional and local government offices provide the most information regarding a facility. Most of these offices can provide technical data and documentation of hearings and interviews involving government, corporate, and community views. Most government offices claim they have inadequate staff to duplicate and send material.

Most large commercial hazardous waste management companies maintain an open-door policy in regard to parties interested in visiting. They recognize the importance of maintaining good communications with consumers and the public. When a problem occurs at a site, however, or if the visiting parties seem adversarial, companies have been known to refuse entry or to be quite uninviting.

Even the inexperienced operator of a plant that produces hazardous wastes should visit RCRA facilities. Compare written information with visual observations. With a quick look and a short conversation, one can often identify inadequate freeboard or collapsing dikes (both deficiencies could result in the release of hazardous liquid wastes) and get a rough idea of management's attitude toward environmental concerns. But don't expect perfection. And don't expect to learn about all weaknesses, either. An on-site inspection can be a showtime for facility employees. Daily observations over time are necessary to detect subtle management strengths and weaknesses, and, in turn, to get a thorough understanding of the quality of a site's operation. U.S. EPA RCRA inspections, for example, require completion of a lengthy checklist that can easily consume a number of days. Reviews of records can take just as long. Also check with neighbors and local officials and review local newspaper articles.

If the investigating party is not a potential customer of a waste facility's services, extensive investigations can be unwelcome and could require legal authority. Local officials may have such authority, but independent researchers may have to rely on written material derived from government investigations. Unfortunately, however, variations in the quality of regional, state, and local regulation can lead to distorted perceptions of the quality of a facility's performance relative to facilities located elsewhere.

Sampling for contamination is prohibitively expensive and is rarely allowed to outsiders. Picture-taking is often controlled; company representatives may develop the film themselves to minimize disclosure of confidential processes or of details that reveal the companies that generate wastes at the facility. All of the firms CEP studied except IT allowed CEP to take photographs.

CEP's visits to the facilities generally lasted only a few hours each. Research and evaluation of the written records, however, took many months.

PART B OF RCRA PERMIT APPLICATIONS

The owner or operator of a facility is required to apply for a permit. These applications are followed by public hearings, reviews, and so forth. Part A of the application identifies general characteristics of the site such as location, wastes accepted, processes and capacities. Part B is a more extensive overview that often necessitates the submission of a dozen volumes and thousands of pages of technical data regarding present and proposed activities.

Because availability of information is an overriding concern in any investigation, submission of Part B was the first criteria CEP used to select the sample sites. As of June 18, 1984 (when CEP's field investigations began), U.S. EPA required the eight major commercial hazardous waste management firms to submit Part B of the permit applications for 55 facilities. Submissions had been made for 46 of these 55. CEP chose ten facilities from these 46 for detailed evaluation. Seven of the ten are among the 25 sites whose Part B submissions had been reviewed in detail by government permit writers (see table 8-1). The voluminous Part B provides much of the information needed to derive hazard scores for topographic and hydrogeologic conditions within a site's boundaries, the containment technologies used at a facility, and, to a lesser degree, the potential targets of contaminant migration.

TABLE 8-1. *Status of Facility Permit Applications as of June 18, 1984*

Company Name	**A**	**B**	**C**	**D**	**E**
American Ecology, Inc.	2	2	1	1	1[a]
Browning–Ferris Industries, Inc.	17	16	9	7	2
Environmental Systems Company, Inc.	2	1	0	0	1[a]
International Technology Corp.	9	6	2	1	1[a]
IU International Corporation	3	4	4	2	1
Rollins Environmental Services[b]	1	1	1	1	1
U.S. Pollution Control, Inc.	2	2	1	1	1
Waste Management, Inc.[c]	20	15	12	12	2
Totals	55	46	29	25	10
Percent of Total	100%	84%	53%	46%	19%

A: Number of facilities with Part B called by U.S. EPA
B: Number of facilities with Part B received by U.S. EPA
C: Number of facilities with Part B reviewed by U.S. EPA
D: Number of Part B reviewed existing facilities
E: Number of CEP evaluated facilities

DERIVED FROM: U.S. EPA, "Hazardous Waste Data Management System: Status of Permit Applications" (Washington, DC: 6/18/84).

NOTE: Companies are listed in alphabetical order.

a. For logistical reasons, the facility CEP evaluated is not the existing facility with a reviewed RCRA Part B application.
b. As of October 28, 1985, U.S. EPA has reviewed four of Rollins Environmental Services subsidiaries' facilities.
c. Figures include facilities acquired from SCA Chemical Services.

Most Part B's, however, are incomplete when they are first submitted, a fact due in part to the complexity of federal requirements. A U.S. EPA or state permit writer is responsible for reviewing the completeness and technical accuracy of Part B. Permit writers use checklists that outline only the general categories that the application should cover but the checklists themselves consist of over 40 pages of questions. Therefore, to evaluate a facility in depth, it is efficient to wait until a permit writer has reviewed the application and the facility has made the required additions and clarifications.

Unfortunately, of the applications reviewed, CEP was able to evaluate only seven revised Part Bs. CEP evaluated IT's Benicia facility because U.S. EPA had only reviewed Part B of EPC's West Side Disposal facility at the time. IT owns only 30 percent of EPC. Furthermore, IT allowed CEP to visit the Benicia facility only, and U.S. EPA was reviewing this facility's application at the time the data gathering took place. None of the Part B applications submitted for ENSCO's facilities had been reviewed. CEP evaluated ENSCO's incinerator at El Dorado, Arkansas, simply because it is the only operating treatment and disposal site ENSCO owns. CEP chose USEC's Texas Ecologists facility in Robstown, Texas, over the Beatty, Nevada, site to achieve better regional distribution of sites.

CORPORATE DISTRIBUTION

CEP evaluated two CWM facilities and two CECOS sites to reflect the magnitude of their hazardous waste management activities relative to the other firms. At that time CEP estimated that these two firms were the largest in the industry. Now, IT and Rollins may be larger than CECOS (see chapter 7).

Of the 12 CWM facilities with reviewed Part B applications, CEP originally chose three to distribute them by location and technology. After CEP evaluated the Evergreen Landfill, however, ownership of the future expansion acreage was transferred to Genstar, Inc. Consequently, the life expectancy of the Evergreen Landfill was reduced to early 1986. Thus, there was little reason to include CEP's evaluation of the Evergreen Landfill.

CECOS requested that CEP not conduct site visits at its two disposal facilities in Louisiana because the company's environmental manager for the southern region was newly appointed. CECOS was concerned that the regional manager would not be able to answer all questions raised concerning the facilities. CEP evaluated the two other active CECOS disposal sites instead.

CEP MODIFIES THE HRS

CEP modified the U.S. EPA's HRS and its normal application to compensate for important and easily remedied shortcomings that result when applying HRS to active facilities. Because of these modifications, CEP's hazard scores are not entirely comparable with the HRS scores used to rank Superfund sites.

Air Releases

U.S. EPA has voiced substantial misgivings regarding the applicability of the air route score to conditions at active facilities. Uncertainties concerning the air route score are threefold:

1. the causes of a release
2. the sampling protocol used to detect a release
3. the definition of an air release.[1]

The HRS users' manual warns that "data based on transitory conditions due to facility disturbance by investigative personnel are not acceptable."[2] At an active facility, however, air releases are likely to be caused by facility personnel, who, by the very nature of hazardous waste management, disturb the wastes on site. Thus, even though the language in the users' manual technically excludes releases caused by investigator disturbances from the HRS air route score, a release caused by any legal human activity somewhat distorts the intended use of the HRS, according to U.S. EPA.

But what if the normal operations of a facility cause air releases that may indeed represent a threat to human health? According to U.S. EPA, contaminants detected at levels "that significantly exceed background levels" do not constitute an observed release if they are recorded from stack monitoring. Otherwise, HRS would automatically categorize many facilities that operate sources of air emissions (such as incinerators) under permit as a hazard. Similarly, at the active face of a landfill, surface impoundment, or landfarm, air quality is likely to be significantly worse than background levels. If personnel working on the active face are wearing proper protective apparatus, this should not cause a problem. Therefore, only contaminants that are detected beyond the immediate vicinity of the active unit qualify as an observed release according to U.S. EPA. Inadequate occupational health and safety practices may result in serious problems from on-site air releases, however.

In addition, inadequate monitoring may understate off-site air emissions. Unless an interim-status incineration facility has received a permit under TSCA to burn PCBs, it may not undergo rigorous stack testing to determine permitted emission levels. Even after RCRA stack tests are initiated, they do not monitor for all potential hazardous releases to the air; rather, they are based upon the concept of destruction efficiencies. According to U.S. EPA's Science Advisory Board

> this approach emphasizes the elimination of several preselected compounds in the waste and does not fully address either partial oxidation or chemical recombinations, which may create new toxic compounds in the incineration process. Thus, to date, only a very small portion of the compounds found in emissions from incinerators has been identified qualitatively or quantitatively.[3]

Once a permit is granted, moreover, full-scale testing is not conducted until permit renewal time—often a year later. Performance tests are usually conducted under optimal burn conditions. During less than optimal conditions sampling is often discontinued.

Amendments in the reauthorized RCRA require U.S. EPA to set standards for monitoring active hazardous waste sites for air emissions. Interim-status facilities in general do not undergo "continuous or frequent monitoring" for air releases, although the preamble to the National Contingency Plan (NCP) states that "continuous or frequent monitoring would be necessary to establish the long-term level of release."[4] Only one landfill (CWM's facility in Model City, New York) out of more than two dozen inspected by CEP during 1984 and 1985 had air-quality devices on site, and the monitor had been installed just prior to CEP's visit. The current lack of data may give a better impression of the air pollution problem associated with RCRA facilities than is realistic.[5]

Even if the contaminants have been detected away from the active portion of a site, what exactly constitutes a release? Can odors detected off site and traced to the facility be interpreted as "data that show levels of a contaminant. . . . that significantly exceed background levels"?[6] The HRS users' manual is vague regarding these issues.

U.S. EPA, which has to defend its decision in court, will not accept any odor, smoke, cloud, spray or other emission as an observed release to the air unless it is detected with instrumentation that can replicate the evidence.[7] Nevertheless, strong odors that are detectable by residents near a site as well as visually identifiable releases of material whose composition is known prior to its emission could be interpreted reasonably as releases. If these releases are not health hazards themselves, they may indicate inadequate containment.

For all these reasons, CEP used a more comprehensive interpretation of what constitutes an observed release to the air than that used by U.S. EPA. The emissions that CEP recognized represent a nuisance and indicate potentially inadequate containment precautions, even though there may be no evidence that the observed releases have actually damaged human health or the environment. This difference between CEP's and U.S. EPA's application of the HRS is the major cause of the incomparability of total migration scores.

Regulated Unit within a Facility over Time

CEP took two other less restrictive approaches to the HRS. First, CEP's application of the HRS considers facilities as a whole, whereas U.S. EPA only considers one regulated unit at a time. One facility, for example, could include an incinerator, a surface impoundment, and a landfill. CEP selected the containment technology that presents the highest potential for migration for each pathway, as well as the characteristics of wastes contained by the method that received the highest score. U.S. EPA would conduct a separate analysis of each unit, whereas CEP scored the units together so that a facility's overall hazard score reflects its most hazardous unit(s).

Second, CEP counted documentation of a past release as an observed release in order to directly account for the historical problems encountered at a facility. This may penalize current management, but many incidents of observed releases detected prior to 1980 also continued after that time. CEP also considered corrective action to account for remediation of problems.

Liner Systems

The HRS was designed prior to the amended RCRA standards for landfill liners. Most of the landfills and surface impoundments on the NPL at the time were barely more than holes in the ground; they seldom had any lining system whatsoever. As a result U.S. EPA did not issue clear guidance on how to rate various lining systems. CEP altered the ground water route containment score for landfills and surface impoundments to differentiate more clearly among technologies that are currently in use.

U.S. EPA's criteria also uses poorly defined phraseology such as "essentially non-permeable" and "moderately permeable" to distinguish among clay liners. When considering the migration of hazardous wastes, numerous studies have indicated that clay liners are never "essentially non-permeable." In fact, clays have varying degrees of permeability, and different wastes have differing capabilities to go through various clays. CEP's alterations are outlined in table 8-2.

Remember, however, that CEP rated facilities by the least effective containment technology on site in order to reflect the potential for ground water contamination. Therefore, even if a facility's operating landfill cell is double lined, if the site also contains closed or inactive cells without liners, the facility's containment score would be 3.

Letter Grades

U.S. EPA's HRS establishes a scale from 0 to 100, with 0 representing a site that poses a relatively insignificant hazard and 100 representing a site that

TABLE 8-2. *Comparison of U.S. EPA and CEP Containment Criteria*

U.S. EPA Criteria	Assigned Value	CEP Criteria	Assigned Value
Essentially nonpermeable liner, liner compatible with wastes, and adequate leachate collection system.	0	RCRA minimum standard double synthetic and/or clay liners, leachate collection and detection systems, and landfill surface precludes ponding.	0
Essentially nonpermeable compatible liner, no leachate collection system, and landfill surface precludes ponding.	1	Single compatible synthetic and/or recompacted, engineered clay liner (10−7 cm/sec) with leachate collection system, and landfill surface precludes ponding.	1
Moderately permeable, compatible liner, and landfill surface precludes ponding.	2	Single recompacted, engineered clay liner (10^{-7} cm/sec) without leachate collection system, and landfill surface precludes ponding.	2
No liner or incompatible liner, moderately permeable compatible liner, landfill surface encourages ponding, no runon control.	3	Natural, in-place clay liner without leachate collection system, and landfill surface does not completely preclude ponding.	3

poses the greatest measurable relative hazard. A site's total migration route score, which is a combination (a root-mean-square) of the ground water, surface water, and air route scores, determines a site's placement on the scale.

Because the total migration scores are ordinal and not cardinal values (that is, they are relative and not absolute), their only significance is their value in relation to the other sites. Thus, a site with a score of 20 poses a greater potential hazard than a site with a score of 10, but not twice as great.

If a site scores higher than the 28.50 threshold, it might qualify for inclusion on the NPL (depending on what accounts for the score; remember, U.S. EPA and CEP score differently). The threshold has no scientific significance, however. U.S. EPA chose it as a matter of policy so that 400 evaluated sites would rank above it and thus be eligible for cleanup in the first Superfund list.

CEP defined its own grading system using the HRS. First, CEP considered each migration route score separately, rather than combining them into a total migration score. Taking a root-mean-square for the total migration score magnifies the impact of a single aberrant route score on the total route score. A site with one very bad release may score higher than a site with two less serious releases. To modify the disproportionate impact of the air route score, CEP disregarded the total migration score and assigned a grade to each route score independently.

It is worth noting that NPL sites are segregated into 11 consecutive groups by U.S. EPA to demarcate general hazard levels. The first group (sites that pose the greatest hazard) consists of those with scores in the range of 58.41 to 100. Below this large range, into which few sites are hazardous enough to fall, most groups comprise only 2 to 4 points each. Group 11 extends from 28.62 to 30.61, Group 10 from 30.77 to 33.66, Group 9 from 33.73 to 35.45, and so forth. The size of each group in terms of the spread in total migration scores varies because the number of sites and not the absolute value of the scores defines the group.

Similar to U.S. EPA's groupings, CEP divided the ten facilities into five groups according to the rank of each of the facility's migration route scores. Each group is represented by a letter grade. For each migration route, the two sites with the lowest (best) score get the grade A (excellent), the next two get Bs (above standard), then Cs (standard), then Ds (below standard), and the two sites with the highest (worst) scores get get Fs (failing). Each group consists of two facilities.

Although there are significant differences among the facilities' scores, these facilities' report cards do not portray a final, unchanging verdict. Many of the facilities CEP evaluated are presently on a schedule to overcome their on-site problems. Others have eliminated some of the old problems that affected the HRS score. If a site scores very well, on the other hand, this in itself does not necessarily mean that there are no problems at all, only that the ranking system considers the hazard relatively low. Also bear in mind that rankings could be greatly affected by inadequate data.

As with school classes this group of ten facilities may be better—or worse—than some other group. Thus a B for one of these facilities, for example, may have received an A or C if it were evaluated in comparison with another group of facilities.

Fire and Explosion and Direct Contact. CEP provided letter grades for the scores for fire and explosion and direct contact. The grades follow the same pattern as the migration route grades, with the top two sites receiving an A and so forth.

According to the HRS user's manual, however, the fire and explosion hazard should not be evaluated unless:

> either a state or local fire marshal has certified that the facility presents a significant threat to the public or to sensitive environments or there is a demonstrated fire and explosion threat based on field observations (e.g., combustible gas indicator readings).[8]

But when CEP called the appropriate state and local fire marshals in reference to the ten facilities, the six who could be reached stated that the matter had not been investigated.[9] These marshals rely on U.S. EPA to mitigate the potential for fire and explosion. The HRS format does not allow for the specificity needed to evaluate a fire prevention and contigency plan, but it can estimate the potential for damage in the event of a serious fire or explosion. CEP, therefore, scored each site using the U.S. EPA fire and explosion worksheet in order to indicate the potential degree of hazard in the event of a fire or explosion, not the relative potential for a fire or explosion to occur in the first place. CEP considered site workers and buildings (not only off-site targets) as potential targets of fire, explosion, and direct contact.

CEP Additions to HRS

CEP graded the sites on a number of parameters not covered by the HRS. Even more than HRS scores, which rely mainly on unchanging site characteristics, these grades will change over time, and thus should be updated regularly.

Corrective Action. One of the most important additional grades rates corrective action taken at a site. CEP's corrective action grades are

A = facility has eliminated the identified problems
C = facility is in the process of correcting or is on a compliance schedule to solve the identified problems and may have eliminated some of them
F = no action has been taken to correct some or all of the identified problems.

Technology. Although the mix of technological options offered by a facility can change quickly, it is an important criterion in selecting a site. Generators of specific waste streams should determine what methods are available to recover, recycle, treat, or destroy their waste, and then identify which sites offer the best of these methods. CEP's grades simply reflect the breadth of technological options each site offers. CEP associated values with general categories of technologies to differentiate their relative effectiveness in eliminating or reducing the hazard. CEP then added values associated with the technologies to provide an overall letter grade. The values were weighted as follows:

0.5 transportation (rail or truck) and proposed technologies
1.0 storage technologies (including landfill)
2.0 physical treatment technologies
3.0 chemical, biological, and thermal treatment technologies.

Again, the sites received grades A through F, with two sites in each group. This method may score sites with more technologies as better. Its principal indication is the degree of hazard destruction provided. Generators may want to ask only whether a specific technology is available and then leave out the technology grade.

Ground Water Monitoring and Financial Instruments. Generators should consider only sites that are either in compliance with RCRA regulations or on a compliance schedule. Remember, however, that a compliance schedule may be nothing more than a permission slip to continue to fail to meet standards or deadlines.

HRS does not score the severity of a violation directly. The scores do, however, account for most incidents of noncompliance at a facility. A citation for improper drum management, for instance, affects the containment score. The fact that a facility did not segregate incompatible wastes would affect the fire and explosion score. Inadequate security would affect the direct contact score.

Compliance in two significant areas, ground water monitoring and financial instruments, does not always affect the HRS scores, however. Compliance with financial instrument requirements does not affect HRS scores at all. CEP added grades for these two items.

Theoretically, the HRS may tend to rank a facility with a deficient ground water-monitoring system better than one with an adequate system because it could be more difficult to detect contamination at the inadequate facility. Five of the six CEP-evaluated facilities that have inadequate ground water-monitoring systems, however, have, in fact, either detected ground water contamination or have indications of contamination. Even so, of the other four, the one facility with an inadequate monitoring system is less likely to detect contamination than one of the three facilities with adequate systems.

CEP used a simple grading system to minimize these potential distortions. All sites that do not meet ground water-monitoring requirements get an F. Since compliance with requirements is an expected minimum, however, facilities that do comply are not graded on a relative basis; rather, the category is excluded from the evaluation when calculating the root-mean-square grade. If CEP had awarded a standard C to complying facilities, the ground water grade would bring down their final grade unfairly. CEP applied the same methodology to the financial instruments grade; all sites that do not meet requirements get an F, and all that do are not graded in this category.

CEP Root-Mean-Square Grades

CEP synthesized the nine grades into an overall grade by assigning values to each and taking their root-mean-square (the square root of the sum of the grade values divided by the number of grades). The grades have values as follows.

A	=	0.0 points	C+	=	1.7
A−	=	0.3	C	=	2.0
B+	=	0.7	C−	=	2.3
B	=	1.0	D+	=	2.7
B−	=	1.3	D	=	3.0
			F	=	any value greater than 3

CEP's synthesizing technique is different from that used in chapter 7 and different from the technique that American schools use to derive grade point averages. Those techniques, which involve averaging, tend to normalize extremes. Because CEP is assessing risks, however, extremes are precisely what the evaluation needs to highlight. The root-mean-square computation is more suitable than a simple average. Like the HRS scores, CEP's facility grade values increase as the grade becomes worse. This weighting scheme mathematically emphasizes extreme hazards.

CHAPTER NINE

Report Cards: Ten Hazardous Waste Management Facilities

FROM BEST TO WORST

Using CEP's grading system, the ten RCRA facilities differed significantly in performance, with overall final grades ranging from B+ to F. Most facilities received an A on at least one of the several criteria, but none of the ten scored a final A. This finding indicates that, according to CEP's selection system, none of the sites is optimally located, optimally managed, and equipped with the best technologies. From the best to the worst, the facilities are:

Facility Owner/Operator	*Facility Location*	*CEP Grade*
U.S. Pollution Control, Inc.	Clive, UT	B+
Chemical Waste Management	Chicago, IL	C
Chemical Waste Management	Carlyss, LA	C
CECOS International, Inc.*	Niagara Falls, NY	C−
ENSCO, Inc.*	El Dorado, AR	C−
Rollins Environmental Services (LA)	Baton Rouge, LA	C−
Texas Ecologists, Inc.	Robstown, TX	D+
IT Corp.	Benicia, CA	D
Envirosafe Services of Idaho, Inc.	Grandview, ID	F
CECOS International, Inc.	Williamsburg, OH	F

* CECOS Niagara Falls and ENSCO El Dorado received identical final scores.

Quite probably, these ten facilities include neither the best nor the worst facility in the United States.

TABLE 9-1. *CEP Grades for Ten RCRA Facilities*

Facility Abbreviation	Ground Water Route Grade	Surface Water Route Grade	Air Route Grade	Fire and Explosion Grade	Direct Contact Grade
Northeast/North Central					
CWM IL	B	B	C	F	A
CECOS NY	A	C	B	B	F
CECOS OH	F	F	D	A	C
South Central					
CWM LA	C	B	F	D	A
ENSCO AR	D	D	B	C	B
RES LA	C	C	F	B	F
TECO TX	F	A	A	A	C
West					
USPCI UT	A	A	A	C	B
IT CA	B	D	D	F	D
ESII ID	D	F	C	D	D

A = 0.0
A− = 0.3
B+ = 0.7
B = 1.0
B− = 1.3
C+ = 1.7
C (Standard Grade) = 2.0
C− = 2.3
D+ = 2.7
D = 3.0
F = >3

For a generator in the Northeast or North Central region of the United States, CWM's incinerator in Chicago or the CECOS facility in Niagara Falls might be the best bet, depending on the waste stream of the generator. Certain wastes can be incinerated, for example, others cannot (see chapter 5). In the South Central region, CWM's Lake Charles facility or ENSCO's El Dorado facility would be the best choices. In the West USPCI's Grassy Mountain facility is the obvious choice since it received the highest grade of all ten facilities.

Once a facility has been chosen, using the treatment technologies that provide the greatest reduction of hazard may be well worth the price. Incinerating or otherwise destroying all wastes that are compatible with such technologies is the best bet for reducing long-term liabilities, although there may be some short-term problems at a particular site.

The RES (LA) facility in Baton Rouge, for example, came in third of the four sites evaluated in the south, even though it has this country's second largest hazardous waste incinerator and ranked best of all ten sites in terms of the quality and variety of technological options offered to its customers. Why? Notice table 9-1, which outlines all of the component grades that make up the final CEP grades for each site. RES (LA) received an F for air route

Corrective Action Grade	Technology Grade	Ground Water Monitoring Grade	Financial Instruments Grade	CEP Route-Mean-Square Grade	Total Value
A	C	*	*	C	1.9
C	B	F	*	C−	2.4
C	F	F	*	F	3.2
A	B	*	*	C	2.1
C	C	F	*	C−	2.4
C	A	*	*	C−	2.5
C	F	*	F	D+	2.7
A	A	*	*	B+	0.8
C	D	F	*	D	3.0
C	D	*	F	F	3.1

Facility abbreviations:
CWM IL = Chemical Waste Management, Inc., Chicago Incinerator, IL
CECOS NY = CECOS International, Inc., Niagara Falls, NY
CECOS OH = CECOS International Inc., Williamsburg, OH
CWM LA = Chemical Waste Management, Inc., Carlyss, LA
ENSCO AR = Energy Systems Company, Inc., El Dorado, AR
TECO TX = Texas Ecologists, Inc., Robstown, TX
RES LA = Rollins Environmental Services of Louisiana, Baton Rouge, LA
USPCI UT = U.S. Pollution Control, Inc., Clive, UT
IT CA = International Technology Corporation, Benicia, CA
ESII ID = Envirosafe Services of Idaho, Inc., Grandview, ID

NOTE: Facility abbreviations are listed by geographical region; within regions, they are listed in order of best to worst root-mean-square grade.

*If a facility is in compliance, category is not factored into final grade.

hazard. In fact, table 9-2A shows that RES (LA) has the worst air route score of all ten facilities. This grade, in conjunction with a number of other average or below-average component grades (RES (LA) also received an F for direct contact) have, at least for now, negated the benefits of RES (LA)'s superior technical capabilities.

THE REPORT CARDS

Table 9-1 outlines the individual grades that make up the grade point average for each site. Table 9-2A through table 9-2E outline every criterion used to

TABLE 9-2A. *CEP Ground Water Route Scores*[a]

Facility Abbreviation	Ground Water Route Score	Observed Release	Route Characteristics: *Depth to Aquifer of Concern*	*Net Precipitation*	*Permeability of Unsaturated Zone*	*Physical State*	Total Route Characteristics
Northeast/North Central							
CECOS NY	00.00	0[b]	6	2	1	1	10
CWM IL	1.63	0	6	2	1	3	12
CECOS OH	45.71	0[b]	6	2	3	3	14
South Central							
CWM LA	5.44	0	6	2	0	0	8
RES LA	9.20	0	2	2	2	1	7
ENSCO AR	14.84	45	4	2	2	3	11
TECO TX	38.78	45	6	3	2	0	11
West							
USPCI UT	00.00	0	6	0	1	3	10
IT CA	00.00	45	6	0	0	3	9
ESII ID	30.61	45	6	0	2	3	11

NOTE: Facility abbreviations are listed by geographical region; within regions, they are ranked from lowest to highest hazard.

a. CEP hazard scores, although based on U.S. EPA's HRS, are not entirely comparable with the scores used tc rank Superfund sites. The scores represent relative categories of hazard. Higher scores indicate greater hazard.

b. There are indications of ground water contamination caused by CECOS NY and CECOS OH. If these indications are verified, the ground water route score for CECOS NY would still be 00.00 because the aquifer is unusable. For CECOS OH the ground water route score would increase to 73.41 if contamination exists.

derive CEP's scores. The greater the value given for any criterion, the higher is the relative hazard.

Facility description forms provide information on each facility concerning:

Site: demography, water use, hydrology, geology, surface water, areas of special concern

Management: observed releases, ground water monitoring, enforcement actions, corrective actions, staff and unionization, safety and health, financial instruments

Technology: processes and capacities, special wastes handled or excluded

History: prior use and prior owners of the site, hazardous waste permits, future processes and uses of the site, citizen concerns and citizen organizations.

	Waste Characteristics			Targets		
Containment	*Toxicity/ Persistence*	*Hazardous Waste Quantity*	**Total Waste Characteristics**	*Ground Water Use*	*Distance to Nearest Well/ Population Served*	**Total Targets Score**
2	18	8	26	0	0	0
1	18	8	26	3	0	3
2	18	8	26	6	30	36
1	18	8	26	9	6	15
1	18	8	26	9	20	29
3	18	3	21	3	6	9
1	18	8	26	9	10	19
2	18	8	26	0	0	0
2	18	8	26	0	0	0
3	18	8	26	9	6	15

A West Coast generator who wants to choose the best commercial facility for its hazardous wastes, will see that CEP graded USPCI UT significantly better than either IT CA or ESII ID. Why? Notice on table 9-1 that USPCI has As for all migration route scores and for corrective action—high scores in four key criteria. IT CA, on the other hand, got four Ds.

Take the ground water score: Table 9-2A shows that IT CA has had an observed release to ground water. But because the ground water under the facility is unused and naturally unuseable (due to a high saline content) its ground water route score still got a B, a relatively good grade. USPCI, however, got an A because there is no indication of ground water contamination at the facility. Even though the unsaturated zone beneath the site is slightly more permeable than at Benicia, the ground water beneath USPCI UT is also unused and naturally unuseable. According to table 9-2A, the depth to aquifer of concern is 6 for all three sites in the Western region. This means that the distance from the lowest point of waste disposal to the uppermost seasonal level of the saturated zone of concern is 0 to 20 feet (a value of 3 with a multiplier of 2 equals a score of 6), the highest hazard score in this category.[1]

Of course, a generator will want to compare the prices each facility charges. Transportation is a factor, too. CEP was usually unable to get pricing schedules

TABLE 9-2B. *CEP Surface Water Route Scores*[a]

Facility Abbreviation	Surface Water Route Score	Observed Release	*Facility Slope & Intervening Terrain*	*1-yr, 24-hr Rainfall*	*Distance to Nearest Surface Water*	*Physical State*	Total Route Characteristics	Containment
			Route Characteristics					
Northeast/North Central								
CWM IL	0.89	0	0	2	6	3	11	1
CECOS NY	1.70	0	0	2	4	1	7	1
CECOS OH	16.36	45	0	2	6	3	11	3
South Central								
TECO TX	00.00	0	0	3	4	3	10	0
WM LA	00.00	0	0	3	4	3	10	0
RES LA	1.62	0	0	3	6	1	10	1
ENSCO AR	9.40	0	0	3	6	3	12	2
West								
USPCI UT	00.00	0	0	0	0	3	3	2
IT CA	14.55	45	3	1	4	3	11	2
ESII ID	13.58	w 0	1	0	4	3	8	3

NOTE: Facility abbreviations are listed by geographical region; within regions, they are ranked from lowest to highest hazard.

a. CEP hazard scores, although based on U.S. EPA's HRS, are not entirely comparable with the scores used to rank Superfund sites. The scores represent relative categories of hazard. Higher scores indicate greater hazard.

because of business confidentiality. Prices change frequently anyway, so CEP's grading system considers factors other than the price tag. Many of these factors have hidden long-term costs associated with them. Potential customers of these facilities, concerned citizens, and government officials should read the CEP grades in conjunction with the background material to weigh the strengths and weakness of a site in relation to a specific waste or other factors of concern.

Generators of single waste streams may want to modify CEP's system by scoring only those units that will actually handle their wastes. CEP has evaluated all units because a generator's liability may not be limited to a particular unit if the site is abandoned. Similarly, rather than grading a site on all of its processes, a generator may want to ask only whether the site has the specific process that the generator wants to use and to limit evaluation to applicable

Waste Characteristics			Targets			
Toxicity/ Persistence	*Hazardous Waste Quantity*	**Total Waste Characteristics**	*Surface Water Use*	*Distance to a Sensitive Environment*	*Population Served/ Distance to Water Intake Downstream*	**Total Targets Score**
18	8	26	0	2	0	2
18	8	26	6	0	0	6
18	8	26	9	0	0	9
0	0	0	0	0	0	0
0	0	0	0	1	0	1
18	8	26	0	4	0	4
18	3	21	6	6	0	12
18	8	26	0	0	0	0
18	8	26	6	2	0	8
18	8	26	6	0	8	14

facilities. CEP also encourages generators to apply CEP's analytical method to the many other commercial hazardous waste management facilities whose services compete with the ten facilities evaluated here. The HSWA requirement that all facilities submit exposure information will greatly facilitate such endeavors.

Finally, generators and other interested persons should update the information contained in these case studies—especially the data used for grading corrective action, technology, ground water monitoring, and financial instruments. The status of this information can change rapidly. CEP is willing to assess additional facilities and to update the evaluations in this book.

DIFFERENCES IN SITE AND TECHNOLOGY

Although the number of facilities CEP evaluated represents a sizeable portion of those for which adequate information is available, they do not constitute a large sample of the commercial hazardous waste management industry as

TABLE 9-2C. *CEP Air Route Scores*[a]

Facility Abbreviation	Air Route Score	Observed Release	Waste Characteristics: *Reactivity and Incompatibility*	*Toxicity*	*Hazardous Waste Quantity*
Northeast/North Central					
CECOS NY	00.00	0	2	9	8
CWM IL	00.00	0	2	9	8
CECOS OH	48.46	45	1	9	8
South Central					
TECO TX	00.00	0	3	9	8
ENSCO AR	00.00	0	1	9	8
CWM LA	58.97	45	3	9	8
RES LA	60.90	45	2	9	8
West					
USPCI UT	00.00	0	3	9	8
IT CA	53.85	45	3	9	8
ESII ID	38.46	45	3	9	8

NOTE: Facility abbreviations are listed by geographical region; within regions, they are ranked from lowest to highest hazard.

a. CEP hazard scores, although based on U.S. EPA's HRS, are not entirely comparable with the scores used to rank Superfund sites. The scores represent relative categories of hazard. Higher scores indicate greater hazard.

a whole. A larger number of facilities owned by the Big Eight are analyzed in less detail in chapter 7.

The ten facilities are located in three broad areas of the United States (see Figure 9-1).

The sites also differ in their demographic environments. One of the three northern sites, two of the four southern sites, and all three western sites are rural (see table 9-3). Four of the facilities are in urban areas. The area surrounding IT's Benicia site, although presently rural, is experiencing significant population growth.

The principal technologies used at the sites are landfills and incinerators. Two incineration facilities are in the south and one in the north. The rest of the sites use landfills for final disposition of hazardous wastes. Of the eight companies, only IT owns an incineration facility west of Texas, the Vine Hill facility in Martinez, California. Seven of the ten facilities perform some form of physical, chemical, or biological treatment to their waste in addition to their principal disposal method (see table 9-4).

	Targets			
Total Waste Charac-teristics	*Population within 4-Mile Radius*	*Distance to Sensitive Environment*	*Land Use*	**Total Targets Score**
19	21	0	3	24
19	21	2	3	26
18	18	0	3	21
20	15	0	3	18
18	21	6	3	30
20	18	2	3	23
19	18	4	3	25
20	9	0	0	9
20	18	2	1	21
20	12	0	3	15

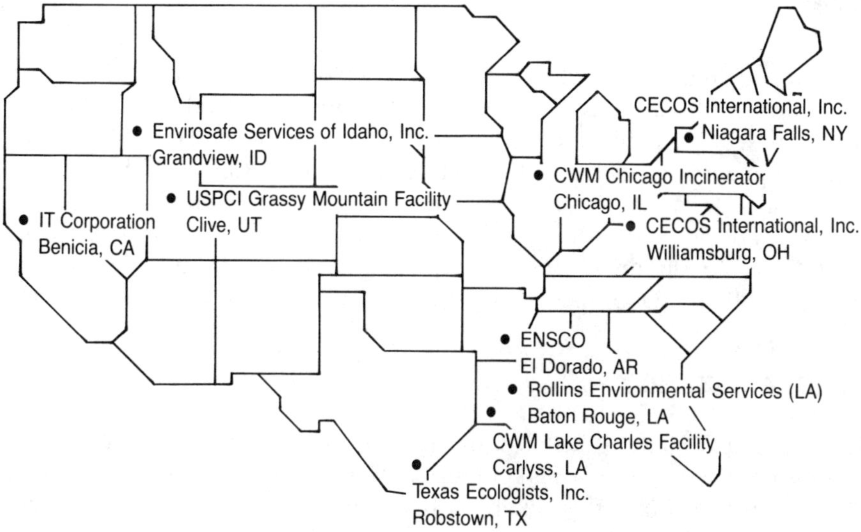

FIGURE 9-1. *Locations of Ten RCRA Facilities*

TABLE 9-2D. *CEP Fire and Explosion Scores*[a]

Facility Abbreviation	Fire and Explosion Score	Containment	Waste Characteristics: *Direct Evidence*	*Ignitability*	*Reactivity*	*Incompatibility*	*Hazardous Waste Quantity*
Northeast/North Central							
CECOS OH	10.63	3	0	3	1	1	3
CECOS NY	17.50	1	0	3	2	1	8
CWM IL	60.21	3	3	3	3	0	8
South Central							
TECO TX	13.60	1	0	3	2	0	8
RES LA	15.35	1	0	2	2	1	8
ENSCO AR	18.89	1	3	2	3	1	7
CWM LA	37.92	3	3	1	3	2	4
West							
USPCI UT	28.13	3	0	3	2	2	8
ESII ID	43.33	3	0	3	3	2	8
IT CA	43.54	3	3	3	2	3	8

NOTE: Facility abbreviations are listed by geographical region; within regions, they are ranked from lowest to highest hazard.

a. CEP hazard scores, although based on U.S. EPA's HRS, are not entirely comparable with the scores used to rank Superfund sites. The scores represent relative categories of hazard. Higher scores indicate greater hazard.

As already mentioned, RES (LA) offers the widest variety of technology to its customers. USPCI's Grassy Mountain facility comes in second and the CECOS facility in New York follows third in terms of the diversity and hazard destruction effectiveness of its technology. TECO TX and the CECOS Williamsburg facility provide landfilling only. The Ohio CECOS facility and IT in Benicia, however, are near company-owned facilities dedicated solely to treatment. CECOS has a treatment center in Cincinnati, 45 minutes from Williamsburg. IT's Vine Hill and Baker in Martinez are just across Suisun Bay from Benicia.

CERCLA HAZARD?

In ranking active RCRA facilities, the first and foremost question is: Could these sites pose a great enough potential hazard to qualify for cleanup under CERCLA? In short, are they likely to become Superfund sites? If they are, generators would be liable for the cleanup bill. The compliance records of the Big Eight indicate that they would probably take corrective action if a serious observed release did occur. CERCLA provisions apply only if a com-

Total Waste Characteristics	Targets: Distance to Nearest Population	Distance to Nearest Building	Distance to Sensitive Environment	Land Use	Population within 2-Mile Radius	Buildings within 2-Mile Radius	Total Targets Score
13	5	3	0	3	3	3	17
14	5	2	0	3	4	4	18
17	5	3	0	3	3	3	17
14	5	3	0	3	2	1	14
13	5	3	0	3	3	3	17
16	4	2	0	3	4	4	17
13	5	2	0	3	2	2	14
15	5	2	0	0	1	1	9
16	5	3	0	3	1	1	13
19	5	3	0	1	1	1	11

pany proved unable (because of inadequate resources) or unwilling to comply with RCRA enforcement actions.

If the worst-case scenario came to pass—if all ten sites leaked contaminants into ground water—CEP believes that only one site, CECOS in Williamsburg, Ohio, could qualify for the NPL. Hydrogeological characteristics and potential exposure targets contribute to a high score for CECOS. At the time of this writing, U.S. EPA and Ohio consider the ground water-monitoring system at the Williamsburg facility to be inadequate and suspect that the facility may be leaking.

Ironically, U.S. EPA chose the Williamsburg facility to receive wastes from at least 19 Superfund sites, more than any of the other facilities we studied, and more than most commercial RCRA facilities in the country. In response to the indications of ground water contamination, Ohio ordered the RCRA facility closed until CECOS submits an adequate contamination assessment. To CEP's knowledge, this is the first time a RCRA facility has been closed—even if temporarily—due to an inadequate ground water-monitoring program. CEP's comparative facility evaluation supports Ohio's action because the performance of the CECOS Williamsburg facility was the worst of the ten sites. Nevertheless, there is nothing in CECOS's history to indicate that it will not comply with the state enforcement action or refuse to clean up the site if necessary.

TABLE 9-2E. *CEP Direct Contact Scores*[a]

Facility Abbreviation	Direct Contact Score	Observed Incident	Accessibility	Containment
Northeast/North Central				
CWM IL	00.00	0	0	0
CECOS OH	00.00	0	0	15
CECOS NY	37.50	0	3	15
South Central				
CWM LA	00.00	0	0	0
ENSCO AR	00.00	0	0	0
TECO TX	00.00	0	0	15
RES LA	37.50	0	3	15
West				
USPCI UT	00.00	0	0	15
IT CA	00.00	0	0	15
ESII ID	12.50	45	0	0

NOTE: Facility abbreviations are listed by geographical region; within regions, they are ranked from lowest to highest hazard.

a. CEP hazard scores, although based on U.S. EPA's HRS, are not entirely comparable with the scores used to rank Superfund sites. The scores represent relative categories of hazard. Higher scores indicate greater hazard.

On August 6, 1985 the RES (LA) incinerator in Baton Rouge was also shut down by state order. Louisiana's Department of Environmental Quality alleged that the incinerator emitted an unusually thick smoke. In addition, workers in the facility's control room were allegedly in a state of confusion with one employee allegedly suffering from dizziness.[2] But only if a laboratory could verify that the samples of toxic emissions collected off site contained contaminants above background levels would U.S. EPA's HRS count this incident as an observed release.

RES (LA) strongly denies the allegations, and the company is contesting the state's order in the courts and through the administrative process. In the meantime, RES (LA) has been authorized by the Court of Appeals for the First Circuit, State of Louisiana, to resume incineration operations. The incinerator went back into operation in October 1985.

CEP's evaluation of the site, which took place months prior to the troubling closure, foreshadowed the incident in a sense. CEP's air route score, which is less restrictive than U.S. EPA's, shows that the RES (LA) facility could pose considerable risks associated with air releases compared with the other nine facilities because of its history of allegedly producing odors and its proximity to residences, workplaces, and sensitive natural environments. In addition,

	Targets		
Toxicity	Population within 1-Mile Radius	Distance to Critical Habitat	Total Targets Score
15	8	0	8
15	12	0	12
15	12	0	12
15	4	0	4
15	12	0	12
15	8	0	8
15	12	0	12
15	4	0	4
15	16	0	16
15	4	0	4

CEP's direct contact score, which accounts for worker exposure, also indicated the potential for serious harm because of the site's accessibility, inadequate containment, and proximity to population.

CEP's evaluation indicates the hazards posed by RES (LA)'s ground water contamination are not severe enough to deserve inclusion on the NPL.

FACILITY DESCRIPTION: USPCI UT

Facility	*CEP Score*		*Grade*
USPCI Grassy Mountain Facility	Ground Water Route	=	A
3 miles east, 7 miles north of Knolls exit, Interstate 80	Surface Water Route	=	A
Clive, UT 84107	Air Route	=	A
	Fire and Explosion	=	C
U.S. EPA ID#: UTD991301748	Direct Contact	=	B
	Corrective Action	=	A
	Technology	=	C
	Ground Water Monitoring	=	*
	Financial Instruments	=	*
	Root-Mean-Square Grade	=	B+

* Facility is in compliance, so this category is not factored into final grade.

TABLE 9-3. *Profile of Ten RCRA Facilities*

Parent Corporation/ Company Name	Facility Name, Location	Region	Rural/ Urban
American Ecology Corporation/ U.S. Ecology, Inc.	Texas Ecologists, Robstown, TX	S	R
Browning-Ferris Industries/ CECOS International, Inc.	CECOS International, Niagara Falls, NY	N	U
	CECOS International, Williamsburg, OH	N	R
Environmental Systems Company, Inc./ Energy Systems Company, Inc.	ENSCO, El Dorado, AR	S	U
International Technology Corporation	IT Corporation, Benicia, CA	W	R[b]
IU International, Inc./ Envirosafe Services, Inc.	Envirosafe Services of Idaho, Grandview, ID	W	R
Rollins Environmental Services, Inc.	Rollins Environmental Services (LA), Baton Rouge, LA	S	U
USPCI, Inc./ U.S. Pollution Control, Inc.	Grassy Mountain Facility, Clive, UT	W	R
Waste Management, Inc./ Chemical Waste Management	Lake Charles Facility, Carlyss, LA	S	R
	Chicago Incinerator, Chicago, IL	N	U

Region:
N = Northeast/North Central
S = South Central
W = West

Rural/Urban:
R = rural
U = urban

Site

Demography. The Grassy Mountain facility is in an isolated area. The Bureau of Land Management owns all of the surrounding land, including mud flats to the west and the Grayback Hills to the east. Neither is presently in use.[3] Interstate 80 leads to Knolls Restaurant 11 miles southwest of the facility, the closest work site. Tooele Air Force Base, 20 miles north, conducts bombing practice between the base and the facility.[4] Salt Lake City is 85 miles east.[5]

Water Use. There is no potable water in the area. Water is hauled in from Wendover, Nevada, 55 miles to the west.[6] The air force base uses reverse osmosis to convert saline ground water into potable water.[7]

Hydrology and Geology. The facility rests upon Quaternary alluvial deposits of silt and silty clay in excess of 120 feet deep. To the south these soils extend for miles; to the north and west, they grade into heavy clay. The average estimated permeability of these sediments is 7×10^{-6} centimeters per second.[8] The ground water has naturally high total dissolved solids (with sodium chloride the main constituent) and low recharge. The uppermost level of the saturated zone is 7 to 40 feet below ground surface. The ground water that does not evaporate generally moves in a west-northwest direction.[9]

Principal Technology	Observed Release/Incident: Ground Water	Surface Water	Air	Direct Contact	CERCLA Wastes Received[a]
L	G				0
L	P				11
L	P	S	A		19
I	G				8
L	G		A		2
L	G		A	D	4
I/L	G[c]		A		6
L					0
L			A		0
I					6

Principal Technology:
L = landfill
I = incinerator
I/L = both

Observed Release/Incident:
G = documented release to ground water
P = indications of possible release to ground water
S = documented release to surface water
A = documented release to air (including odors)
D = documented incident of direct contact

a. As of October 1984, facilities received hazardous wastes removed from the given number of Superfund sites, according to U.S. EPA, "Disposition of CERCLA Wastes Data Base" (10/84).
b. Benicia, California is steadily becoming urban.
c. RES (LA) has had documented release to ground water, but not to the aquifer of concern.

Surface Water. Great Salt Lake is 22 miles to the east.[10] There is no other surface water within the vicinity.

Areas of Special Concern. There are no sensitive areas, parklands, wetlands, critical habitats, or known archaeological sites within the immediate area. A large population of bald eagles winters approximately 60 miles to the southeast. A breeding site for peregrine falcons lies approximately 25 miles east.[11]

Management

Observed Releases. No releases from this facility have been observed.

Ground Water Monitoring. As of January 9, 1985, the facility was conducting detection monitoring. U.S. EPA judged the well system to be adequate and the sampling and analysis to be inadequate. Ground water-monitoring records are adequate.[12]

TABLE 9-4. *Hazardous Waste Management Technologies at Ten RCRA Facilities*

	CECOS NY	CECOS OH	CWM IL	CWM LA	ENSCO AR	ESII ID	IT CA	RESLA LA	TECO TX	USPCI UT
CEP Grade	B	F	C	B	C	D	D	A	F	A
ACCESS (0.5 points each)										
Rail			X		X			X		
Truck	X	X	X	X	X	X	X	X	X	X
Storage (1 point each)										
Container	X		X	X	X			X	X	X
Tank		P	X	X	X			X	X	X
Waste Pile						P				X
Surface Impoundment						P	X	X		X
Landfill		X		X		X	X	X	X	X
Physical treatment (2 points each)										
Clarification								X		
Coagulation								X		
Decanting				X		X		X		
Distillation					X					
Flocculation								X		X
Solar Evaporation						P	X			X
Filtration	X									
Spray Irrigation							X			
Stabilization	X			X		X		X		

Biological treatment (3 points each)										
Sequencing Batch Reaction	X									
Land Treatment										X
Chemical treatment (3 points each)										
Adsorption	X								P	
Degradation				X						X
Fixation				X		P				X
Neutralization	X			X		P		X		X
Oxidation	X			X				X		X
Precipitation	X									X
Reduction	X									P
Thermal destruction (3 points each)										
Liquid Injection			X		X			X		
Rotary Kiln			X		X			X		
Thermal Oxidation								P		
Supercritical Water Reactor	P									
Plasma Arc Reactor										
Total points	24	2	9	19.5	11	6	6.5	30.5	4	28

DERIVED FROM: Environmental Information, Ltd., *Industrial and Hazardous Waste Management Firms, 1985* (Minnesota: 1984); CEP site visits; and company data.

NOTE: Facility abbreviations are listed in alphabetical order; technologies are listed in order of increasing hazard destruction effectiveness.

X = Existing on-site technology
P = Proposed technology (0.5 points each)

Enforcement Actions. *1984:* State Notice of Violation and Compliance Order. In June, the Utah Solid and Hazardous Waste Commission (USHWC) cited the facility for 18 violations, including (1) improper storage of primary oil/solids/water separation sludge from the petroleum refining industry (K051), (2) failure to manifest shipments of listed hazardous wastes, (3) failure to ship hazardous wastes to an approved TSDF, (4) failure to obtain proper waste analysis, (5) acceptance and management of wastes not specified in the current Part A permit application, (6) exceeding storage capacity at the drum storage dock, (7) failure to segregate incompatible wastes in containers on the drum dock, (8) inadequate drum dock aisle space, (9) failure to develop, install, and maintain fire-fighting capabilities.[13] *1984:* State Notice of Violation. In September USHWC cited the facility for improper storage of K051.[14] *1984:* In October USHWC cited the facility for failure to minimize the possibility of any unplanned release of contaminated pieces of plastic to air, soil, or surface water.[15] *1984:* Notice of Violation. In November USHWC cited the facility for placing noncontainerized or bulk waste containing free liquids into the landfill cell.[16]

Corrective Actions. (1) Separation sludge that the generator had shipped in illegal containers was temporarily placed in an inappropriate "frac tank" because it was the only available container on site; returning the sludge in the original containers would have violated DOT regulations. Although USPCI explained the circumstances surrounding this situation to the state and corrected the violations, a similar violation occurred 3 months later. (2–5) USPCI UT disagrees that violations 2–5 occurred. (6–8) USPCI claims that no incompatible wastes were stored at the facility but admits to insufficient storage space for which Part B requests an expansion.[17] (9) Facility installed a 20,000-gallon tank in May 1984 to be used in the event of a fire at the office building on site. There is one portable fire extinguisher to be used in the event of a fire at the active area of the landfill. USPCI UT no longer uses plastic to line its vehicles that transport bulk solids for landfarming because USPCI UT was unable to prevent the liner from becoming sliced and airborne after tilling.[18]

Staff and Unionization. Six full-time and three temporary secretarial employees work on site. None is unionized.[19]

Safety and Health. Winds scattered plastic liner strips that were contaminated with oil refinery waste across the facility and off site. USHWC was concerned that the temporary personnel engaged in removing the contaminated plastic might have been improperly trained and might not have adequate personal protection.[20]

Financial Instruments. As of January 14, 1983 U.S. EPA had not evaluated the facility's $484,200 closure cost or $333,000 postclosure cost estimates. The facility carries insurance of $5 million per occurrence and annual aggregate sudden occurrence. Other insurance includes $3 million per occurrence and $6 million annual aggregate non-sudden occurrence. U.S. EPA judged that the insurance was adequate.[21]

Technology

Processes and Capacities. The facility has a capacity of 631.5 acre-feet of landfill and 206 acres for land treatment. Other capacities include: 41-acre feet of surface impoundment for storage and disposal, covered drum dock with 250-drum capacity, five tanks for blending and storage with 390,000-gallon storage capacity, and treatment capacity of 10,000 gallons per day.[22] Neutralization performed as of 1984; chemical oxidation or reduction planned for 1985 or 1986.[23]

Special Wastes Handled or Excluded. Facility does not accept toxic P wastes, acutely hazardous U wastes, or chlorinated organics.[24]

History

Prior Use and Owners. Property was acquired from the Bureau of Land Management by an individual landowner for waste management activities.[25]

Construction and Permits. Hazardous waste management activities began in 1980. In 1983 U.S. EPA called for Part B of the RCRA application, which the department received and reviewed.

Future Processes and Site Uses. Land treatment could continue for an estimated 60 to 80 years. The site can only be used for industrial activities similar to hazardous waste management in the future.[26]

Citizen Concerns and Organizations. There are no local residents, and CEP is not aware of any citizen organizations concerned about the facility. The Utah Environment Center had not even heard of the facility.

FACILITY DESCRIPTION: CWM IL

Facility	*CEP Score*		*Grade*
SCA Chemical Services,	Ground Water Route	=	B
Inc., Chicago Incinerator	Surface Water Route	=	B
11700 Stony Island Ave.	Air Route	=	C
Chicago, IL 60617	Fire and Explosion	=	F
	Direct Contact	=	A
U.S. EPA ID#:	Corrective Action	=	A
ILD000672121	Technology	=	C
	Ground Water Monitoring	=	*
	Financial Instruments	=	*
	Root-Mean-Square Grade	=	C+

* Facility is in compliance, so this category is not factored into final grade.

Site

Demography. This facility is located in the highly industrialized Lake Calumet area of southeast Chicago. Almost half of all the land area zoned for manufacturing in Chicago is in the Lake Calumet area.[27] The closest residential area is approximately a mile west of the facility; all land in the immediate vicinity of the site is used industrially.[28] According to the 1980 census, there are 107,600 persons within the 60617 ZIP code area. The neighborhoods of Burnham, Burnside, East Side, Hegewisch, Pullman, Riverdale, Roseland, and South Deering, as well as the towns of Calumet City, Dolton, and Thorton are all within 3 to 4 miles of the site.[29]

Water Use. The public water-supply system serves the entire metropolitan area. Its major Lake Michigan intakes are located considerably more than 3 miles north of the site; shore intakes are used during peak demand, however. A filtration plant is 4 miles from SCA, but the distance is over 8 miles via the Calumet River, which flows away from Lake Michigan.[30] The high mineral content of area ground water is acceptable for use only by a few industries.[31] Lake Calumet and Calumet River are designated as secondary contact and indigenous aquatic life waters; there is no direct human contact with these surface waters via recreation or supply.[32]

Hydrology and Geology. The facility rests upon a fill composed of slag, clay, silt, stones, and rubble that is from 7 to 18 feet thick. Below this lies gray silty clay that ranges from stiff to hard with isolated layers of silt. Most soil samples show permeabilities of 10^{-7} to 10^{-8}; however, a permeability of 2.6×10^{-6} was detected in one sample.[33] Upper ground water is contained in the fill and in organic marsh deposits and is generally found 6 to 10 feet below ground surface at the site.[34] The shallow aquifer is separated from the next low yielding aquifer by approximately 400 feet of relatively impermeable clay glacial deposits.[35] The high mineral content Ironton-Galesville aquifer, 1500 feet below ground surface, is capable of supplying substantial yields.[36]

Surface Water. Water is a dominant feature in Southeast Chicago. The facility's surface impoundments are built upon a fill that extends west into Lake Calumet about 10 to 12 feet above the surface of the lake. There are approximately 50 to 70 feet of fill between the impoundments and the lake. The Calumet River flows south approximately 2 miles east/southeast of the facility. Lake Michigan is 4 miles northeast of the site. The facility is not within the 100-year floodplain.[37]

Areas of Special Concern. A fresh-water wetland is one-half mile northeast of the site.[38] Calumet Park, Eggers Woods, Wolf Lane State Park, Burnham Woods, and Beaubien Woods are all within 2 miles of the site.[39] In addition, Lake Calumet is a stopping place for migratory birds.[40]

Management

Observed Releases. There are no reported releases to ground water, surface water, or air. A 1980 U.S. EPA inspection and three 1984 Illinois EPA (IL EPA) inspections found insufficient freeboard in the surface impoundments, but there were no observed resulting releases.[41] The facility's aqueous waste water treatment area discharges into the sewer system of the Metropolitan Sanitary District of Greater Chicago.[42] *1983:* On two occasions high-BTU waste liquids were fed (in drums) into the kiln. Once inside the kiln, the drums broke and the liquid "flashed," causing flames to come out the ash door and around the kiln seals.[43] *1984:* In January air lines froze and the fans that maintain negative pressure failed causing the emergency stack to remain open for 20 minutes. This failure could have allowed an emission of partially combusted wastes. Although the waste feed did shut off, the emergency stack normally stays closed until combustion has completed. It then opens to release heat and closes again. Wipe samples were taken, but no significant releases were detectable.[44]

Ground Water Monitoring. As of January 20, 1984, U.S. EPA found that the facility's ground water-monitoring well system, sampling, analysis, evaluation program, and records met regulatory requirements; U.S. EPA may request installation of more wells, however.[45]

Enforcement Actions. *1980:* Complaint and Findings of Violation. In December U.S. EPA alleged SCA violated RCRA by failing to post danger signs, to maintain adequate freeboard on a surface impoundment and to provide protective cover on earthen dikes.[46] *1981:* Consent Agreement and Final Order. In March SCA agreed to correct the alleged violations.[47]

Corrective Actions. *1981:* Danger signs were posted at appropriate locations and an appropriate cover was planted on all earthen dikes.[48] *1983:* The surface impoundments were reconstructed and the lime additive was changed to caustic scrubbing in order to reduce formation of sludges in these ponds.[49] The kiln seal design has been changed and new seals installed. Drums with 100 percent liquids are no longer fed into the kiln.[50] *1984:* A manual override with compressed air has been installed to close the emergency stack if the air lines freeze.[51]

Staff and Unionization. There are 75 full-time employees on site. None is unionized.

Safety and Health. In 1979 the facility was cited for violations of U.S. OSHA respiratory protection and sanitation requirements.[52]

Financial Instruments. As of November 8, 1985 estimated closure cost would be $385,550. Sudden occurrences ($1,000,000 per occurrence and $2,000,000 annual amount) insurance coverage was adequate.[53]

Technology

Processes and Capacities. The facility has liquid injection incineration capacity of 6012 pounds per hour and 2910 pounds per hour for solids. It has storage capacity of 55,000 gallons, 112,000 gallons of tank storage capacity, four surface impoundments with a total capacity of 2,000,000 gallons, and a rail car area that can store two 20,000-gallon cars.[54]

Special Wastes Handled or Excluded. The facility handles PCBs. From 1983 to 1984 the facility accepted wastes from the following Superfund sites: Illegal Drum Disposal, North Dartmouth, Massachusetts; Liquid Disposal, Utica, Michigan; Kin-Buc Landfill, Edison, New Jersey; Northwest Dust Control, White City, Oregon; Alburn; and Liquid Dynamics.[55] Most of the wastes that the facility accepts are organic.

History

Prior Use and Owners. Hyon, Inc. and Envirotherm, Inc. operated the site as a hazardous waste facility.

Construction and Permits. In 1981 SCA Chemical Services bought the facility and completely rebuilt it. In 1982 SCA conducted test burns and U.S. EPA called in Part B of the RCRA application. In 1983 U.S. EPA reviewed Part B, and the facility received Chicago, state, and TSCA permits to burn PCBs. The first annual trial burn was completed in July 1980.[56]

Future Processes and Site Uses. Management plans to substitute above-ground storage tanks for surface impoundments in the near future.[57] The site is expandable from 5 to 30 acres.[58]

Citizen Concerns and Organizations. Avalon Trails Improvement Association, Irondalers Against the Chemical Threat (I-ACT), and the Chicago branch of Citizens for a Better Environment (CBE) opposed SCA during the initial permit application period. CBE questioned the results of SCA's two test burns, which showed PCB concentrations in kiln ash that differed by a factor of nearly 500 (16 parts per million in Run 11 versus 7600 parts per million in Run 13) and concluded that "the approval conditions proposed by U.S. EPA, Region 5 are woefully inadequate and significantly less stringent and/or comprehensive than previous PCB incinerator approvals issued by the Agency in other regions of the country."[59] Avalon Trails wanted SCA to spend $50,000 to review the test burn data, but SCA refused to do so unless it received a permit without delay.[60] The concerns of these groups, especially I-ACT, prompted IL EPA to conduct a multimedia environmental assessment of Southeast Chicago, which has long served as a disposal area for a wide variety of hazardous wastes. In fact there are 31 waste-handling facilities within 3 miles of the SCA facility.[61] U.S. EPA denied a 1984 petition filed by I-ACT and CBE that pointed out a number of alleged shortcomings regarding the IL EPA study and requested more impact analyses of industrial pollution in Southeast Chicago.[62] Jim Foley, the president of Avalon Trails Improvement Association, sums up the citizens' concerns and frustrations by asking, "Why should Southeast Chicago be made into a hellhole? Whether or not there is currently a health risk. . . . I don't think there is any way to find out unless people start dropping in the streets."[63]

FACILITY DESCRIPTION: CWM LA

Facility	*CEP Score*		*Grade*
Lake Charles Facility	Ground Water Route	=	C
Chemical Waste Management, Inc.	Surface Water Route	=	B
	Air Route	=	F
John Brannon Rd.	Fire and Explosion	=	D
Carlyss, LA 70663	Direct Contact	=	A
	Corrective Action	=	A
U.S. EPA ID#:	Technology	=	B
LAD000777201	Ground Water Monitoring	=	*
	Financial Instruments	=	*
	Root-Mean-Square Grade	=	C+

* Facility is in compliance, so this category is not factored into final grade.

Site

Demography. The site covers an area of 280 acres in Calcasieu Parish (167,048 persons in 1980), approximately 7 miles south of Sulpher (19,709 persons in 1980), and 10 miles southwest of Lake Charles (75,051 persons in 1980). The entire facility is surrounded by prime agricultural land that is used primarily for soybeans and cattle.[64] Five buildings are within a one-quarter mile of the site; 26 are within one mile. Approximately 100 people live within a mile of the facility and nearly 400 permanent residents live within 2 miles. The closest residences are 3000 feet north and 1400 feet south of the site.[65]

Water Use. Reportedly, the nearest water well that taps the shallowest aquifers is less than 1 mile southeast of the landfill area; however, the closest documented well that taps an aquifer as shallow as 27 feet is 1.5 miles due north.[66] Within 3 miles of the site in the eastern part of Calcasieu Parish, the 200-foot sand of the Chicot Aquifer is the principle source of water for domestic and agricultural use. The 500-foot sand is the major aquifer in the west of the parish, and it serves the towns of Sulpher, Edgewater, and Vinton. The city of Lake Charles uses the 700-foot sand.[67] In all, approximately 47 wells screened at depths from 27 to 600 feet are located within 3 miles of the facility.[68] Farming in the immediate area of the facility has shifted from rice, which needs irrigation, to soybeans, which do not need irrigation; consequently, irrigation ditches have reportedly not been used for at least 5 years.[69]

Hydrology and Geology. Numerous geological formations lie beneath the facility. At approximately 8 to 19 feet below ground surface is a clayey silt and sandy upper pervious zone that flows south/southwest. At 21 to 25 feet below ground surface is a silty, clayey sand—lower pervious zone that flows northeast. An underground river channel under the northeast corner of the site 42 to 80 feet down flows northeast. A third pervious zone of silty, clayey sands is beneath the western section of the site 58 to 62 feet down. The Chicot Aquifer is in three sands beneath the site at approximately 200, 500, and 700 feet.[70] The pervious zones within the upper 100 feet of the Prairie Formation are likely to be hydraulically connected with streams, lakes, and marshlands of the area on a localized basis.[71] The clay layers found between the pervious zones decrease in permeability from 10^{-7} to 10^{-7} centimeters per second as they get deeper, averaging at 4.4×10^{-8} centimeters per second.[72] The site has been completely surrounded by a 40-foot-deep and 3-foot-wide bentonite clay slurry wall that has an estimated permeability of 10^{-7} centimeters per second in order to prevent ground water from flooding the landfill area.[73]

Surface Water. Approximately 2500 feet west of the facility, Bayou Chopique flows south/southeast toward the Intra Coastal Waterway, which is less than 2 miles south of the facility, and ultimately into the Calcasieu River.[74]

Areas of Special Concern. Bayou Chopique, a fresh-water wetland, is about 2500 feet to the west and curves around the southern edge of the property at a greater distance.

Management

Observed Releases. In February 1984 the facility accepted 562 truckloads of wastes from the army's Pine Bluff Arsenal in Arkansas. On February 28 drums from two of the loads that contained spent smoke cannisters and a dye and smoke mixture containing hexachloroethane, granular aluminum, and potassium chlorate activated the ignited liquids that were being solidified in a nearby mixing pit. A resultant fire lasted 30 minutes. The follow-up inspection conducted by the Louisiana Department of Environmental Quality (LADEQ) revealed that the manifest documenting what the drums contained was completed improperly; they contained improper U.S. EPA identification numbers. ENRAC (a Chemical Waste Management subsidiary which was contracted to do the clean-up at Pine Bluff) and army personnel provided conflicting information. They said the smoke mix was nonreactive except when in contact with free oils. Some of the army's loads contained cans of oil-based paints; others appeared to contain free liquids including solvents and waste oils that had spilled onto the active landfill area. Allegedly, the detonating devices of the cannisters were removed, but seven smoke bombs were subsequently found with handles and detonators intact. Rick Hunter, the army representative at the inspection, theorized that, "due to the excess freeboard [empty space] within the drum and the lack of filler material, a spark was formed."[75] CWM had been compacting drums that contained hazardous wastes by driving over them with bulldozers. The friction caused by this practice may have ignited a smoke cannister inside a drum as it was being crushed.[76] During a second occurrence in August, a bulldozer hooked a landfilled drum containing Pine Bluff materials and ignited a smoke cannister and caused another fire. CWM mixed liquid and sludge wastes with kiln dust and fly ash in mixing pits that are located within the landfill trench area. After solidifying the wastes in this manner, CWM reexcavated the materials and pushed them onto the working face of the landfill. On the night of August 27, 1984, a dozer operator was doing just that—except he underestimated the depth of his blade. He uncovered a Pine Bluff drum and by backing up over it apparently caused a smoke cannister to ignite and the drum to vent. The dozer operator quickly pushed the solidified wastes on top of the fire to put it out.[77] Ground water monitoring data for the first quarter of 1984 show statistically significant increases of pH, specific conductants, and total organic halides in more than a half dozen down-gradient wells.[78] CWM claims, however, that "the significant changes shown from the Student's T-Tests are due primarily to the nature of the statistical evaluation and partly to experimental interferences."[79] CWM had similar problems with its monitoring results for the fourth quarter of 1983 and throughout 1982.[80]

Ground Water Monitoring. As of December 19, 1984, U.S. EPA found the facility's ground water-monitoring well system, sampling, analysis and evaluation program, and monitoring records to be adequate.[81]

Enforcement Actions. In August 1984 CWM received a state compliance order requesting a full investigation of the fires and a report showing how such incidents could be prevented in the future. The site also prohibited the receipt of any more Pine Bluff wastes without special state approval for each shipment.[82]

Corrective Actions. The facility eliminated all open mixing pits. It complied with a state recommendation to stop crushing drums unless a drum crusher is used, and in addition has installed a drum decant system and storage tanks. Liquid organic materials that might contribute to a fire will be separated in the tanks and removed to an off-site recovery or disposal facility. Treatment to stabilize sludges is conducted in an enclosed building equipped with air emission control scrubbers.[83] The facility's former practice of inspecting only 10 percent of incoming drums in each vehicle said to contain identical wastes has been revised; the current policy is to inspect l00 percent of drums on all incoming loads as recommended by the state.[84] Although there was no evidence of leakage, CWM exhumed all older trenches and redisposed them into trench #5.[85]

Staff and Unionization. There are 70 to 75 workers on site. None is unionized.[86]

Safety and Health. There have been no reported incidents of health and safety problems other than the two fires discussed above.

Financial Instruments. As of December 19, 1984, U.S. EPA found the facility's insurance to be adequate. Coverage included a $372,000 closure cost estimate, $2,604,000 postclosure cost estimate, $1,000,000 per occurrence and $2,000,000 annual aggregate sudden occurrence insurance ($1,000,000 per occurrence and $2,000,000 annual amount), and $3,000,000 per occurrence/$6,000,000 annual aggregate non-sudden occurrence insurance ($3,000,000 per occurrence and $6,000,000 annual amount).[87]

Technology

Processes and Capacities. The facility has approximately 5656 acre-feet of landfilling capacity, which is to be divided into about 20 independent trenches, the first of which (trench #5) is partially full and as of October 15, 1984 can hold some 474,074 cubic yards of waste.[88] The trenches are operated with a "moving" or "modified" procedure that allows for technological upgrading over time. The facility also has 480 current and 960 proposed drum storage capacity; drum decanting capacity of 320 drums per day; a 1,000,000-gallon leachate tank; as well as treatment for cyanide and sulfide destruction, neutralization, and fixation; and chemical oxidation of water-reactive wastes.[89]

Special Wastes Handled or Excluded. This facility is not reported to have accepted any wastes removed from Superfund sites.

History

Prior Use and Owners. In 1975 Sediment Removers, Inc. began landfarming petroleum wastes at the site.

Construction and Permits. In 1978 CWM acquired the facility. In 1982 the state issued the facility a permit. During 1983 U.S. EPA called and received Part B of the facility's RCRA application.[90]

Future Processes and Site Uses. Proposed for the site are a deep well for the injection of aqueous wastes, an incinerator for organics and combustibles, and a solvent reclamation center. The facility is expected to be in operation for another 22 years.[91]

Citizen Concerns and Organization. The Calcasieu League for Environmental Action Now (CLEAN) is an active citizen organization located in the parish. This facility, however, is not its principal concern; rather, CLEAN is involved in a number of environmental issues in the Lake Charles area, including protests against the CECOS hazardous waste facility in Willow Springs and ocean incineration.[92] Marvin Hargar is a local resident and member of CLEAN who has kept close watch of and is outspoken about the Lake Charles facility. He is concerned that the detection of bis (2-ethylhexyl) phthalate (BEHP) in the range of 1500 to 3400 parts per billion in an irrigation well approximately 15 feet from the north end of the landfill area and 2000 feet from the then-active cell may indicate ground water contamination from the facility.[93] In addition he alleges that an irrigation well 514 feet deep at the southwest corner of the landfill area was backfilled with gravel, that an abandoned water well has not been capped, and that there are two abandoned gas wells at the site. This could, he says, create a "swiss cheese effect," allowing vertical migration of contaminants between aquifers if the wells are not properly sealed. According to Hargar, CWM has compacted an area of 25 feet surrounding only one gas well that is within the landfill area. Hargar also claims that selenium and thalium were found in surface water southeast from the site in 1981.[94] CWM denies Hargar's claims, saying no BEHP was detected in CWM monitoring wells, and that the company "is unaware of any study indicating selenium and thalium were found in surface waters surrounding the site."[95] Currently, no federal or state drinking water standards regulate BEHP.[96]

FACILITY DESCRIPTION: CECOS NY

Facility	*CEP Score*		*Grade*
CECOS International Inc.	Ground Water Route	=	A
56th & Pine Ave.	Surface Water Route	=	C
Niagara Falls, NY 14302	Air Route	=	B
	Fire and Explosion	=	B
U.S. EPA ID#:	Direct Contact	=	F
NYD080336241	Corrective Action	=	C
	Technology	=	B
	Ground Water Monitoring	=	F
	Financial Instruments	=	*
	Root-Mean-Square Grade	=	C−

* Facility is in compliance, so this category is not factored into final grade.

Site

Demography. All of the city of Niagara Falls (population 71,384 in 1980) and much of the town of Niagara (population 9648 in 1980) are within 4 miles of this upstate New York facility. It is in a busy industrial and residential area with Interstate Highway 190 flanking its eastern boundary and State Highway 62 (Pine Avenue) to the south. Five of the facility's 385 acres are within the city limits of Niagara Falls, the rest are inside the town of Niagara.[97]

Water Use. The entire Niagara Falls area gets its water from the Niagara River at intakes below Buffalo.[98] There are no drinking water intakes within 3 miles downstream of CECOS's discharge points, but the area above and below Niagara Falls is widely used for fishing, bathing, boating, and other water sports.[99] One-half mile downstream from CECOS's discharge into the Niagara a number of plants within the city limits draw water from the river for industrial purposes.[100] The river is not used for drinking until Lake Ontario, where Canada draws municipal supplies. Natural salinity and contaminants from other industrial sources have made the aquifers of concern unuseable for domestic, agricultural, or industrial purposes.[101]

Hydrology and Geology. The topography of the site has been completely altered by over 30 years of industrial deposits and excavations. Three hydrologically connected aquifers, the uppermost beginning just 1 foot below ground surface, lie under the site.[102] Ground water flows generally south southeast, but because of cracks and crevices in the bedrock and pumping at the adjacent E. I. DuPont de Nemours & Co. Necco Park site, it also flows northwest in certain places.[103] The local hydrogeology contains several layers. Uppermost are industrial deposits 30 to 55 feet thick in some places. Topsoil (removed in some places) can be up to 1 foot thick. A 1- to 5.5-foot thick unconfined aquifer that emerges as surface runoff lies under the topsoil. Continuing downward are 2 to 16 feet of lacustrine clays with an estimated permeability of 10^{-7} centimeters per second, although joints raise the permeability of the clays in certain places. Next there is up to 1 foot of Lake Lundy deposits of sandy, silty clay with 2.4×10^{-6} permeability. Under that are 5 to 10 feet of glacial till with 2.4×10^{-3} permeability. Finally, the 10-foot thick Lockport Dolomite aquifer lies where the bedrock is cracked and has a permeability of 10^{-3} centimeters per second.[104] Although the lacustrine clays were originally thought to act as secondary liners, joints and springs discovered during excavation increase their permeability by a factor of 10 or more.[105] Effluents from the Lockport Dolomite enter the Niagara River.[106]

Surface Water. The Niagara River is approximately 1.5 miles south of the facility. There are also two perennial streams near the facility, one a mile east and the other a mile west.[107] Drainage from the site's active area is stored in underground tanks. One city and two state permits specify the allowable levels of contaminants that CECOS can discharge into the Niagara Falls waste-water treatment system. Since July 1978 the city's carbon filtration system has been out of service, but CECOS has an extensive activated carbon filtration and pretreatment system. Storm water runoff from the site periphery is released directly into the 60th Street sewer system from where it flows into the river a mile south.[108]

Areas of Special Concern. A 5-acre fresh-water wetland, which had been 13 acres prior to the expansion of a BFI sanitary landfill, is directly northeast of the facility.[109]

Management

Observed Releases. *1981:* Leachate from the hazardous waste landfills was found seeping into wetlands northeast of the site.[110] *1982:* In August the state demanded cessation of leaching into a construction area at the site.[111] In November containers with RCRA wastes were found leaking and open.[112] *1983:* In May the facility began assessment monitoring after reporting detection of significant increases in pesticides, phenols, lindane, benzene, and other hazardous chemicals in down-gradient wells.[113] Also in May surface impoundments were found in violation of RCRA because of "severe berm erosion in many areas" on two and inadequate freeboard on three of them; however, there was no observed release.[114] During the entire month of December, CECOS was unable to meet the standards required by its discharge permit, so the state granted the company a variance to discharge into the Niagara 3 pounds of toxic phenols per day.[115] Also in December ponding, flooding, leachate collection system failures, and leachate levels in standpipes that exceeded allowable limits were reported at the site on several occasions, and some runoff was found continuing past the property line.[116] *1984:* In January a state inspector noted broken drums in the heavy metal cell, one of which had spilled approximately 25 percent of its contents.[117] In February CECOS exceeded its permitted discharge levels for chemical oxygen demand.[118]

During CEP's site visit of CECOS in October, a whitish plume was seen rising from a tank identified as containing hydrochloric acid for the neutralization basin. The plant official said the emission was steam, and that this event occurred whenever the tank was overfilled.[119] Toxic emissions, however, have never been observed coming from the facility. Also in October U.S. EPA released a draft study indicating that "there is circumstantial evidence that the two units in question, 3 & 4, are leaking and probably were leaking at the time Superfund wastes were deposited in them" and that ground water contaminants may be "migrating off site and could endanger the health of the public living down stream."[120] *1985:* U.S. EPA alleges that "evidence exists to show that the ground water beneath and around [CECOS's] SCMF [Secure Chemical Management Facilities] Nos. 1–3 and Necco Park is contaminated with hazardous wastes" and that CECOS needs to upgrade its ground water-monitoring program to determine whether SCMF Nos. 1–3 have leaked.[121]

Ground Water Monitoring. As of February 22, 1985, EPA found the facility's ground water-monitoring well system and sampling, analysis, and evaluation program to be inadequate.[122]

Enforcement Actions. *1978:* Consent Agreement. U.S. EPA assessed a fine of $5500 for the facility's failure to record high leachate levels.[123] *1980:* Compliance Complaint and Notice of Opportunity for Hearing. In May U.S. EPA warned the facility that it was in violation of TSCA because landfilled PCB containers were found opened, damaged, and without absorbant material.[124] *1981:* State Consent Order. In August NYSDEC cited the facility after it found leachate from the hazardous waste landfills seeping into wetlands northeast of the site.[125] Final Administrative Order. U.S. EPA cited the facility for violating TSCA on three counts including inadequate absorbant material in a PCB drum, lack of warning signs, and exceeding leachate level limits. U.S. EPA assessed a $30,000 fine.[126] *1982:* Compliance Complaint. In March U.S. EPA alleged malfunctions of the automatic leachate collection system were in violation of TSCA.[127] State Consent Order. In April NYSDEC ordered the facility to stop allowing leachate to seep into a construction area at the site.[128] *1983:* 3008 Warning Letters. In February U.S. EPA notified facility that it had not submitted closure and postclosure financial instruments assurance and had not supplied missing ground water-monitoring reports.[129] In August U.S. EPA warned facility of violating 40 CFR 265.223.[130] State Consent Order. In November NYSDEC ordered the evacuation of one landfill cell because of a torn liner and the closure of another because of overfilling.[131] *1984:* State Warning Letter. In June NYSDEC warned facility to cover PCB materials that could become airborne and leave the site.[132] In November U.S. EPA directed facility to conduct off-site ground water monitoring, claiming that current sampling and analysis procedures were inadequate.[133] *1985:* Administrative Order on Consent. In February U.S. EPA ordered the reevaluation of hydrology and existing ground water wells, a plan for an initial monitoring network, a plan for installation of new wells, a plan for enlarging the monitoring network, as well as compliance with additional and miscellaneous reporting provisions relating to a study area surrounding closed landfill units 1, 2, and 3.[134] Consent Agreement and Consent Order. On the same day U.S. EPA directed the facility to begin assessment monitoring, to install more wells, and to develop and conduct more extensive monitoring for landfill units 4 and 5.[135]

Corrective Actions. The facility complied with past enforcement actions and is on a compliance schedule to upgrade its ground water-monitoring program. Prior to the 1985 consent agreements, the facility had 50 regularly sampled ground water-monitoring wells to detect contamination and three full-time NYSDEC hazardous waste inspectors who are stationed at the site.[136] Volatile organic chemicals leaking from the adjacent DuPont Necco Park site and Union Carbide Corporation's 30- to 50-foot-high lime deposits cause such high and erratic ground water contamination levels that it is difficult to statistically verify leakage from the CECOS facility. For over 5 years, state and regional U.S. EPA officials have been negotiating a cleanup strategy with DuPont. DuPont is currently pumping chemicals out of the ground water.[137] In addition, the fractured bedrock beneath the facility increases the difficulty in determining the hydraulic gradient of the ground water.[138] Off-site monitoring conducted in 1985 did not detect migration of significant contamination in the unsaturated zone beyond the facility boundaries.[139]

Staff and Unionization. One hundred and two full-time employees work on site. None is unionized.[140]

Safety and Health. In 1981 U.S. OSHA inspected the facility in response to a worker's complaint.[141] During CEP's 1984 visit of the site, CEP saw that adjacent to Pine Avenue, trucks park on CECOS property to await manifest verification. There does not appear to be a gate to prevent outsiders from entering the parking area even though trucks have been found leaking hazardous wastes on at least one occasion.[142] Nevertheless, there have been no reported incidents of direct contact.

Financial Instruments. As of December 1983 EPA had not evaluated facility's closure assurance of $3,000,000 and postclosure assurance of $2,008,570. The facility's sudden ($1,000,000 per occurrence and $2,000,000 annual amount) and non-sudden ($3,000,000 per occurrence and $6,000,000 annual amount) insurance policies effective as of November and June 1982 (respectively) meet federal requirements.[143]

Technology

Processes and Capacities. The facility has 1,314,867 cubic yards of closed pre-RCRA landfill, 930,000 cubic yards of RCRA landfill capacity, tank treatment capacity of 160,000 gallons per day, 8,000,000 gallons of surface impoundment storage capacity, and 250,000 gallons of container storage capacity.[144] Treatment offered at the facility includes: chemical oxidation, chemical reduction, neutralization, stabilization, filtration, activated carbon, sand filter, and a Sequencing Batch Reactor (SBR) that uses microorganisms to degrade organic contaminants such as phenols in industrial waste water.[145] The SBR also performs equalization, aeration, and sedimentation processes in a timed sequence using a conically shaped batch tank. The SBR is reported to reduce biological oxygen demand by 99 percent and phenol by 95 percent to 98 percent and to save 720 billion BTUs per year in comparison to conventional waste water treatment systems.[146]

Special Wastes Handled or Excluded. The facility is authorized to accept PCB wastes. From 1982 to 1984, the facility accepted wastes removed from the following Superfund sites: W. R. Grace, Acton, Massachusetts; Ottattie & Goss, Kingston, New Hampshire; Keefe Environmental Services, Epping, New Hampshire; Syncon Resins, Kearny, New Jersey; Summit National, Deerfield Township, Ohio; Dracup Warehouse/Youngstown PC, Youngsville, Pennsylvania; Lehigh Electric Phase I, Pennsylvania; Lehigh Electric Phase II, Pennsylvania; Picillo/RI, Coventry, Rhode Island; Big John Salvage, Fairmont, West Virginia.[147]

History

Prior Use and Owners. Union Carbide Corp. was the former owner of the site and used it as a dumping area for its industrial wastes. Approximately 25 to 30 acres of the site are covered with 30 to 55 feet of lime as a result of Union Carbide's activities.[148]

Construction and Permits. In 1976 CECOS began hazardous waste disposal activities. In 1983 U.S. EPA called for, received, and reviewed Part B of the facility's RCRA application.

Future Processes and Site Uses. During either fall 1985 or spring 1986, CECOS plans to construct a Durachem II chemical treatment process facility that will separate, encapsulate, and recover organic residues from still bottom, paint, and solvent wastes. With Modar, Inc. CECOS has developed a supercritical water thermal oxidation reactor that it would like to bring on line at the facility by 1989.[149]

Citizen Concerns and Organizations. A citizens' organization called Society to Oppose Pollution in Towns (STOP IT), which formed during the 1978 to 1979 state permit hearings, is no longer active. Many residents, however, have remained concerned about the facility's impact, especially after the potential for contaminants to migrate offsite became front-page national news in 1984.[150] Tony Bassili, for example, runs a motel directly across Pine Avenue, less than 1000 feet south of the site. He remembers when the government rented his rooms to house Love Canal evacuees and does not enjoy the idea of being displaced himself if Love Canal wastes or materials removed from ten other Superfund sites leak from the CECOS facility.[151] In addition to potential hazards to the environment and human health, other residents are concerned about what they see as excessive local political power given to CECOS.[152] Three citizen groups, Niagara Environmental Action, Lasalle and Niagara Demand (LAND), and New York Community Action Network (NY CAN), have joined together to create the Necco Park–CECOS Committee. Their principal fear is that contaminants will percolate into basements.[153] To date, there has been no evidence of such exposure.

FACILITY DESCRIPTION: ENSCO AR

Facility	*CEP Score*		*Grade*
ENSCO, Inc.	Ground Water Route	=	D
American Rd.	Surface Water Route	=	D
El Dorado, AR 71730	Air Route	=	B
	Fire and Explosion	=	C
U.S. EPA ID#:	Direct Contact	=	B
ARD069748102	Corrective Action	=	C
	Technology	=	C
	Ground Water Monitoring	=	F
	Financial Instruments	=	*
	Root-Mean-Square Grade	=	C−

* Facility is in compliance, so this category is not factored into final grade.

Site

Demography. Almost 27,000 residents of the city of El Dorado, near the southern border of Arkansas, live within 4 miles of the facility.[154] More than 1000 people live or work within a mile. The immediate vicinity is industrial. A largely abandoned oil field is located northeast of the site, with residences primarily to the north, west, and south.[155]

Water Use. The Sparta Sand Formation, approximately 600 to 800 feet below ground surface, is essentially the sole source of water for most of southern Arkansas. Within a 3-mile radius of ENSCO, nine wells tapping into the Sparta serve 28,000 people, and the 48,769 people in Union County rely on ground water for domestic use.[156] Many years ago El Dorado pumped water from the Cockfield Formation, using wells screened typically from 80 feet or lower.[157] Columbian Carbon operates the closest well, which taps the Cockfield 8200 feet east of ENSCO for industrial use.[158] Three other wells, including one used by ENSCO for process water, are in the immediate vicinity of the site, but all have been shut in.[159] In the areas outside of the El Dorado city limits but within 3 miles of ENSCO, up to a dozen families could have wells that tap the Cockfield for domestic use.[160]

Hydrology and Geology. Sandy, silty clay lies from ground surface to 10 feet below. The Cockfield Formation is the uppermost aquifer at 10 to 200 feet below ground surface.[161] Although the Cockfield is undifferentiated on a regional basis, at the site it is divided at 90 feet by a clay layer less than 10 feet thick.[162] The upper Cockfield is hydrologically connected to Boggy Creek.[163] The lower Cockfield's lignitic sands at 100 to 140 feet below ground surface are under greater artesian pressure than the upper Cockfield at the southeast of the site. At the northwest, the opposite is true.[164] From 150 to 500 feet below ground surface, the Cook Mountain Formation acts as an aquiclude (that is, it prevents water from migrating vertically).[165] The upper Cockfield flows toward the facility from all directions due to ENSCO's ground water cleanup-program.[166]

Surface Water. Boggy Creek runs north to south through the eastern edge of the site, with drainage ditches entering it from the southern portion of the site. Boggy Creek is a class B waterway; it serves as a recreational or agricultural water supply, warm-water fishery or public water supply, or an industrial water supply.[167]

Areas of Special Concern. At approximately 1 mile south of the site, Boggy Creek is a wetland. Bayou Loutre at about 3 miles southeast is also a wetland. According to ENSCO's Part B, wetlands are assumed to exist within the facility perimeter.[168]

Management

Observed Releases. *Mid 1970s:* ENSCO had problems handling some odorous sulfur-containing compounds. The problem has since been solved.[169] *1975:* The company spilled hazardous wastes being transported from Kansas City, Missouri. The shipper mislabeled the load, and the spill resulted in wildlife kills along the stream where the spill took place.[170] The shipper (not ENSCO) accepted financial responsibility for cleanup. *1981:* Deteriorated drums on site spilled contents onto the soil.[171] Scrubber lagoons 1 and 2 leaked chlorides, iron, manganese, sodium, and sulfate.[172] Lead was detected above drinking-water standards in monitoring wells B-2, B-2a, and B-2b where ENSCO could have contributed to the contamination.[173] The company suspected a leak in an API lagoon that treats storm water. Endrin and dieldrin were detected at twice the level of drinking-water standards.[174] *1983:* Tanks 509, 510, and 511 were suspected of having leaked chlorinated organic chemicals detected at levels of 10,000 parts per billion.[175] Tanks 152 and 109 leaked, and tank 109 spilled 169,000 gallons of what ENSCO said was rainwater into a diked area.[176] Also during 1983, one of the legs on a pressure sphere gave way, causing the sphere to rupture on impact and spill 180,000 gallons of chlorinated hydrocarbon solvents into a diked area.[177] The company believes a leak in tank 122 caused a rise in the level of total organic halogens from 110 parts per billion to 40,100 parts per billion between May 1983 and January 1984. The background level was 10 to 20 parts per billion. Key contaminants detected in lignitic sands are 1-chloro, 3-bromopropane; 1,3-dichloropropane; and either 1-chloro, 2-bromopane or 1-bromo, 2-chloropropane.[178] *1984:* The PCB storage tank was overfilled and approximately 3000 gallons spilled into a diked area.[179] API separator collector overflowed and crude oil spilled into a ditch that leads to Boggy Creek.[180] The waste-fired boiler failed the first performance test by exceeding allowable hydrochloric acid emissions.[181] Fire occurred in a tank suspected of containing sludge from diesel or crude left by Amoco. ENSCO took the proper contingency measures, and there was no observed release "except smoke."[182]

Ground Water Monitoring. As of December 1984 the facility was conducting detection monitoring. U.S. EPA evaluated well system and sampling, analysis, and evaluation program to be inadequate. Monitoring records have not been evaluated.[183] ENSCO believes its ground water-monitoring program is adequate and maintains that the state agrees with the company.

Enforcement Actions. *1981: United States* v. *Energy Systems Company,* Complaint and Consent Decree. Among the charges were: improper waste drum management, periodic overflow and leaking scrubber effluent lagoons, inadequate fire protection and response measures, bulk waste storage facilities at 12 lots in disrepair, inadequate runon and runoff prevention systems, and inadequate containment of transferred wastes in loading areas. Over 5000 drums were stored in various locations throughout the site; some leaked onto the soil. According to the decree, the wastes stored in these drums include dieldrin (a carcinogen, mutagen, and teratogen), chromium (the hexavalent form is a carcinogen and a possible teratogen), lead compounds (carcinogens, which also cause anemia, neurological disorders, and kidney damage), and cyanide (a highly toxic compound). The decree required that ENSCO construct a permanent barrel storage area for non-PCB wastes, incinerate or recontainerize all leaking barrels, clean up the barrel storage area, and reduce the inventory to less than 1000 barrels.[184] All items of the 1981 decree have been corrected and ENSCO has been released by the court. *1982:* U.S. EPA Administrative Complaint and State Compliance Order. In March and May of 1982, inspection reports documented the failure to complete the program for non-PCB waste barrel management on schedule. State and U.S. EPA inspectors found leaking and deteriorating drums outside designated storage areas. Drums in the proper area were standing in water even though it had not rained 48 hours prior to the inspection, and the 1000-drum limit had been exceeded. The scrubber lagoons continued to have inadequate freeboard, and an unlined sump used to trap runoff was within the water table adjacent to a creek.[185] All items of the 1982 complaint and order have been corrected and settled. Also in 1982 U.S. EPA filed an administrative complaint alleging ENSCO had violated provisions of the TSCA. Although ENSCO considers the complaint to be without merit, it settled for a fine of $25,000.[186] *1983:* On December 6 a state inspector noted four Class I and five Class III violations.[187] The scrubber lagoons still had inadequate freeboard.[188] *1984:* Arkansas Department of Pollution Control and Ecology Consent Administrative Order. The order alleges that ENSCO violated state laws by causing contamination of ground water at or emanating from the facility, constructing and operating the calcium chloride processing facility without a water permit, and operating a source capable of emitting air contaminants contrary to the provisions of ENSCO's permit.[189]

Corrective Actions. TSCA authorization in 1981 required that an independent consultant, reporting directly to U.S. EPA Region 6, conduct a minimum of five unannounced inspections each month. Quarterly summaries covering a minimum of 15 inspections each quarter state that ENSCO has operated and is operating in full compliance with TSCA requirements for PCB treatment. ENSCO cleaned up leaking drums. A water permit has been issued for the calcium chloride recycling unit. In fact, ENSCO had applied for a federal NPDES permit one year prior to the order. The waste-fired boiler that failed its first performance test has met air emissions standards ever since.[190] A ground water-monitoring and hydrogeological reassessment program is intended in part to evaluate the success of ENSCO's remedial actions. All of the problematic tanks have been brought up to RCRA standards; ENSCO incinerated the contents of over half the tanks or disposed of them in landfills in RCRA or TSCA facilities. The company demolished many of the tanks, too, and disposed of them appropriately. ENSCO has constructed barrels of hazardous waste, excavated contaminated soils, installed synthetic liners, and closed the two brine scrubber lagoons. ENSCO has constructed a french drain to collect contamination. The reassessment will try to determine whether the closed lagoons are leaking and whether the french drain is containing the contamination. The company also implemented a system of surface water control. Only one violation of the NPDES permit has occurred since the full implementation of the surface water collection and treatment system. As required by the consent decree, ENSCO has achieved control of wastes left by the refinery operations prior to ENSCO's acquisition of the site. ENSCO paved, diked, and sumped loading stations for trucks and rail cars with concrete.[191]

Staff and Unionization. There are approximately 300 full-time, on-site employees. Some workers are members of the Oil Chemical and Atomic Workers Union.[192]

Safety and Health. *1975.* ENSCO violated OSHA standards for walking and working surfaces; hazardous materials; and welding, cutting, and brazing. *1977:* ENSCO committed two serious OSHA violations and nine other violations. The company was fined $180, a reduction of the initial penalty of $245. *1978:* ENSCO committed 15 serious violations of OSHA standards and 12 other OSHA violations. It was fined a total of $2000, a reduction of the initial penalties of $30,670.[193]

Financial Instruments. As of September 1984 U.S. EPA considered ENSCO's $2,961,300 closure assurance to be adequate. The facility's sudden ($1,000,000 per occurrence and $2,000,000 annual amount) and non-sudden insurance ($3,000,000 per occurrence and $6,000,000 annual amount) also met U.S. EPA standards.[194]

Technology

Processes and Capacities. The facility's capacities include incineration capacity of 1800 gallons per hour, 40,000 gallons per day of calcium chloride reclamation, solvent reclamation and pretreatment of 10,000 gallons per day, 4000 gallons per day of fuels, and 1.5 tons per hour of ferrous metals, 408,000 gallons of drum storage, 11,651,400 gallons of tank storage, and 26,500 cubic yards of waste pile. A Pyrotech modular incinerator is based at the site.[195]

Special Wastes Handled or Excluded. The facility has a license to incinerate PCBs. Wastes with high heavy-metal content are not accepted. Waste from the following eight Superfund sites were accepted between 1982 and 1983: Arrcom Corp., Rathdrum, Idaho; Pacific Recycling, Pocatello, Idaho; Keefe Environmental Services, Epping, New Hampshire; Chem-Dyne 2, Hamilton, Ohio; Vulcanized Rubber and Plastics Co., Morrisville, Metal Bank, and Lehigh Electric Phase I, Pennsylvania; and Bio Ecology, Grand Prairie, Texas.[196]

History

Prior Use and Owners. American Oil Company (Amoco) operated an oil refinery on the site for more than 20 years.

Construction and Permits. *1973:* ENSCO, at the time called Pollution Control, Inc., made a lease-purchase agreement with Union County and the El Dorado Industrial Development Corp. to operate an incinerator for hazardous waste. *1974:* Arkansas granted a construction permit. *1977:* ENSCO completed construction and started test burns. *1978:* ENSCO received an Arkansas permit for incineration. *1981:* ENSCO received U.S. EPA TSCA permit for PCB incineration. *1983:* U.S. EPA called for submission of Part B of the facility's RCRA permit application.

Future Processes and Uses. ENSCO plans to base two additional mobile rotary kilns at the site. The site will never be useable for purposes other than industrial waste disposal.

Citizen Concerns and Organizations. *1978:* In August Union County passed an ordinance banning incineration, shipment, or storage of PCBs in the county. In September U.S. EPA held public hearings to consider the proposal to conduct test burns of PCBs at the site. A 4600-signature petition introduced to the hearing record stated citizen opposition to the test burn. A poll showed that of 1256 people contacted, 1200 were opposed to the importation of wastes from out of state or county. *1979:* In January Citizens for a Safe Environment (CASE) was incorporated to monitor ENSCO's operations.[197] *1981:* In January U.S. EPA approved PCB incineration at the facility, a move that preempted local restrictions.[198] *1985:* CASE, which once had a membership of some 250 households, now has only 10 households involved, and they are no longer very active. Rachel McKinney claims the group was treated shabbily by the state and U.S. EPA, so they gave up.[199] The Dioxin Task Group, which reportedly has approximately 1000 households involved within 50 miles of ENSCO is active. Both groups are concerned about air pollution as well as ground water contamination. The Dioxin Task Group strongly supported a county ordinance to ban storage of acutely hazardous wastes at ENSCO. ENSCO has appealed this local ordinance.[200]

FACILITY DESCRIPTION: RES (LA)

Facility	*CEP Score*		*Grade*
Rollins Environmental Services (LA), Inc.	Ground Water Route	=	C
13351 Scenic Highway	Surface Water Route	=	C
Baton Rouge, LA 70807	Air Route	=	F
	Fire and Explosion	=	B
	Direct Contact	=	F
U.S. EPA ID#:	Corrective Action	=	C
LAD010395127	Technology	=	A
	Ground Water Monitoring	=	*
	Financial Instruments	=	*
	Root-Mean-Square Grade	=	C−

* Facility is in compliance, so this category is not factored into final grade.

Site

Demography. RES (LA) is just northeast of the Baton Rouge city limits. Over 800 persons live within 1 mile in the residential community of Alsen, which is across State Highway 61 from the site.[201] In addition, a large portion of the town of Baker (population 12,865 in 1980), 1.5 miles northeast of the site, is within 4 miles of the facility. Adjacent land is used industrially, including an Allied Corp. plant flanking the southern boundary, Barnes Chemical one-third of a mile north, and the Petro-Processors Superfund site in between RES (LA) and Barnes.

Water Use. Within 3 miles of the facility, approximately 32,000 residents and workers use water tapped from up-gradient wells in aquifers 400 to 2800 feet below ground surface.[202] The 400-foot sands serve over 200 people. The largest well is an up-gradient municipal utility well a mile east of the site that serves approximately 140 Alsen residents.[203] The shallowest well screen is 232 feet below ground surface; shallower ground water is not currently in use.[204] According to RES (LA), no down-gradient withdrawal ground water is used for drinking use.

Hydrology and Geology. Soils beneath the facility consist of clay layers with permeabilities ranging from 10^{-7} to 10^{-10} centimeters per second and intermittent layers of clayey silt, clayey sand, silty sand, and sand and silt with permeabilities of 10^{-3} to 10^{-6} centimeters per second.[205] Ground water lies in silty and sandy layers 25 to 45 feet and 50 to 78 feet below ground surface.[206] These aquifers flow southwest at an estimated rate of 0.1 feet per year.[207] The 400-foot sands contain a major aquifer at 180 to 600 feet below the facility.[208] No continuity or hydraulic interconnection between the three shallow aquifers has been established.[209] Other major aquifers are found at 1000 and 1300 feet below ground surface.[210]

Surface Water. Water drains southwest from the site into Bayou Baton Rouge, Devil's Swamp (one-quarter mile west of the site), Baton Rouge Harbor (one-half mile west of the site), then into the Mississippi River, and ultimately into the Gulf of Mexico.[211] The facility is 40 feet above the flood plain and approximately 6000 feet hydraulically up-gradient from the river.[212]

Areas of Special Concern. There is no critical habitat in the area but the fresh water wetland called Devil's Swamp is a sensitive environment.[213]

Management

Observed Releases. *1980 to 1983:* Engineering contractors of the city-parish alleged that 113 out of 132 odor complaints were attributable to releases from the facility and that the facility caused a nuisance to persons living and working in the vicinity. Allegedly, the odors sometimes caused nausea and respiratory irritation.[214] RES (LA), on the other hand, alleges that the engineering contractors of the city-parish were not qualified to quantify odor, to identify its source, or to make findings on the causes of any alleged respiratory irritation and nausea. Also according to RES (LA), the air-quality inspectors of the state of Louisiana ascribe validity only to a small number of these complaints, particularly those after RES (LA) closed down its landfarming operations in 1981. From 1981 to 1983 contaminants were first detected in seven ground water-monitoring wells. These contaminants are believed to have leaked from closed landfill cells. RES (LA) asserts that all of the contaminated ground water is retained within the current boundaries of the RES (LA) site. Compounds detected include 1,1-dichloroethylene, trichloroethylene, 1,1,2-trichloroethane, 1,1,2,2-tetrachloroethane (reported only once at a level of 15 parts per billion), 1,1-dichloroethane, chlorobenzene, benzene, ethylbenzene, toluene, chloroform (the last five chemicals were detected only in one well), and carbon tetrachloride, all of which have been shown in scientific studies to have toxic, carcinogenic, mutagenic, or teratogenic effects on humans and other life forms.[215] However, since RES (LA) believes that the contaminated ground water is limited to the RES (LA) site, that it is not in a useable aquifer, and that it is being removed and treated, the company does not believe it poses a threat to human health. *1983:* A U.S. EPA investigator reportedly found an open culvert gateway leading to the Mississippi River from RW5, a surface impoundment without a permit for hazardous waste. RW5 is within the shell of an exhumed landfill. Facility management denies that it uses RW5 for hazardous waste.[216] According to RES (LA), U.S. EPA has accepted the company's report on this matter and considers the matter closed. Moreover, RES (LA) states that the open culvert was, in fact, a small pipe that went to a landfill containment ditch, not to an outfall. The outfall is sampled and reported per the facility's NPDES discharge permit. A contractor inspection found surface erosion on a bio-equalization pond and a truck-wash basin, the potential for erosion on another pond, and an 8-foot manmade breach in another. The inspector saw no release.[217] *1983 to 1984:* Of seven confirmed odor complaints, two were attributed to the incinerator, although Louisiana state engineers indicate the incinerator is not a likely source, according to RES (LA).[218]

Ground Water Monitoring. As of September 27, 1984, U.S. EPA alleged that the facility's ground water-monitoring well system, sampling, analysis, and evaluation program, and assessment report were inadequate.[219] However, detailed hydrogeological analyses prepared by Geraghty and Miller, a national ground water consulting firm, demonstrated that the ground water program at the facility is functioning effectively. According to RES (LA), U.S. EPA agrees. In early October 1985 the matter was resolved successfully.

Enforcement Actions. *1980:* In April the Louisiana Environmental Control Commission (ECC) ordered RES (LA) to "clean up its operation." The City–Parish Council adopted a resolution directing the parish attorney to take legal action against RES (LA) and for the parish technical representative to continue investigating complaints.[220] *1981:* In June the parish attorney general held proceedings against RES (LA) before the ECC. The ECC ordered the RES (LA) landfarm closed by December.[221] *1983:* The Louisiana Department of Natural Resources referred two odor complaints to the state attorney general for action and ordered RES (LA) to cover and add scrubbers to the decanting basins, to enclose the tank farm servicing area, and to install a separation unit to keep organic materials out of the aeration basin.[222] *1984:* State Notice of Violation. Pursuant to an April RCRA inspection, the Louisiana Department of Environmental Quality issued a citation alleging eight violations at the facility including open containers and insufficient aisle space in the drum storage area, insufficient waste analysis prior to incineration, inadequate freeboard on surface impoundments, four improper recordkeeping violations, and failure to fence the facility's NPDES ditch (a deficiency also noted in the previous RCRA inspection).[223] Compliance Order and Notice of Opportunity for Hearing. Pursuant to a June RCRA inspection, U.S. EPA cited the facility for many of the same violations noted earlier by the state and additional violations including failure to provide a barrier that completely surrounded the active portion of the facility; failure to develop and follow an inspection schedule for quality control of monitoring, safety, security, operating, and structural equipment; failure to obtain approval or permits for increases in design capacities (U.S. EPA subsequently withdrew this allegation); operation of hazardous waste surface impoundment without approval; inadequate ground water-monitoring system and assessment program; inadequate freeboard on scrubber basin surface impoundment; failure to perform daily inspections of the incinerator emergency shut-down controls and system alarms; failure to transfer wastes from bulging and leaking containers to containers in proper condition. U.S. EPA proposed to require the facility to take 13 measures to correct these deficiencies and proposed a civil penalty of $88,400.[224] RES (LA) vigorously contested U.S. EPA's allegations. The company particularly objected to allegations that the facility's inspection and ground water-management programs were inadequate and that any unpermitted increases in design capacity had occurred. U.S. EPA subsequently withdrew the allegation of unauthorized increases in design capacity and accepted RES (LA)'s explanation that the surface impoundment at issue had not been used to manage hazardous waste. The proposed civil penalty was reduced to $40,000, which RES (LA) paid, and the matter was closed in early October 1985.[225] Order Requiring Submission and Implementation of Proposal for Monitoring, Testing, Analysis, and Reporting. In October U.S. EPA evaluated the facility's pollution-abatement program and alleged that concentrations of contaminants had actually increased in the most down-gradient (southwestern) ground water-monitoring wells. U.S. EPA ordered that RES (LA) provide a new geohydrologic summary and that the company schedule excavation and complete closure of old landfill, propose a new, detailed ground water-monitoring, -testing, -analysis, and -reporting plan, and appoint a facility coordinator to oversee implementation of the order.[226]

Corrective Actions. There have only been a few relatively minor surface-water discharge violations since 1981.[227] RES (LA) ceased land application in 1981 and covered an open tank in 1983. But according to the city-parish inspector, "everyone is surprised that the odors continue to be there."[228] RES (LA) maintains that, "the subjective sense of smell relied on in their "inspections" simply cannot measure levels of emissions. The level of emissions and so-called "nuisance" odors has not been and probably *cannot* be quantitatively detected and documented. . . . Our review indicates that VOC [volatile organic compounds] emissions . . . have been nearly eliminated. . . . Any airborne VOC emissions that might remain are very limited in lateral and vertical extent and are contained within the RES (LA) facility."[229] In 1981 two pollution-abatement wells were installed and in 1983 three additional abatement wells were installed in the southwestern portion of the site to reverse the southwest ground water flow and to remove contaminated ground water for treatment.[230] All known landfill cells have been excavated, and, where necessary, the excavated wastes have been stabilized and replaced in lined cells.[231] The concentration of total VOCs in the 50- to 75-foot-deep monitoring well 4B has declined from more than 4.5 parts per million in 1982 to less than 0.25 parts per million by the end of 1984; however, in well 4A, readings for chlorobenzene show an increase from 11 parts per billion (parts per billion) to

22 parts per billion during the last months of 1984.[232] During 1985 additional monitoring wells were installed in the north-central area of the facility, and RES (LA) has agreed to install an abatement well at that location. Volatile organic pollutants were detected there in 1983.[233] U.S. EPA believed that monitoring wells at the site were situated at such a distance that the extent of the plume could not be determined empirically.[234] RES (LA) and its ground water consultants disagreed with U.S. EPA. Nevertheless, RES (LA) did install additional monitoring wells at the locations in question. All the wells, with one exception, have since been determined to be clean. RES (LA) has acquired the property of Allied Corp., so if contaminants were to migrate to this area, it would not be off-site migration. Erosion on surface impoundment dikes in 1983 has been repaired, and continuing action aims at maintaining adequate freeboard. RES (LA) added 2 feet to dikes around the scrubber effluent basin and basin 302.[235] RES (LA) strongly objected to U.S. EPA's October 1985 allegations but commissioned Geraghty and Miller to review the hydrogeology of the Baton Rouge site and the effectiveness of the ground water-monitoring and -management program. The review demonstrated to U.S. EPA's satisfaction that the increase in contaminants in the southwestern monitoring wells was evidence that the facility's ground water-pumping and -treatment program was functioning effectively by reversing the ground water gradient in areas of concern and recapturing contaminated ground water for treatment. U.S. EPA has subsequently determined that RES (LA) has fully responded to the agency's concerns and declared the proceeding closed as of October 23, 1985.[236]

Staff and Unionization. Approximately 105 full-time employees work at the site. Forty-five plant operations workers are members of Teamsters Union Local 270.

Safety and Health. From 1979 to 1982 there were three U.S. OSHA inspections pursuant to worker complaints but no recorded violations.[237]

Financial Instruments. U.S. EPA considers facility's closure cost estimate of $1,178,341 and postclosure cost estimate of $2,286,690 and its sudden ($1,000,000 per occurrence and $2,000,000 annual amount) and non-sudden occurrences ($3,000,000 per occurrence and $6,000,000 annual amount) insurance to be adequate.[238]

Technology

Processes and Capacities. The site covers approximately 260 acres. RES (LA)'s rotary kiln and liquid injection incinerator can process 13 metric tons per hour. The facility has a surface impoundment treatment capacity of 1,008,000 gallons per day, 576,000 gallons per day of tank treatment capacity, 180,000 gallons per day of encapsulation treatment capacity, 14,400 acre-feet of landfill capacity, and 9,981,060 gallons of surface impoundment, tank, and container storage capacity.[239] The facility performs clarification, coagulation, decanting, flocculation, stabilization, neutralization and oxidation treatment.[240]

Special Wastes Handled or Excluded. From 1983 to 1984 the facility received wastes from the following Superfund sites: Cleve Reber, Sorrento, Louisiana; Petro Processors, Scotlandville, Louisiana; W. J. Oil, Larose, Louisiana; Victor Products,

St. Louis, Missouri; Reichold Chemical, Columbia, Mississippi; and Bio Ecology, Grand Prairie, Texas.[241]

History

Prior Use and Owners. Prior to the purchase of most of the site by Rollins–Purle in 1969, the land was undeveloped pasture.

Construction and Permits. Facility began operations on January 17, 1971.[242] U.S. EPA called for and received Part B of RES (LA)'s application during 1983. The application is still undergoing review as of October 1985.

Future Processes and Site Uses. RES (LA) plans to use thermal oxidation at facility.[243]

Citizen Concerns and Organizations. Alsen residents, led by Mary McCastle, have initiated a class-action suit against the facility. The class of some 4000 area residents seeks damages allegedly sustained from the landfarming of nonhazardous wastes from an Exxon facility. The plaintiffs allege that the wastes caused illness and discomfort with burning, tearing eyes, headaches, sore throats, and upset stomachs.[244] The company is vigorously contesting these allegations and no damages have been awarded by the courts as yet. Mary McCastle claims to have suffered from emphysema since the plant began operations in 1971. The medication she needs for her condition "fills a shopping bag." Most people in the community have very low incomes and used to depend on their own gardens for vegetables and fruit, according to McCastle. After the waste facility moved in, McCastle claims, nothing would grow—and what would, most people were afraid to eat. She says local residents have to keep their windows shut, even during hot, humid Louisiana summers, for fear of fainting. McCastle claims that, as a result of off-site emissions of hazardous waste, she passed out once when she opened her door to go to her car.[245]

FACILITY DESCRIPTION: TECO TX

Facility	*CEP Score*		*Grade*
Texas Ecologists, Inc.	Ground Water Route	=	F
3½ miles south on Petronilla Rd.	Surface Water Route	=	A
	Air Route	=	A
Robstown, TX 78380	Fire and Explosion	=	A
	Direct Contact	=	C
U.S. EPA ID#:	Corrective Action	=	C
TXD069452340	Technology	=	F
	Ground Water Monitoring	=	*
	Financial Instruments	=	F
	Root-Mean-Square Grade	=	D+

* Facility is in compliance, so this category is not factored into final grade.

Site

Demography. TECO TX is in the prime agricultural area of Nueces County, Texas (population of 268,999 in 1980), approximately 4 miles south of Robstown (population 12,100) and 16 miles west of Corpus Christi (population 231,999).[246] All areas surrounding the facility are farmland; cotton is the principal crop. There are 13 houses within one-half mile of the facility.[247]

Water Use. The Gulf Coast Aquifer, at 150 to 500 feet deep, is the major source of drinking water in the area.[248] On average, water wells that tap it are 300 feet deep.[249] At least 31 domestic water wells are within a 3-mile area of TECO TX.[250] The depth at which these wells are screened is unknown. At the facility, a shallow aquifer 25 feet below ground surface is not used due to high chloride and sulfate content, but in areas beyond the site, this aquifer is used by some local residents. Drinking-quality water can be pumped from it only with low yield. Some local residents also use the aquifer for irrigation and livestock.[251]

Hydrology and Geology. Below ground surface is a surficial clay layer 8 to 22 feet thick, thinning in the northwest and southeast with interspersed sand lenses. The next layer is 10 to 39 feet of sand, which contains the uppermost aquifer. Below this sand layer is 6 to 15 feet of clay with an estimated permeability of 3.3×10^{-3} to 3.3×10^{-5} centimeters per second.[252] The Gulf Coast Aquifer is below this moderately permeable clay layer in a second sand layer that is 75 feet thick at the site. Chloride content in the Gulf Coast Aquifer ranges from 670 to 890 parts per million.[253] The highest level of the ground water in this zone, which ranges from fresh to saline, varies over the region from 200 feet below ground surface in the north to 60 feet in the south with a thickness varying from 0 to 450 feet.[254] The higher levels of salinity in shallow and deeper sands indicate little migration into the major drinking-water aquifer, which exhibits lateral movement primarily to the east.[255] If gravitational pressure from the upper aquifer exceeds the artesian pressure gradient of the lower aquifer, however, vertical migration between the two saturated zones is possible. This event would be most likely to occur during the occasional tropical disturbances in the summer and fall that cause torrential rains.[256]

Surface Water. There is no surface water within a 3-mile radius of the site, except for two shallow ponds located approximately 1500 feet to the east of the facility; the Nueces County Drainage Ditch, which run north to south through the site; and possibly some drainage ditches in the surrounding farmland.[257] Reportedly, the neighborhood farmers do not irrigate their fields, and thus do not use surface water.[258]

Management

Observed Releases. *1970 to 1980:* Throughout the decade, there were numerous odor problems resulting from activities at the facility.[259] *1977:* Contaminants were detected in the shallow aquifer beneath the site. Total organics were found at levels approximately 1000 times greater than background. Phenols were detected in the range of 300 to 600 parts per million.[260] Reportedly, TECO TX contaminated approximately 14 million gallons of shallow ground water with phenols, which are suspected carcinogens.[261] In addition, analytical results for other toxic compounds followed the pattern set by phenols.[262] Results of ground water-monitoring tests showed the highest concentration of contaminants below pre-RCRA landfills and ponds and reduced concentrations near the site periphery. *1979:* Wind dispersed herbicides into a maize field north of the site.[263] *1981 to 1983:* TECO TX reported 10 spills of hazardous waste to U.S. DOT. Although this is more spills than reported for any other facility owned by the Big Eight, only 7 gallons and 5 pounds of hazardous wastes were released, with no penalties assessed for the damages.[264] *1982 to 1984:* Six odor complaints and one dust complaint were filed with the Texas Air Control Board (TACB), but none resulted in violations because none could be confirmed.[265] Results of sampling three ground water-monitoring wells at the site's periphery reported in February 1984 showed a sudden and unexpected rise in TOC contamination.[266]

Ground Water Monitoring. As of February 14, 1984, U.S. EPA found the facility's ground water-monitoring well system, sampling, analysis, evaluation program, and records to be adequate.[267]

Enforcement Actions. *1981:* On January 26 the Texas Department of Water Resources (TDWR) negotiated a compliance agreement with TECO TX to remove the contamination and to arrest additional aquifer contamination. TDWR did not recommend direct U.S. EPA enforcement so long as TECO TX complied with the remedial plan. U.S. EPA action has been limited to site inspections, and discussion with TDWR on state actions.[268]

Corrective Actions. The suspected cause of ground water contamination is a damaged pond, the liner of which was ruptured during cleaning. The pond has since been exhumed. A second plume of contamination is suspected to have resulted from the operation of a mixing pit where hazardous waste liquids were poured to be solidified with soil. The mixing pit was closed in 1980.[269] TECO TX conducts ground water monitoring and reclamation in response to the terms of the 1981 compliance agreement with the state. Monitoring utilizes approximately 80 wells for monthly sampling, 14 pumping wells, a french drain below trench #19 to recover contaminated ground water, and an activated carbon waste-water treatment plant to adsorb contaminants.[270] Trench #19 is the only interim-status cell built over the contaminated area. According to Paul Lewis, a hydrologist with TDWR, "it would be impossible to tell if any of the cells over existing contamination are leaking," because the contamination is so severe that only a massive leak would be detectable, and "seepage out of an active landfill would not be that great."[271] TECO TX plans to build future landfill trenches over the area which is now contaminated. In areas not affected by the contamination, however, no leakage has been detected below closed and operating trenches. TECO TX purchased the 15-acre crop after a release of herbicides from the facility contaminated the produce.[272] Since 1983 odor problems associated with the facility have reportedly improved. TECO TX is accepting fewer materials with high volatility and is now handling odor-causing wastes with better techniques. Local residents still complain about smells, but there have been no substantiated odor releases since 1982.[273]

Staff and Unionization. The facility employs 27 full-time and 4 or 5 part-time workers, none of whom is unionized.[274]

Safety and Health. There have been no reported occupational incidents at the facility due to waste materials. A few smashed fingers, smashed toes, and one hernia have occurred due to construction activities, as well as the murder of a night guard by a thief in early 1985.[275]

Financial Instruments. As of February 14, 1984, the facility's closure cost is estimated at $1,311,294 and its postclosure cost at $1,787,582. Neither estimate had been evaluated by the state, nor had the site's sudden and non-sudden occurrences instruments been reviewed.[276]

Technology

Processes and Capacities. The facility has a 3150 acre-feet landfill capacity, a 27,500 gallon container storage capacity, and an evaporation pond for reclaimed ground water and runon.[277] The total site area of 240 acres can be used for landfill storage with the exception of a 200-foot-wide peripheral buffer zone, interior buffer zones around buried telephone cables, natural gas pipelines, and the Nueces County drainage ditch.[278]

Special Wastes Handled or Excluded. The site reportedly received state permission to accept PCB wastes from Huntsville Utilities of Huntsville, Alabama, in 1977 and allegedly received illegal shipments of PCBs from Diamond Shamrock Corp. of Deer Park, Texas.[279] U.S. EPA assessed a fine of $2.3 million against Diamond Shamrock for the alleged infraction but took no action against TECO TX.[280] TECO TX has not received wastes from Superfund sites.[281]

History

Prior Use and Owners. Prior to hazardous waste activities, the site was used for farming.[282]

Construction and Permits. *1973:* On April 10 the facility began hazardous waste management operations. The facility had a pilot incinerator, but it was removed in 1979 because it could not meet regulatory requirements.[283] *1983:* EPA called for and received Part B of TECO TX's RCRA application.

Future Processes and Site Uses. The carbon adsorption unit that was first used for ground water reclamation may be used for waste-water treatment in the future. The facility is expected to continue operating 15 to 20 more years, after which TECO anticipates no future use for the site.[284]

Citizen Concerns and Organizations. A formal citizens' organization of area residents and farmers called Citizens Against Chemical Dump Sites was formed in an attempt to close the facility.[285] Although this formal group is no longer active, area residents still remain opposed to the facility. There is a strong distrust of facility management and state regulatory efforts. They are concerned about too many unanswered questions regarding "what's going on under the ground." They are suspicious of erosion trenches TECO TX claims result from runoff from inactive portions of the site. Kenneth Ahlrich, who lives less than one-half mile west of the site, claims TDWR will not install a ground water-monitoring well on his property adjacent to the site "because if [the contamination] is outside the property, they'd have to shut them down. If we shut them down, we would have to clean up their mess. It could take 10 years to clean up the ground water. It's a continuous battle from here to eternity."[286] Mrs. Douglas Salyer is identified as the closest neighbor. Her house is less than 2000 feet from the landfill area. Salyer has a well that she thinks is deeper than 100 feet but does not know for sure. "The place [TECO TX] has made a nervous wreck of me," she says, "they dig and dig and dig. Now there's a mountain of the stuff out there. It's gonna seep through." Because of her concern, she now buys all of her drinking water.[287]

FACILITY DESCRIPTION: IT CA

Facility	*CEP Score*		*Grade*
IT Corp., Benicia Facility	Ground Water Route	=	B
Lake Herman Rd.	Surface Water Route	=	D
Benicia, CA 94510	Air Route	=	D
	Fire and Explosion	=	F
U.S. EPA ID#:	Direct Contact	=	D
CAD000060012	Corrective Action	=	C
	Technology	=	D
	Ground Water Monitoring	=	F
	Financial Instruments	=	*
	Root-Mean-Square Grade	=	D

* Facility is in compliance, so this category is not factored into final grade.

Site

Demography. The facility is in Southern Solano County, 15 miles northeast of the San Francisco–Oakland area. The town of Benicia is between 3 and 3.5 miles southwest.[288] This predominantly rural area is steadily becoming suburban.[289] Between 1970 and 1980 Benicia's population doubled to 15,376. In 1984 the closest dwelling was a mile from the facility and just under 4000 people live within 4 miles of the site.[290] The surrounding area is used for livestock grazing.[291]

Water Use. Lake Herman is an emergency source of water for Benicia and has never been used, except for recreational purposes.[292] Intakes from the Sacramento River are 30 miles upstream from the site and serve much of Solano and Contra Costa counties.[293]

Hydrology and Geology. The hilly site varies in elevation from 260 to 850 feet above sea level.[294] At least five major faults and numerous minor faults cross it; however, none are active (Holocene faults).[295] Six active faults lie within 33 miles. The closest is three-quarters of a mile away.[296] Sedimentary beds of shaley claystone with sandstone and limestone lenses lie under the site and dip northeast at 20 to 40 degrees.[297] Earlier studies show the permeabilities associated with these soils to be in the range of 10^{-8} centimeters per second; however, later studies indicate permeabilities as great as 6×10^{-6} to 3×10^{-4} associated with the sandstones. Fractured and jointed bedrock, moreover, may provide conduits with even faster rates of permeability.[298] Springs have been noted coming from the fault zones, indicating the potential for rapid migration; however, it is often the case that the same features will create a tighter barrier. Readings taken during the winter and spring of 1982 and 1983 from six wells showed ground water lenses from 0.5 to 25 feet below ground surface.[299] September 1984 readings also detected water-bearing layers from approximately 50 to 200 feet below ground surface.[300] The ground water is generally not useable due to low yield and high salinity. Brackish (saline) ground water lies 400 feet below ground surface.[301]

Surface Water. Natural drainage from the site is toward Suisun Bay, 1.5 miles to the east. Paddy Lake, approximately 1 mile west of the active treatment and disposal area, is a valley of riparian growth in an intermittent stream.[302] Less than 0.5 miles to the southeast of the active area, an intermittent unnamed creek flows toward Suisun Bay. IT discharges rainwater into the creek. The creek is also used to water stock downstream.[303] Suisun Bay is used for recreation, fish and bird habitat, navigation, and industrial cooling.[304]

Areas of Special Concern. At 1.5 miles east of the facility, a saltwater wetland borders Suisun Bay.[305]

Management

Observed Releases: *1978:* Leaks in ponds O, P, and Q were reported.[306] *1979 to 1980:* During the winter a slide occurred on the exterior dike of pond #12 at a time when the liquid level in the pond was near its maximum; fortunately, the dike did not collapse and a release was prevented.[307] *1981:* Evidence of seepage from pond #13, including discoloration of soils, was observed.[308] Leaks from ponds O, P, and Q were reported.[309] *1983:* On September 2 a fire occurred in the unloading area, indicating that incompatible wastes commingled prior to being landfilled.[310] *1984:* On February 16 IT notified the state that contaminants had migrated off site and initiated remedial action.[311] On March 7 a fire occurred in the landfill, indicating that ignitable wastes were landfilled.[312] During a March 9 and 12 U.S. EPA inspection, dikes at impoundments 12, 13, 14, 15, and 16 were found to have inadequate protective cover and were leaking. Significant increases in electrical conductance and chloride were detected by ground water-monitoring well C-3.[313] Material identified as nickel catalyst was observed being dispersed by wind, and the inspector saw containers of the same material leaking in the storage area.[314] IT maintains that it had neither stored nor transported the leaking drums, which were part of a bulk shipment from a site cleanup.[315] The drums were accepted as solid wastes (not as drums); they were hauled and stored as bulk solids and contained no liquids. IT maintains that no contamination or pollution has occurred off site. They do have nuisance concentrations of chloride (a nonhazardous constituent) off site that are under a continuing monitoring program. On April 3 and March 25, IT notified the state that contaminants had again migrated off site and that the company had initiated remedial action.[316] On November 19 and 27, water samples collected below impoundments 12 and 13 indicated that wastes or contaminated water continued to flow off site to surface waters. High levels of lead, copper, and total organic carbons were present at a seep area down gradient from IT CA's recovery well.[317] The lead, copper, and total organic carbon (TOC) detected above background levels were samples taken on IT property. There has been no off-site movement of these materials. *1985:* Samples from 11 of 27 (40 percent) ground water-monitoring wells indicated contamination. Contaminants detected included: vinyl chloride at 10 parts per billion (a concentration that is five times greater than that allowed in drinking water), cadmium at 100 to 270 parts per billion, trichloroethylene (TCE) at 6 to 260 parts per billion, methyl chloride at 100 parts per billion, and lead. The organics (vinyl chloride, TCE, and methyl chloride) have all been verified as contaminants in the sample process. They are not present in the water samples (that is, they are a result of well installation and sampling practice, not migration).[318]

Ground Water Monitoring. As of March 12, 1984, the facility's ground water monitoring-well system, sampling, analysis, evaluation, and ground water records are inadequate.[319]

Enforcement Actions. *1971 to 1973:* California issued two cleanup and abatement orders and one cease and desist order for pond overflows and illegal surface discharges.[320] *1984:* Notice Letter. On May 23 U.S. EPA notified the Department of Health Services of the State of California (CA DOHS) of alleged violations at the Benicia facility. The violations included failure to implement an adequate ground water-monitoring program, to maintain adequate containment systems at impoundments, to prevent wind dispersal of waste, to manage wastes to prevent fire, to maintain containers in good condition, to maintain required freeboard, and to develop a complete waste-analysis plan.[321] Determination of Violation, Compliance Order, and Notice of Right to Request a Hearing. On September 27 U.S. EPA reiterated the findings of the notice and, in addition, cited IT CA for land application of inappropriate wastes. U.S. EPA proposed a civil penalty of $81,500 for the violations and ordered IT CA to comply with 14 provisions to either discontinue, repair, or certify implementation of corrective measures that addressed all of the cited violations.[322] Warning Letter. On October 4, U.S. EPA provided IT CA with a 42-page outline of deficiencies in Part B of its RCRA application.[323] Clean-up and Abatement Order. On January 16 the San Francisco Bay Regional Water Quality Control Board outlined 22 tasks to correct violations at impoundments 12, 13, 14, 15, 16, O, P, and violations of ground water-monitoring requirements.[324] *1985:* Consent Decree. On February 21 IT CA agreed to comply with the abatement order without admitting guilt and to pay a $30,000 civil penalty for alleged violations at Benicia, Vine Hill, and Baker.[325]

Corrective Actions. The O, P, Q releases have been retained on site. IT CA drained impoundments 12 through 16, constructed a recovery trench drain and installed a recovery well. New ground water-monitoring wells have been bored. IT CA is undergoing assessment to determine the extent of ground water contamination. IT CA has consented to comply with the state cleanup and abatement order.

Staff and Unionization. Thirty-two full-time workers are employed at the disposal site. Twenty-three are truck and heavy equipment operators and are members of the teamsters' union. IT CA employs 56 transporters to service the site.

Safety and Health. In 1984 truck drivers and site workers without any protective clothing were seen walking adjacent to the landfill where there is no barrier.[326]

Financial Instruments. As of March 12, 1984, U.S. EPA found the facility's closure and postclosure cost estimates to be inadequate. The status of sudden and non-sudden occurrences instruments is unknown.[327]

Technology

Processes and Capacities. The site has 300 acre-feet of landfill capacity, 80,000,000 gallons of surface impoundment storage capacity, and 3,000,000 gallons per day of surface impoundment treatment capacity.[328]

Special Wastes Handled or Excluded. The facility has received wastes from the following three Superfund sites during the period of 1983 through 1984: Celtor Chemical Works, Hoopa, California; McClellan Air Force Base, Sacramento, California; and Gene Murren, Dire, Colorado.[329]

History

Prior Use and Owners. Howard Jenkins began hazardous waste management activities at the site on December 28, 1968.[330] Before IT purchased the site in 1974, and ownership changed hands twice.[331]

Construction and Permits. On March 30, 1981 California issued the facility an interim-status document. During 1983 U.S. EPA called for Part B of the RCRA application and received it. In 1984 U.S. EPA reviewed Part B for completeness.

Future Processes and Site Uses. IT has applied for permission to operate a tank storage and treatment facility with a capacity of 33,100 gallons per day.[332] The site has no future use other than hazardous waste management.

Citizen Concerns and Organizations. CEP is unaware of any local citizen groups or households concerned about IT CA's facility. Six national public-interest organizations—Environmental Defense Fund, Sierra Club, Citizens for a Better Environment, Campaign for Economic Democracy, Concerned Neighbors in Action, and Coalition on Environmental and Occupational Health Hazards—sponsored a study, *Nowhere to Go*. The study investigated all Class I landfills (those that can receive extremely hazardous wastes) in California, of which IT's Benicia facility is one. The study, whose research is quoted in this section, alleges that every one of these landfills is leaking. IT characterizes the report as "... a political document created for political purposes and ... intended to support predetermined conclusions."[333]

FACILITY DESCRIPTION: ESII ID

Facility	*CEP Score*		*Grade*
Envirosafe Services, of Idaho, Inc., Site B	Ground Water Route	=	D
Missile Base Rd.	Surface Water Route	=	F
Grandview, ID 83624	Air Route	=	C
	Fire and Explosion	=	D
U.S. EPA ID#:	Direct Contact	=	D
IDD073114654	Corrective Action	=	C
	Technology	=	D
	Ground Water Monitoring	=	*
	Financial Instruments	=	F
	Root-Mean-Square Grade	=	F

* Facility is in compliance, so this category is not factored into final grade.

Site

Demography. The facility is on a desert plateau. All land within 100 feet of the site is owned by the Bureau of Land Management and is either used for grazing or is left open.[334] Irrigated pastureland lies approximately one-quarter mile to the north, a farm is approximately 2 miles to the east, and a cattle feedlot is 8 miles to the east.[335] No residences or cropland are within 1000 feet of the facility. Approximately 30 homes are within 4 miles of the site; 8272 people live in Owyhee County in 1980.[336]

Water Use. Area residents have no alternative to ground water for domestic use. Other than a 3000-foot well used for ground water monitoring and nonpotable uses on site, there are no producing wells within 1000 feet of the facility.[337] The closest off-site well is 7000 feet southwest. Houses to the east of the site have wells sunk into the 25-foot-deep alluvium of the Snake River. Residences to the north have wells that are screened 800 to 1000 feet deep. Wells to the southwest of the facility are reportedly shallower.[338] The majority of the irrigation supply wells within a 3-mile radius draw from the 3000-foot-deep aquifer.[339] Castle Creek and the Snake River are also used for stock watering and irrigation.

Hydrology and Geology. The soil down to 10 feet deep comprises air fall, water-laid ash, cinders, tuffaceous silt, silty and gravelly sands, and some clay at the north end of the site. From approximately 10 to 35 feet deep, well to poorly graded, dense, and silty to gravelly sand has an estimated permeability from 10^{-2} to 5×10^{-4} centimeters per second or greater. From approximately 35 to 125 feet deep are light brown, dense, fine, or silty sands stratified with a few layers of clay and silt. From approximately 120 feet deep to 140 feet, there is an estimated permeability of 3×10^{-6} centimeters per second. The highest seasonal level of the saturated zone lies approximately 145 to 160 feet below ground surface and approximately 32 feet below the lowest point of waste disposal. This cold-water upper aquifer flows approximately 5 degrees north of east at a rate of 10^{-2} centimeters per second and yields approximately 1 to 6 gallons per minute. Below these sand and silt zones is a clay layer with an average permeabilty of 2 to 3×10^{-6} centimeters per second that extends to approximately 430 feet deep. Below the clay are alternating layers of shale and basalt to a depth of 2500 to 3000 feet wherein lies a layer of sandstone. The sandstone contains the hot-water Castle Creek Aquifer, which may produce up to 1500 gallons per minute. Within the region is a warm-water aquifer at 1000 feet that results from the mixing of the cold- and hot-water aquifers; however, this warm-water aquifer is not below the facility.[340]

Surface Water. Castle Creek, a small tributary of the Snake River, is located approximately 0.5 miles west of the facility. The Snake River is approximately 2.5 miles north and east of the site. The site slopes generally toward the northeast although northwest drainage into Castle Creek is possible. The Grandview Irrigation District Canal, which obtains its water from the C. J. Strike Reservoir 12 miles to the southeast of the site, parallels the Snake River about 2 miles east of the facility.[341]

Areas of Special Concern. A sanctuary for birds of prey is approximately 3 miles from the site.[342]

Management

Observed Releases: *1979:* Contamination of soil with PCBs in the range of 290 to 2150 parts per million was documented.[343] *1983:* On August 20 to 21, a storm and flash flood resulted in erosion and other evidence of runoff from the site.[344] A U.S. EPA inspection on October 27 and November 1 found leaking drums containing wastes contaminated by cadmium, nonhalogenated solvents, phosphoric acid, and ignitable wastes stored directly on the ground in areas of poor drainage. Runoff from active portions of the facility was not collected. Wastes formed ponds along the ailing grade and were allowed to evaporate or percolate into the soil.[345] A December 19 inspection found corrosive wastes, lead-contaminated wastes, and spent nonhalogenated solvent wastes leaking from 17 drums.[346] *1984:* On March 26 Charles Scott, Owyhee County safety officer, reported: "As I approached the mixing area . . . strong odors quite similar to those detected earlier at the Ellsworth's residence were coming from the pond, strong enough, in fact, that they were able to pass through the cartridges of my respirator. . . . Though I did not notice any odors upon arrival, as I was preparing to leave I did detect an odor of a phenolic nature varying in intensity."[347] On June 5 a vapor cloud of unknown material formed due to activity in trench 11.[348] On June 6 and 14, drums in storage, the storage pallets, and the soil beneath the drums were contaminated with the hazardous waste contents of the drums due to the opening of drums for sampling. The sampling released 3200 parts per million of cyanide solution.[349] On October 11 leaks and spills occurred in storage areas with no containment or other means to minimize releases.[350] *1985:* According to U.S. EPA Region 10, "a variety of toxic organic solvents have been released into soil and ground water. These substances are suspected to have leaked from old Titan missile silos as ESII. . . . "

Ground Water Monitoring. As of 1984 ESII ID claimed a waiver which EPA had not reviewed.[351]

Enforcement Actions. *1983:* Notice of Violation and Warning. U.S. EPA alleged ESII violated requirements for the waste analysis, general inspections, personnel training, contingency plan, ground water monitoring, and closure and postclosure. U.S EPA also alleged that ESII ID illegally disposed of drums containing liquids and failed to provide adequate diversion for runon at the active trenches.[352] Compliance Order. U.S. EPA ordered ESII ID to meet 16 conditions: stepped-up inspections of storage areas, excavation of barrels in one trench, construction of a truck decontamination facility, establishment of a ground water-monitoring program, sealing of an artesian well near the silos, building of berms and a holding pond to contain future flash floods, and cleanup of the 8000 to 10,000 leaking drums on site. In addition, U.S. EPA assessed a civil penalty of $150,000.[353] *1984:* On March 6 U.S. EPA issued ESII ID a compliance order for violations of TSCA including outside storage of PCBs for over 30 days in leaking and improperly labeled drums. ESII ID accepted a $43,500 civil penalty to settle the U.S. EPA administrative complaint of PCB storage violations and agreed to upgrade the PCB operations manual.[354] Complaint, Compliance Order and Notice of Opportunity for Hearing. In August U.S. EPA alleged that ESII ID violated the terms of the 1983 compliance order by continued improper drum management, inadequate precautions to prevent disposal of liquids, deficient inspection, training, waste analysis, closure and postclosure plans, and deficient procedures to minimize the possibility of unplanned sudden or non-sudden releases. U.S. EPA increased the assessed penalty to $156,100.[355] Consent Agreement and Final Order. After negotiations with ESII ID, U.S. EPA presented an amended enforcement action in October. Several earlier findings were deleted or altered in response to ESII ID's defense; U.S. EPA remained concerned, however. On June 20 and 22, storm water was found diverted into active areas of the site including drum storage areas, mixing areas, and even into an active trench. U.S. EPA reduced the assessed penalty to $67,500 and ordered ESII ID to take 19 corrective measures to address drum management, trench disposal, and other problems.[356] *1985:* Corrective Action Order. On October 31, 1985, U.S. EPA ordered ESII ID to perform monthly monitoring around silos, and if contamination equals over 1 part per million, to correct the problem immediately. The second part of the order requires ESII ID to develop a remedial design, install engineered measures and equipment, and prepare a plan for long-term monitoring and maintenance. Consent Order. On August 15, 1985, U.S. EPA cited ESII ID for manifest violations, an inadequate closure plan and closure cost estimate, and an incomplete contingency plan. U.S. EPA assessed a penalty of $49,000.

Corrective Actions. After the original 1983 compliance order, ESII ID voluntarily stopped receiving all wastes, even though U.S. EPA withdrew this portion of the order after 5 days. Plans for ground water monitoring, storm-water control, truck washing, removal of some buried drums, and sealing an uncapped well in the missile silo complex also resulted from this order.[357] The inventory of drums in storage was reduced from approximately 17,000 in late 1983 to less than 100 one year later. U.S. EPA verified ESII ID's defense of its use of waste analysis and contingency plans, its maintenance of emergency equipment, and its manifest discrepancy reporting.[358] During the summer of 1984, ESII improved the site's runoff collection system at a cost of nearly $1 million.[359] ESII ID is on a compliance schedule to correct outstanding deficiencies identified in the final order of 1984.

Staff and Unionization. Approximately 40 full-time employees work at the facility. Some workers are members of the Service Employees International Union of Idaho.[360]

Safety and Health. Two employees were taken to the hospital by ambulance on April 5, 1984 after complaining of health problems associated with pesticide exposure. Both had been nauseous after working near the volatile toxins area. Similar problems had occurred in the same area on January 18, when employees were sent to the hospital after experiencing almost identical symptoms.[361] On November 1 as much as 200 gallons of PCBs spilled in a workshop, contaminating a 39-year-old employee. Reportedly, the man was not told that the truck still contained PCB-contaminated wastes when he opened a valve to repair it. The employee was not wearing required splash gear. PCB concentrations of 7642 parts per billion were detected in his fat tissues.[362]

Financial Instruments. In 1982 the amount needed to close ESII ID was estimated to be $215,591. ESII has a trust fund worth $2,021,000 to cover closure and postclosure requirements at a number of its facilities. ESII ID had adequate sudden ($1,000,000 per occurrence and $2,000,000 annual amount) and non-sudden occurrences insurance ($3,000,000 per occurrence and $6,000,000 annual amount) policies for 1984.[363] On August 15, 1985 U.S. EPA cited ESII ID for inadequate closure plan and closure cost estimate.

Technology

Processes and Capacities. 1700 acre-feet of landfill capacity. One TSCA and one RCRA trench are in operation. Three 15,000-gallon storage tanks and 187,220 gallons of container storage capacity are on site. Decanting and stabilization are performed on site.[364]

Special Wastes Handled or Excluded. ESII ID accepts PCBs and during 1983 accepted wastes from the following Superfund sites: Aarcom Corporation in Rathdrum, Idaho; Pacific Recycling in Pocatello, Idaho; American Surplus in Tacoma, Washington; and Western Processing in Kent, Washington.[365]

History

Prior Use and Owners. From 1954 to 1972 the site was used as a U.S. Air Force Titan Missile Complex. From 1973 to 1981, Wes-Con, Inc. operated a hazardous waste facility at the site, filling the three abandoned missile silos and 18 landfill trenches with hazardous wastes.[366]

Construction and Permits. ESII ID has operated the facility since 1981. U.S. EPA called for and received Part B of ESII ID's RCRA application in 1983. As of 1985 it had been reviewed, amended, and resubmitted twice.

Future Processes and Site Uses. ESII expects to continue operating the facility for 25 years.[367] Installation of a sludge-handling system, drum storage expansion, drum-thawing building, evaporation pond (excluding runoff units), physical-chemical reaction systems, remote tank storage, waste pile storage building, and a chemical fixation unit is planned between 1985 and 1987.[368]

Citizen Concerns and Organizations. No active formal citizen organization focuses on ESII ID's operation. Neighboring households are concerned that runoff in the event of a flash flood will not be properly contained. They also voice concern about their children playing in Castle Creek downstream of the site as well as worker safety and fair employment practices. Some residents complain of smells. They feel that ESII ID does not notify authorities of problems and incidents promptly enough for an accurate outside assessment, and that ESII ID does not follow through on agreements.[369] ESII ID has brought Owyhee County to court, questioning its right to tax the facility.

FACILITY DESCRIPTION: CECOS OH

Facility	*CEP Score*		*Grade*
CECOS International, Inc.	Ground Water Route	=	F
5092 Aber Rd.	Surface Water Route	=	F
Williamsburg, OH 45176	Air Route	=	D
	Fire and Explosion	=	A
U.S. EPA ID#:	Direct Contact	=	C
OHD087433744	Corrective Action	=	C
	Technology	=	F
	Ground Water Monitoring	=	F
	Financial Instruments	=	*
	Root-Mean-Square Grade	=	F

* Facility is in compliance, so this category is not factored into final grade.

Site

Demography. The facility is in an agricultural area. The eastern end of the site is bordered by a large farm.[370] On the western boundary there are a few scattered homes. The closest are less than 1000 feet from the landfill operations.[371] More than 1000 people live within 1 mile of the facility and 2460 live in Jackson Township.[372] In 1980 the Village of Williamsburg approximately 5 miles south, had a population of 2054.

Water Use. Much of the Williamsburg area is served by the Harlan–East Fork Water System, which takes water from the East Fork of the Little Miami River 9 miles downstream.[373] Within 3 miles of the facility, however, 1000 to 3000 persons rely on ground water for domestic or agricultural use. During 1984 three new wells were being constructed or were completed within 3 miles of the facility.[374]

Hydrology and Geology. Underlying a thin veneer of topsoil and extending to a depth of 6 to 8 feet is a silty gray clay with a permeability of 10^{-5} to 10^{-7} centimeters per second. Below this a 30- to 60-foot layer of sandy, silty clay and glacial till with gravel and rock fragments has a permeability of 10^{-8} centimeters per second. Below this layer is several hundred feet of relatively impervious shale and thin hard limestone.[375] Sandy and gravelly seams are 1 to 9 feet thick throughout the site from 20 to 30 feet below ground. These seams have a permeability as great as 10^{-2}.[376] The highest level of the saturated zone in Clermont County can be less than 1 foot from ground surface during late winter and spring, according to the Soil Conservation Service.[377] CECOS OH's pumping of 28,000 to 30,000 gallons per day has lowered the water table at the facility to 5 feet below the lowest level of waste disposal at an active cell.[378] Ground water is confined to the upper portions of the bedrock and the discontinuous sand and gravel seams. Wells generally yield 2 to 4 gallons per minute. The primary source of ground water is rainfall percolation.[379]

Surface Water. Pleasant Run and its intermittent East Branch tributary are the natural ground water discharge areas of the site. Pleasant Run flows south along the western portion of the property and its East Branch flows south along the southern border.[380] Landfill cells are allowed within 100 feet of these streams. Pleasant Run and its tributary join the East Fork of the Little Miami River 1.3 miles south of the site.[381]

Areas of Special Concern. CEP did not identify any areas of special concern.

Management

Observed Releases. *1982:* On August 24 placement of incompatible materials in a heavy metal subcell caused a chemical reaction resulting in white smoke and spurts of flame from 2 to 3 drums. A truckload of paint sludge containing 45 percent pigments and resins, 20 to 30 percent water, and miscellaneous paint debris reacted with 42 drums of hair bleach containing 60 percent potassium and sodium persulfate, and 40 percent titanium dioxide, sodium metasilicate, magnesium carbonate, actosil 300, pharasorb, and diammonium phosphate.[382] On December 16 state inspectors found the contents of drums and contaminated rainwater leaking from an uncovered waste load in a truck.[383] *1983:* Landfill-related odor complaints were received approximately once a month in groups of two to five. The odors seldom travelled more than 1 mile but sometimes were detected as far as Marathon, which is 3 miles away.[384] *1984:* Water from fire pond 4/5 was spray-evaporated and irrigated. The spray, which contained twice the allowable levels of chromium, as well as arsenic, lead, and pesticides, was seen drifting onto adjacent properties and houses.[385] Odor complaints continued during 1984. In one case, a neighbor complained of having vertigo and headaches as a result of emissions from the site.[386] There have reportedly been continuous problems with ill-managed trucks that leak onto roads enroute to the site, and Aber Road was found to be contaminated with PCBs.[387] On November 4 and 7, runoff contaminated with phenols at concentrations greater than permissible under drinking water standards was discharged into the east branch of Pleasant Run Creek. An Ohio Environmental Protection Agency (OEPA) inspector found a pipe leading to the tributary from a landfill cell that was under construction but had not received wastes at the time.[388] *1985:* U.S. EPA data and a state-commissioned draft report indicate potential ground water contamination resulting from the site. Certain wells detected elevated levels of conductivity, and total organic halogenated materials.[389]

Ground Water Monitoring. As of May 1985, the facility was under assessment. The well system was inadequate; sampling and analysis and records have not been fully evaluated.[390]

Enforcement Actions. *1983:* State Warning Letter. In April OEPA served a formal warning to CECOS OH which ordered the facility to comply with ground water-monitoring requirements.[391] *1984:* In February OEPA ordered CECOS OH to halt construction of hazardous waste cell #8, because ground water pressure collapsed a portion of the cell wall.[392] On May 31 OEPA required CECOS OH to put up a $300,000 surety bond to guarantee against similar failures.[393] U.S. EPA suspended CECOS OH's TSCA permit for PCB disposal from February 22 to April 13 as a result of the slope failure.[394] Complaint and Finding of Violations and Order. On September 24 U.S. EPA alleged that Part B of CECOS OH's RCRA application was incomplete and late. CECOS OH has failed to correct deficiencies in its permit application. U.S. EPA assessed a civil penalty of $11,000.[395] On November 9 OEPA shut down the facility after receiving reports that its operators pumped phenol-contaminated water from a landfill cell into a tributary that joins the East Fork of the Little Miami River upstream of public drinking water intakes.[396] Director's Final Findings and Orders. On November 26 OEPA ordered CECOS OH to comply with 25 provisions including specification and monitoring of truck routes to the facility, construction of a truck wash, limitation of operating hours, submission of revised surface water management plan, retention of an independent environmental auditor, sampling of off-site water wells, submission of particulate emissions-control and personnel-decontamination plans, and holding monthly meetings with a citizen committee.[397] On November 27 OEPA allowed the facility to reopen.[398] *1985:* In May U.S. EPA ordered CECOS to submit a ground water-quality assessment plan after indications of ground water contamination were detected and determined that the newly constructed cell #9 did not meet minimum technology standards. OEPA closed the facility again pending development of an adequate ground water-monitoring program.[399] In June a county grand jury indicted BFI, CECOS, and two former site employees on 96 counts. The criminal indictment alleges that site employees deliberately pumped contaminated rainwater into a nearby creek. The charges carry a maximum fine of $25,000 and a maximum prison sentence of 4 years on each of 20 felony counts.[400]

Corrective Actions. CECOS ceased spray irrigation in October 1984. Ground water is pumped to prevent slips of cell walls. The facility assessed its ground water-monitoring program in response to the OEPA order of May 9, 1985.

Staff and Unionization. There are 45 full-time workers on site; none is a member of any union.[401]

Safety and Health. *1979:* U.S. OSHA fined the site $180 for two serious violations; three other serious violations and five less serious violations were identified.[402] *1982:* "There have been several incidents which the agency [OEPA] feels were potential hazards to the employees at the landfill, such as a drum of reacting material in a secure cell and a spill of toxic wastes which caused a bad odor problem within a cell."[403] OEPA cited inadequate personnel training at the facility.[404] *1983:* "[David C.] Proud, 28, claimed he saw drums containing liquid wastes backed off the back of trucks into the disposal cell and covered with kitty litter to absorb the liquid. Bulldozers sometimes split drums and once splashed the contents on him."[405]

Financial Instruments. As of May 4, 1984, the closure cost of the facility was estimated to be $564,900 and post-closure maintenance was estimated to cost $1,549,860. CECOS passed a financial test to cover these costs. As of July 5, 1983, CECOS had a $1,000,000 insurance policy to cover sudden occurrences.[406]

Technology

Processes and Capacities. The site has a landfill capacity of 2,200,000 acre-feet. It has 100,000,000 gallons of container storage capacity. It has 500,000,000 gallons of surface impoundment capacity, which is used for the storage of runoff and water that would be used in case of fire.[407]

Special Wastes Handled or Excluded. The facility has received PCBs. From 1981 through 1984, the facility received wastes from the following 18 Superfund sites: A & F Materials, A. L. Taylor Site (Valley of the Drums), Berlin & Farro, Chem Dyne Inc., Danville Plating, Dennett's Quarry, Gebhart, Harrison County Site, Hixon Plating, Liquid Disposal, Northernaire Plating, Old Mill Site, Peerless Plating, Rason Tanning, Riomple Site, Summit National, Syncon Resins, and the Wade Park Chemical Facility.[408]

History

Prior Use and Site Owners. Since 1972 Clermont Environmental Reclamation Company and LEW, Inc. operated the site as a solid waste disposal facility.

Construction and Permits. In November 1976 OEPA granted facility permission to dispose of hazardous wastes. In 1983 U.S. EPA called for, received, and reviewed Part B of the RCRA application.

Future Process and Site Uses. CECOS estimates another 20 years of landfill capacity at the facility. After closure the site has no future use.[409]

Citizen Concerns and Organizations. No active formal citizen organization focuses on CECOS OH. Local residents, however, have a number of concerns regarding the facility. They dislike the trucks driving—and sometimes seen speeding—at all hours on roads they consider poor. They disapprove of trucks driving past the local school and on the same road that school buses use. Residents complain that "the trucks are damn near running wild" and have been seen on occasion to spill. They are angered by the limited control over the placement and operation of hazardous waste facilities afforded to local government by state law. They feel the politicians listen more to manufacturers who need to use commercial landfill capacity than to ordinary constituents who live near the landfills. They are annoyed by "atrocious" odors. They fear there are inadequate local supplies of foam and other fire prevention materials. They are frustrated by falling property values. They resent CECOS for suing local residents for slander for comments made at a public hearing.[410]

CHAPTER 10
Information Management: Ranking the Regulators

AN INFORMATION SYSTEM'S THREE-FOLD FUNCTION

Regulatory agencies nationwide are having difficulty performing the three essential functions of a competent public information system:

- gathering data accurately and comprehensively
- processing the information to reveal its significance
- communicating the findings effectively to industry, the public, and other government users.

The problem is not only limited funds but also a fragmenting of the funds and other resources that are available.

Gathering Information

U.S. DOT, for instance, relies on motor carriers themselves to report both accidents and unintentional releases of hazardous wastes. One U.S. DOT official claimed "75 percent get reported."[1] Without random inspections, however, an agency cannot possibly know what percent of releases is reported, and it certainly cannot uncover criminal activities. In the past many transporters with motive and opportunity were found to have perpetrated "midnight dumping" and illegal spiking of fuel oils used in residential and commercial boilers.

U.S. EPA's record for inspecting stationary handlers of hazardous waste is nearly as bad: As of November 1984 90 percent of New York City's hazardous waste facilities registered in the U.S. EPA database had never been inspected. Only 216 inspections took place between 1981 and the beginning of 1984.[2] One part-time employee averaging one and a half inspections per week could have accomplished this many inspections in those years. At this rate it will take almost 20 more years to make only one inspection of all the New York City facilities that are listed as generating and managing hazardous wastes.

Processing Information

Until the reauthorization of RCRA in 1984, generators of up to 1 ton of hazardous waste per month were not required to notify U.S. EPA of their activities. Prior to 1984 New York was one of only ten states that had required these small generators to describe their activities. Over twice as many small generators as large ones reported in New York. U.S. EPA estimates that 75 percent of the notifications of hazardous waste activities in its database are nonregulated or "protective" notifiers that were not required to inform U.S. EPA of their activities.[3]

With the lowered limit requiring small generators to meet certain reporting requirements, U.S. EPA's database will have to accommodate the new entries. U.S. EPA will have difficulty determining the universe of small generators unless it spends the time and money to modify the existing system to differentiate generators of small volumes of waste from larger ones. Ironically, small generators that voluntarily notified U.S. EPA of their activities may have to do so again because the original information system cleared by the Office of Management and Budget (OMB) did not request information concerning the quantity of waste generated at a site. The impact of OMB's attempt to reduce reporting requirements will cost industry money in this case.

Communicating Information

In general, CEP found U.S. EPA to be relatively competent in terms of ability to answer requests for information. Although U.S. EPA headquarters in Washington, D.C., has little detailed information, the ten U.S. EPA regional offices (see figure 10-1) responded partially or completely to approximately three-quarters of CEP's FOIA requests. Only two regions (Regions 2 and 7), however, were able to provide the counties in which facilities are located. The regional offices decided to waive all fees for their responses because the information would be used in a manner that "can be considered as primarily benefiting the general public" as stipulated by the amended Freedom of Information Act (5 USC § 552).

GRADES FOR U.S. EPA AND THE STATES

CEP rated U.S. EPA's regional offices on three criteria: inspection performance (gathering data), enforcement performance (processing data), and the quality of their responses to a series of FOIA requests (communication). The regions ranked as follows:

U.S. EPA Region	*Location*	*CEP Grade*
3	Mid-Atlantic States (DC, MD, PA, VA, WV)	B+
2	Northeastern Corridor & Caribbean Islands (NY, NJ, PR, VI)	B−
9	Southwestern States & Pacific Islands (AZ, CA, GU, HH, NV)	C+
5[a]	Great Lakes States (IL, IN, MI, MN, OH, WI)	C+
6[a]	South Central States (AR, LA, NM, OK, TX)	C+
8[a]	Rocky Mountain States (CO, MT, ND, SD, UT, WY)	C+
4	Southeastern States (AL, FL, GA, KY, MS, NC, SC, TN)	C−
7	Midwestern States (IA, KS, MO, NE)	D+
10	Northwestern States (AK, ID, OR, WA)	D−
1	New England States (CT, ME, MA, NH, RI, VT)	D−

[a] U.S. EPA Regions 5, 6, and 8 received identical grades (Region 9 is slightly better). Regions are shown in figure 10-1.

None of the regions received a final A because their performance varied in the three grading categories. For example, even though CEP gave Region 3 an A+ for inspection performance (see table 10-1), and an A for enforcement (see table 10-2), the region received only a C for its communication (see table 10-3), which brought its average down to a B+.

Other federal agencies, such as the U.S. departments of transportation and labor, were less willing to waive fees associated with information requests and were less able to provide complete information. And state agencies, to which the federal FOIA does not apply, were, in comparison, poorly equipped to respond to information requests.

CEP rated 27 states on their ability to communicate information to the public. Illinois, Michigan, Nevada, and Texas did best; California, Idaho, New York, and Ohio showed the most room for improvement. For inspection and enforcement performance during fiscal year 1984, CEP ranks Maryland's hazardous waste regulatory program the best. On average, Maryland inspected each major facility ten times and took seven enforcement actions against every serious violator. Among the largest waste-generating states, New Jersey and California also scored excellently. Hawaii and Wyoming, on the other hand, were ranked the worst, because they did not inspect any major facilities during

US EPA, Region 1
Solid Waste Program
John F. Kennedy Bldg.
Boston, MA 02203
617-223-5775

US EPA, Region 2
Solid Waste Section
26 Federal Plaza
New York, NY 10007
212-264-0503

US EPA, Region 3
Solid Waste Program
6th and Walnut Sts.
Philadelphia, PA 19106
215-597-9377

US EPA, Region 4
Solid Waste Program
345 Courtland St., NE
Atlanta, GA 30308
404-881-3015

US EPA, Region 5
Solid Waste Program
230 South Dearborn St.
Chicago, IL 60604
312-353-2197

US EPA, Region 6
Solid Waste Section
1201 Elm St.
Dallas, TX 75270
214-767-2734

US EPA, Region 7
Solid Waste Section
1735 Baltimore Ave.
Kansas City, MO 64108
816-374-3307

US EPA, Region 8
Solid Waste Section
1860 Lincoln St.
Denver, CO 80295
303-837-2221

US EPA, Region 9
Solid Waste Program
215 Fremont St.
San Francisco, CA 94105
415-556-4606

US EPA, Region 10
Solid Waste Program
1200 6th Ave.
Seattle, WA 98101
206-442-1260

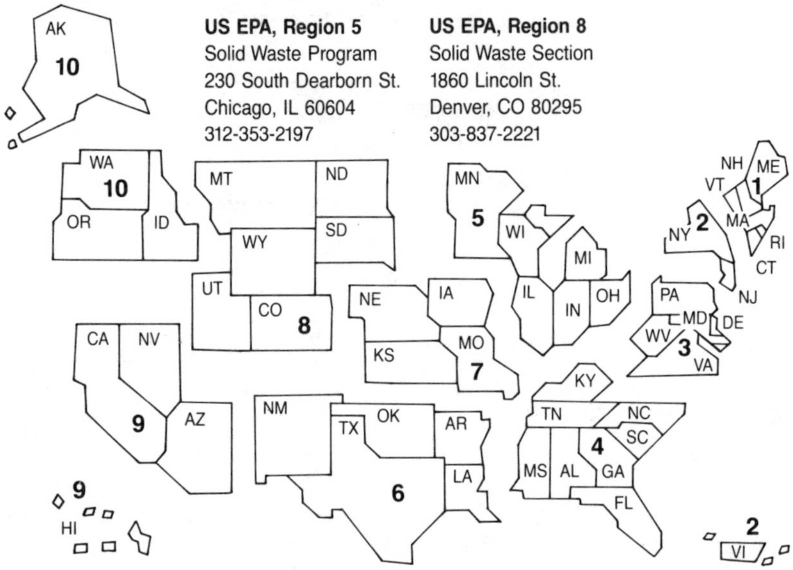

FIGURE 10-1. *The U.S. EPA Regions Identified*

the fiscal year 1984. Louisiana and New York also have among the poorest records (see table 10-4).

GATHERING INFORMATION ABOUT GENERAL AREAS OF CONCERN

Transportation

U.S. EPA has reported that the odds of hazardous waste being released during shipment by truck are as great as during treatment, storage, and disposal.[4] But it is virtually impossible to find out what or how many traffic accidents involve trucks carrying hazardous wastes. And there is almost no inspection of these trucks. U.S. DOT inspects only vehicles that carry hazardous products, like gasoline. It will not inspect truckloads of hazardous waste except at weigh stations. Its inspection reports often do not distinguish such vehicles anyway. According to Bill Wilson of U.S. DOT, "We don't have the manpower to go out and investigate incidents involving hazardous wastes."[5]

U.S. DOT has two databases from which one can obtain limited information concerning transportation incidents involving hazardous wastes. The data are derived from reports that transporters submit voluntarily to the department after an accident or spill. One of these databases is the Hazardous Materials Information System (HMIS), operated by Wilson Hill Associates, a contractor to the Materials Transportation Bureau of U.S. DOT's Research and Special Programs Administration. Using this system, it is possible to segregate transportation incidents involving hazardous waste from those involving other hazardous materials. From 1977 through 1983, however, only 242 incidents involving hazardous wastes were recorded in the database.

The U.S. Coast Guard operates another U.S. DOT database that contains far more extensive information on transportation incidents than the HMIS. Unfortunately, this system, which is administered by the National Spill Response Center in Washington, D.C., cannot easily segregate incidents involving wastes from those involving other hazardous materials. Because the Coast Guard's database can be sorted by the name of the transporter, however, one can at least identify the portion of transportation incidents that involve trucks driven by the eight major hazardous waste management companies.

Discrepancies between the two transportation information systems reveal the inadequacy of the system. For the years 1982 and 1983, the Coast Guard data show 64 incidents involving vehicles of the eight major companies, whereas for the entire period from 1977 through 1983, the HMIS reports only 35 incidents involving vehicles of these firms. Furthermore, during the overlapping years, not a single incident is recorded in both databases.

Local inspection is no better. The New Jersey Public Interest Research Group spent 6 months visiting every police headquarters in the state to discover that during every day of 1980, an average of six traffic accidents resulted in spills of hazardous materials (mainly chemicals and fuels) in New Jersey alone. Perhaps current reporting is better; RCRA's 1984 amendments make it a criminal offense not to report, and the penalty for ignoring the law can be fines of up to $10,000 and a year in jail.

U.S. EPA's Hazardous Waste Data Management System (HWDMS) does not show violations of transport rules occurring in New York City. This appears encouraging. But it does not actually indicate a perfect record. Only a special unit of the city police department inspects vehicles transporting hazardous wastes. This information is not communicated between regulatory agencies. U.S. EPA depends on NYSDEC to perform inspections. But NYSDEC inspects transportation terminals only; it does not inspect moving vehicles. The U.S. DOT's databases record only those voluntarily reported violations or accidents that result in spills, not the number of improperly loaded or labeled vehicles, though such infractions significantly increase potential risks.

TABLE 10-1. *Inspection Performance of the U.S. EPA Regions*[a]

U.S. EPA Region	A	B	C	D	E	F	G
Region 1	442	15.4	130	464	275	103	215
Region 2	411	23.4	108	473	481	318	27
Region 3	198	19.2	91	819	396	100	103
Region 4	331	21.1	193	832	444	344	256
Region 5	490	39.0	69	596	601	973	165
Region 6	423	18.2	122	458	411	399	317
Region 7	151	12.0	60	177	181	193	54
Region 8	116	8.6	71	211	148	53	11
Region 9	168	14.2	93	400	221	396	93
Region 10	133	5.4	44	149	143	53	34
Totals	2863	176.5	981	4579	3301	2932	1275

Columns A to G: Statistics for fiscal year 1984
A: Major facilities
B: Federal inspectors (FTE)
C: U.S. EPA inspections
D: Total inspections
E: U.S. EPA total inspections target
F: Record reviews
G: D and F with Class 1 violations

a. Excluding federal inspectors (FTE = full time equivalents), all statistics pertain to major facilities only.

TABLE 10-2. *Enforcement Performance of the U.S. EPA Regions*[a]

U.S. EPA Region	A	B	C	D[b]	E	F	G	H
Region 1	442	49	51	166	161	47	22	146
Region 2	411	24	5	18	215	8	50	–16[c]
Region 3	198	25	26	100	339	25	100	0
Region 4	331	100	97	239	658	3	46	290
Region 5	490	125	38	118	265	53	88	102
Region 6	423	138	195	253	156	71	72	248
Region 7	151	25	23	52	156	8	1	68
Region 8	116	12	5	19	137	2	16	13
Region 9	168	30	29	50	315	14	6	60
Region 10	133	7	8	26	51	7	4	22
Totals	2863	535	477	1041	2453	238	405	933

Columns A to H: Statistics for fiscal year 1984
A: Major facilities
B: Class 1 violators at beginning of year
C: Class 1 violators during fourth quarter
D: Class 1 violators during entire year
E: Enforcement actions
F: Pursuant compliers at beginning of year
G: Pursuant compliers during entire year
H: Remaining Class 1 violators (B – F) + (D – G)

a. Excluding enforcement actions, which include EPA actions on all facilities, statistics pertain to Major Facilities only.
b. Column D is not the sum of columns B and C. Some beginning-year violators were still in violation at the end of 1984.
c. Negative number due to U.S. EPA data gathering error.

U.S. EPA Region	H	I	J	K	L	M	N	CEP Regional Performance Grades
Region 1	104.97%	168.72%	23.30%	8.44	3.48	37.91%	271.01	D−
Region 2	115.08	98.33	77.37	4.61	5.69	3.41	297.68	D
Region 3	413.63	206.81	50.50	4.73	9.69	11.20	674.18	A+
Region 4	251.35	187.38	103.92	9.14	6.37	21.76	536.42	B+
Region 5	121.63	99.16	198.57	1.76	7.95	10.51	418.58	C
Region 6	108.27	111.43	94.32	6.70	4.30	36.98	288.05	D
Region 7	117.21	97.79	127.81	5.00	7.94	14.59	341.17	D+
Region 8	181.89	142.56	45.68	8.25	7.41	4.16	381.65	C
Region 9	238.09	180.99	235.71	6.54	8.45	11.68	658.12	A+
Region 10	112.03	104.19	39.84	8.14	4.06	16.83	251.45	D−
Averages	159.93	138.71	102.41	5.55	6.16	16.97	395.81	C

DERIVED FROM: U. S. Environmental Protection Agency, "Quarterly Report—Regional Data" (November 1984).

Columns H to N: Ratios and grades for fiscal year 1984

H: Total inspections/majors
I: Total inspections/target
J: Record reviews/majors
K: U.S. EPA inspections/federal inspector
L: Federal inspectors/100 majors
M: D and F with Class 1 violations/D and F
N: Summation of ratios for grades (I + J + K + L − M)

SCALE: 250 275 325 350 375 425 450 475 525 550 575 625 650

| − | D | + | − | C | + | − | B | + | − | A | + |

U.S. EPA Region	I	J	K	L	M	CEP Regional Performance Grades
Region 1	48.64%	96.98%	42.07%	197.95%	−204.52	D−
Region 2	10.21	1194.44	156.75	−166.66	313.20	A+
Region 3	63.13	339.00	126.26	−100.00	163.13	A
Region 4	102.41	275.31	20.24	190.00	−272.16	D−
Region 5	49.59	224.57	68.78	−18.40	37.58	C
Region 6	92.43	61.66	72.95	79.71	99.18	C+
Region 7	50.99	300.00	16.66	172.00	−206.32	D−
Region 8	26.72	721.05	69.23	8.33	34.17	C
Region 9	47.61	630.00	39.21	100.00	−108.40	D
Region 10	24.81	196.15	44.00	214.28	−195.09	D
Averages	55.04	235.63	58.50	74.39	70.93	C

DERIVED FROM: U.S. Environmental Protection Agency, "Quarterly Report—Regional Data" (November 1984).

Columns I to M: Ratios and grades for fiscal year 1984

I: Class 1 violators/majors
J: Enforcement actions/violators
K: Correction rate with one quarter lag allowed
L: Rate of change in overall noncompliance
M: Summation of ratios for grades K − (I + L)

SCALE: −300 −200 −100 0 25 75 100 200 300

| − | D | + | − |C| + | − |B| + | − |A| + |

A = Excellent; B = Above Average; C = Standard; D = Below Standard.

TABLE 10-3. *Ranked Responses of the EPA Regional Offices to CEP Requests*

EPA Region	A	B	C	D	E	F	G	H	I	J	Response Quality Score[a]
CEP Grade A (Excellent/ > or = 3.50)											
Region 6	C	C	C	C	C	C	C	C	PC	NC	3.70
CEP Grade B (Above Average/ > or = 3.00)											
Region 8	C	C	C	C	C	C	C	C	—	PNC	3.30
Region 2	C	C	C	C	C	C	N/A	C	PNC	—	3.22
Region 5	C	C	C	C	—	C	C	C	—	C	3.20
Region 10	C	C	C	C	C	C	N/A	C	PNC	PNC	3.00
CEP Grade C (Standard/ > = 2.00)											
Region 3	C	PC	C	C	C	C	N/A	NC	PNC	—	2.89
Region 7	C	PC	C	C	—	C	N/A	C	NC	—	2.78
Region 9	C	—	C	C	—	—	C	PNC	NC	PC	2.20
CEP Grade D (Below Standard > or = 1.00)											
Region 1	—	—	NC	NC	C	—	C	PNC	—	NC	1.70
Region 4	C	PNC	—	C	—	—	C	C	—	—	1.70

A: Standard industrial classification code (11/10/83)[b]
B: Financial instruments (11/10/83)[b]
C: Inspections (11/10/83)[b]
D: Enforcements (11/10/83)[b]
E: Process capacity (11/10/83)[b]
F: Estimated amount of waste (11/10/83)[b]
G: RCRA Part A application (7/9/84)
H: Inspections/enforcements (10/22/84)
I: Owners/operators of CERCLA sites (11/2/84)
J: Financial instruments (12/11/84)

NOTE 1: EPA regions are listed from highest to lowest quality of responses.

NOTE 2: C (value = 4): computerized complete response; NC (value = 2): computerized partial response; PNC (value = 1): noncomputerized partial response; — (value = 0): region unable to respond with information; N/A = information request not sent to region (average does not include this category).

a. Response quality score = average response value for the EPA region.
b. Date information request was mailed to the EPA region.

TABLE 10-4. *State Inspection and Enforcement Performance for Fiscal Year 1984*[a]

State/Territory	CEP Performance Grades	State/Territory	CEP Performance Grades
Alabama	D+	Montana	B+
Alaska	D−[b]	Nebraska	C
Arizona	D+	Nevada	C
Arkansas	D	New Hampshire	D
California	A−	New Jersey	A
Colorado	D	New Mexico	D
Connecticut	D	New York	D
Delaware	C	North Carolina	C−
Florida	C−	North Dakota	D+
Georgia	B+	Ohio	B−
Guam	C	Oklahoma	D+
Hawaii	D−[b]	Oregon	D+
Idaho	D−[b]	Pennsylvania	B
Illinois	C−	Puerto Rico	D+
Indiana	D+	Rhode Island	D
Iowa	C−	South Carolina	C+
Kansas	C−	South Dakota	D−[b]
Kentucky	B−	Tennessee	C
Louisiana	D	Texas	D+
Maine	D	Utah	D+
Maryland	A+	Vermont	D
Massachusetts	D	Virginia	C−
Michigan	D+	Washington	C−
Minnesota	C−	West Virginia	B−
Mississippi	C−	Wisconsin	D+
Missouri	D+	Wyoming	D−[b]

A = Excellent; B = Above average; C = Standard; D = Below standard.

a. Grades based on key regulatory statistics from EPA'S "Quarterly Report—Regional Data" (November 1984) regarding major hazardous waste treatment, storage, and disposal facilities. The ratios used include: the average number of inspections, record reviews, and Class 1 violators per major facility and the average number of enforcement actions taken against each Class 1 violator. The District of Columbia is not included because EPA has not identified any major facilities there.

b. These states are graded on the basis of only the first two ratios, because no Class 1 violators were found there.

Air

A 1985 U.S. EPA Science Advisory Board draft report on the impact of hazardous waste incinerators on air quality states: "The monitoring programs used to date were few and narrow in scope."[6] While U.S. EPA monitors what happens to certain chemicals as they are burned, most compounds released during incineration are neither monitored nor analyzed for toxicity. That's so, even though U.S. EPA has had the authority to set federal emissions standards since the CAA amendments of 1970. For the great number of combustion products and the volatiles associated with the handling of compounds, EPA has yet to set such limits. The Science Advisory Board also found existing studies of the impacts of meteorological conditions and fugitive emissions to be inadequate.

Incineration is not the only form of hazardous waste management that can significantly harm air quality. But of more than two dozen landfills CEP inspected, only CWM's Model City, New York, facility had on-site air-monitoring devices, and these devices were just being installed at the time of the visit (see chapter 8).

HSWA requires that U.S. EPA develop air-monitoring standards at hazardous waste facilities by May 1989. U.S. EPA has already announced that it will not meet this deadline. Key personnel say this failure is due to severely limited resources.[7]

Ground Water

A national strategy has emerged for the protection of groundwater. In 1984 U.S. EPA established a special Office of Groundwater Protection, which in August of that year outlined a program to address the following needs:

- short-term building of institutions at the state level
- assessing the problems that may exist from unaddressed sources of contamination (sources that weren't covered directly by federal regulations)—in particular, leaking storage tanks, surface impoundments, and landfills
- issuing guidelines for U.S. EPA decisions affecting ground water protection and cleanup
- strengthening U.S. EPA's organization for ground water management at the headquarters and regional levels and strengthening U.S. EPA's cooperation with federal and state agencies.[8]

The reauthorization and strengthening of RCRA in late 1984 brought underground storage tanks, one of the unaddressed sources of contamination that the new ground water office was designated to investigate, under the jurisdiction of federal regulation.

Municipal and privately owned solid waste landfills that are authorized to accept only nonhazardous wastes in any sizeable quantity, however, are still not routinely checked to see if hazardous wastes are present in dangerous amounts. Solid waste landfills are still allowed to receive "small" quantities of hazardous waste from households and other sources. At the time of this writing, only four states (Maine, Ohio, Oregon, and Wisconsin) prohibit small quantities of hazardous waste from being disposed in solid waste landfills.[9]

These small quantities can accumulate. Of the uncontrolled sites on the Superfund NPL, a fifth are landfills that were not originally designed to accept hazardous wastes. It will cost billions of dollars to clean these sites (see table 10-5).[10]

Municipal, county, and privately owned solid waste landfills are prohibited from accepting large shipments of hazardous wastes, but such deliveries took place regularly before they were outlawed by RCRA in late 1976. They still take place, sometimes with the connivance of landfill employees. For example, an illegal PCB-dumping scheme at the Fountain Avenue Landfill in Brooklyn, New York, was uncovered in 1982. Because of inadequate monitoring, escaping contamination was not detected for another 2 years, until 1984.

These problems are certainly not limited to New York. Kent Stoddard, former director of California's OAT, says the lack of information on contami-

TABLE 10-5. *Summary Data on Solid Waste Facilities*

Percent of uncontrolled sites that are solid waste facilities:	
Of 1389 sites with actual or presumed problems of releases of hazardous substances	18%
Of 550 sites on the NPL	20%
Two most prevalent effects at problem solid waste sites:	
Leachate migration, ground water pollution:	at 89% of sites
Drinking water contamination:	at 49% of sites
Mean size of problem solid waste sites	67.4 acres
Median hazard ranking score[a]:	
Solid waste sites on the NPL	40.8
All NPL sites	42.2
Estimates for national number of solid waste sites:	
Operating sanitary, municipal landfills	14,000
Closed sanitary, municipal landfills	42,000
Operating industrial landfills	75,000
Closed industrial landfills	150,000
Operating surface impoundments	170,000
Closed surface impoundments	170,000
Total	621,000
Estimated number of sites needing future cleanup:	
Low: 5% landfills, 1% impoundments likely to release toxic substances	17,400
High: 10% landfills, 2% impoundments likely to release toxic substances	34,800
Conservative figure used for cleanup by Superfund	5,000

SOURCE: Office of Technology Assessment, "Superfund Strategy—Summary" (1985).

a. 28.5 required for placement on the NPL; current highest site score on NPL is 75.6.

nation in California (a state with a well-regarded program) "is shocking" and that "compliance with ground water monitoring requirements has been terrible."[11]

A report of the California State Legislature concludes, "State and federal standards do not ensure that required ground water monitoring will detect migrating pollution before ground water supplies are contaminated." Two-thirds of the toxic ponds investigated in California are located next to or above domestic and municipal drinking-water supplies. Of these, 72 percent had no liners to prevent their contents from seeping into the ground; 77 percent did not conduct any ground water monitoring, and many of the 23 percent that did monitor, monitored inadequately.[12]

Nationally, U.S. EPA's 1983 SIA found that "in general, impoundments have historically been sited and constructed without apparent regard for the protection of ground water quality." Furthermore, only six states regulated sites specifically for ground water contamination from surface impoundments at the time the report was conducted, and less than 1 percent of the surface impoundment sites were known to be monitored for ground water contamination.[13]

Once again, however, poor execution of the data compilation phase has provided a weak foundation. When U.S. EPA entered the data on monitoring wells into a computer, the difference between "unknown" and "no monitoring wells" was lost; neither U.S. EPA nor the states can use the information to target their few enforcement dollars.

U.S. EPA has also attempted to gather information on ground water monitoring by facilities that have been granted interim status prior to being issued a final RCRA operating permit. The third and final phase of U.S. EPA's "Interim Status Ground Water Monitoring Implementation Study" requested by OMB in 1982 has findings that are just as disheartening as the SIA:

- Several states with a substantial number of RCRA ground water facilities have bifurcated organizational structure [parts of programs are handled by different agencies] that has resulted in very poor implementation.
- Many U.S. EPA regions and most states have severe technical staffing problems due to insufficient fiscal resources and an inability to attract experienced personnel.
- U.S. EPA's policy of turning the RCRA program over to the states as soon as possible has resulted in authorizing many states that were ill prepared to implement the program. This, in part, accounts for the program's poor compliance record.
- A number of authorized states have a law enforcement policy toward monitoring violations.
- U.S. EPA, for several administrative reasons, has declined to enforce or pursue cases involving the major performance standards that make up much of the 40 CFR part 265 subpart F regulations.[14]

The principal source of information for U.S. EPA's ground water-monitoring study is the Major Facility Status Sheet, a form that U.S. EPA and states must fill out for all major facilities. These sheets outline the status of both the ground water-monitoring program at a site and a facility's compliance with financial requirements.

CEP requested copies of these status sheets for 84 major facilities owned by the eight publicly held U.S. commercial hazardous waste management companies. U.S. EPA reported that only 74 of these facilities were considered "major," even though its own computer system had identified 84 as such. Furthermore, the agency claimed it was able to locate only 37 status sheets. Internal memos sent by Lee M. Thomas (now U.S. EPA administrator) to U.S. EPA's ten regional administrators, however, highlight 15 of the missing 47 facilities as deficient (see chapter 4).

U.S. EPA headquarters claims to have paper copies of the status sheets for all 1749 major facilities. Why, then, can it not retrieve them even if requested under FOIA? And how can the administrator quote some that don't seem to be in the files?

Ohio was the only state to follow U.S. EPA headquarters' instructions to enter the status reports into U.S. EPA's computerized HWDMS. The other states sent paper copies that were tabulated by hand. The central computerized data system has since been found to have severe technical problems. "In other words," asserts the study, "the only state to follow U.S. EPA headquarter's direction cannot have its data accessed because the system does not work."[15]

Furthermore, from most of the status sheets one cannot determine whether a facility's monitoring system is adequate or whether a site is leaking, simply because evaluations have not taken place.

Financial Instruments

The information-gathering activities necessary for U.S. EPA's HWDMS are eroding. The agency's management of financial data, the RCRA 40 CFR part 265 subparts G and H requirements, exemplifies this problem.

CEP filed FOIA requests during November 1983 at ten regional U.S. EPA offices. Only half were able to provide this information— a year and a half after the regulations requiring them were issued. These five regional offices said the data were available only from state agencies. Thus, for 21 states, U.S. EPA's database does not indicate the ability of facilities to pay for remedial action in the event of an emergency.

U.S. EPA Region 4 (the Southeastern states) temporarily withheld the requested financial data, uncertain whether it might be considered confidential business information. When U.S. EPA's Office of General Counsel in Washington, D.C., was asked for an opinion, its legal staff could not give a definite answer—even though five regions had already provided the data.

States with more advanced authorization to handle hazardous waste programs manage the financial information themselves. How well? When CEP called eight such states (Alaska, Arizona, California, Maryland, Nevada, Pennsylvania, and Virginia), all but Arizona and Nevada reported that they lacked adequate staffing to process the requests. The information for these two states was, in effect, unavailable. Furthermore, some states possessed only interim status when the financial instruments regulation was promulgated. Thus neither the state nor U.S. EPA had access to this information! In Massachusetts, for example, TSDFs were not required to submit financial instruments until February 13, 1984.

CEP filed another FOIA request to the regional offices regarding financial instruments a year later. Only one of the ten regions— Region 5—was again able to provide a computerized printout of the data. Region 8, which had provided a computerized response the year before, responded this time that, "a computer printout on the information you desire does not exist."[16]

U.S. EPA's Region 2 office, which also was able to provide the information in response to the previous request, noted that New York and Puerto Rico had since taken over jurisdiction of the financial instruments requirements. Region 3 staff responded that Pennsylvania had responsibility for the financial instruments, but when the state was called, it at first said U.S. EPA was responsible for the information. After checking into the matter more closely, a state official found that Pennsylvania had been putting all of its financial instrument records into a box, and was waiting to figure out what to do with them.

Inspections

Due to the enormous universe of industrial facilities that generate, transport, treat, store, or dispose of hazardous wastes, U.S. EPA and the states have had to focus their inspection activities on only those sites that are considered particularly important. They call these sites *major handlers* and they meet the following federal criteria:

- all facilities subject to ground water-monitoring and/or protection requirements
- all incinerators
- up to 10 percent of remaining treatment, storage, and disposal facilities (TSDFs)
- up to 3 percent of generators and transporters.[17]

On average, all major handlers were inspected more than once during fiscal year 1984, according to U.S. EPA's RCRA Enforcement Management System, a database derived from two questionnaires that U.S. EPA sends to the regional offices. Federal inspectors average only five and a half inspections per year, however. They inspected only a third of all major handlers during 1984. The rest of the inspections (almost 80 percent of them) were conducted by state personnel.

When one considers that these 2863 major handlers constitute only 4.3 percent of the estimated 66,428 facilities that handle large quantities of hazardous wastes, one begins to understand the difficulty of inspecting the entire universe of hazardous waste facilities. As the Internal Revenue Service causes fear of audits, facility operators must at least be wary of the possibility of a RCRA inspection for the regulatory program to be effective.

Using the RCRA Enforcement Management System, it is possible to judge the quality of U.S. EPA's inspection program on a regional basis (see table 10-1). Region 3 (the Mid-Atlantic states) performed the best of all regions during fiscal year 1984, by completing twice as many inspections as U.S. EPA Headquarters targeted for the region by the end of the year. On average, major handlers in the region were each inspected four times, and only 11 percent of all inspections and record reviews identified Class 1 violations or "those that pose direct and immediate harm or threats of harm to public health and the environment."

Nationally, however, the correlation between the number of inspections performed by a region and compliance with federal requirements does not appear to be very significant. Some regions with comparably less vigorous inspection programs than Region 3 had lower rates of Class 1 violations. The rates may possibly be due to local variations in population, geography, and types of waste being handled. In fact, a more rigorous inspection program is likely to detect more violations.

Occupational Safety and Health

According to CEP's analysis of federal OSHA data, U.S. OSHA has not focused on hazards that are specific to hazardous waste facilities. Only 3 percent of all U.S. OSHA violations at facilities in related industries from July 1972 through April 1984 involved toxic and hazardous substances, the so-called Subpart Z Standards.[18]

CEP obtained data from U.S. OSHA's computerized Inspection History Listing from 1972 through 1984 for the following four-digit SIC categories:

2842: specialty cleaning, polishing, and sanitation preparations
4952: sewerage systems
4953: refuse systems
4959: sanitary services, not elsewhere classified
9511: air, water, and solid waste management (public).

This universe includes much of the hazardous waste industry, but only a fraction of these industrial classifications represent the hazardous waste management industry.

There were 1771 inspections of 1055 establishments identified in these five SIC categories. Over 12 years, therefore, an establishment was likely to be inspected, on average, less than twice. On average, each inspection found three violations (5122 during the 12-year period). Only one inspection in ten was likely to result in violations related specifically to hazardous substances (175 subpart Z violations found during 1771 inspections).

CEP focused on five of the Big Eight commercial hazardous waste firms that were identified by one or more of these five SIC codes. The five Big Eight companies listed in the U.S. OSHA database are BFI, ENSCO, ESI, RES, and WMI. From 1972 to 1984 U.S. OSHA made 118 inspections of 58 facilities owned by these five companies. Again, that's an average of only two inspections per establishment. The five companies were cited for 250 violations or an average of two violations per inspection. There were no subpart Z violations and no willful violations cited.

The data indicate that these companies had fewer violations than the average for all companies in these four-digit SIC categories. Could it be, however, that there were actually no instances of hazardous and toxic substances violations to be uncovered? While possible, these results indicate that there is a serious gap between the risks to workers that exist at hazardous waste sites and the performance of U.S. OSHA to ensure safe and healthful workplaces.

Penalties for violations have been very low. Total penalties for the five companies were $25,719 for the 12-year period. That's an average of $103 per violation, even though the database indicates that six people were injured and two died as a result of violations at these establishments.

Both the data in U.S. OSHA's Inspection History Listing and the inspections themselves vary greatly in quality. For example, an explosion caused by contractor personnel at the RES (NJ) plant in 1977 that resulted in the deaths of six people and numerous injuries was not even recorded in the database. According to the data, U.S. OSHA inspected ENSCO's one facility 11 times but made only 31 inspections of 20 WMI facilities. That's an average of 11 inspections per establishment for ENSCO and only 1.5 per establishment for WMI. Inspections for these five companies more than tripled during the years 1977 to 1980 (from 23 to 80) and dropped precipitously since 1980. While total inspections for all establishments in these five SIC categories fell 22 percent since 1981 (from 632 to 491), for the five specific companies, they fell by over 81 percent (down to 15 inspections). In addition, there were no repeat inspections among the five companies since 1980.

The fact the definition of an inspection changed in October 1981 complicates data analysis. Now the review of only an establishment's paperwork constitutes an inspection. Such an inspection is less likely to lead to discovery of serious hazards than an on-site wall-to-wall inspection with worker walk-around and serious monitoring measurements.

There is no data in U.S. OSHA's information system for states with approved state OSHA programs. Authorized states have changed over the past 12 years, so there may be data for some of these states but for only a portion of the time.

California's Department of Industrial Relations (CADIR) is one of 23 state agencies which, as of 1984, operated the safety and health program under section 18b of the OSHA of 1970. CEP asked California for its equivalent of the federal OSHA's Inspection History Listing.

CADIR's Division of Occupation Safety and Health (CALOSHA) is split up into five regional and 27 district offices. CALOSHA has a centralized computer data system, but the system cannot provide data on a facility-specific basis. CALOSHA's information specialist responded to CEP's information request by stating that CEP would have to contact each of the 27 offices. The district offices use at least a dozen independent forms to record what the federal OSHA's inspection history listing contains. Furthermore, even though most forms are identical, if CALOSHA attempts to merge the files (and that would be a prohibitively expensive procedure), "the resultant data becomes questionable."[19]

As regulations require more complete containment of hazardous wastes within controlled units, occupational safety and health precautions will become even more important. Risks of worker exposure increase, for example, as more waste-handling is conducted indoors to prevent toxic emissions to the air. It will become even more important in the future for regulators to conduct thorough and repeated inspections to ensure compliance with OSHA requirements. A good data system with full reporting is critical.

PROCESSING INFORMATION

Determining Scale

Of equal importance to gathering information is manipulating and evaluating it to reveal its significance. Regulators must organize data better to estimate the scale and seriousness of a problem, identify underlying patterns of behavior, and determine the needed corrective measures.

U.S. EPA's database lists approximately 1100 large hazardous waste generators in New York City alone, yet NYSDEC claims there are only 1400 generators of all sizes, large and small, in the whole state.[20] U.S. EPA's figure has little resemblance to reality, because New York City generates less than 3 percent of the state total of hazardous waste.

Inaccuracies may unnecessarily hurt companies' images. Just one industrial plant, E. I. DuPont de Nemours's Chambers Works in Deepwater, New Jersey, accounts for 15 percent of U.S. EPA's estimate of the volume of waste generated annually. But plant manager Joseph Piacenti claims, "99 percent was water used in the advanced treatment process to make the waste safe."[21]

According to one U.S. EPA computer printout received by CEP, DuPont's Chambers Works annually produces 1.9 million tons of paint solvent wastes. A subsequent printout from the same database indicates the Chambers Works generates only 0.9 million tons of waste. Surely DuPont did not reduce its production rate by a million tons of paint solvent wastes during the interval between the two printouts. All quantities on the two printouts are identical, except for the single, but quite substantial, missing digit. Such routine data errors can make it impossible for U.S. EPA—let alone the public—to determine the magnitude of a problem.

Equitable corporate taxation to pay for Superfund requires reliable estimates of waste generation. Companies have been taxed on the basis of the amount of chemical and petrochemical materials they use because Congress was unable to determine how to collect data that would enable taxation according to the amount of waste generated. U.S. EPA still cannot provide accurate estimates. The survey estimates that 85 percent of hazardous wastes are produced by chemical and petroleum industries. Given the disregard of dilution during treatment and unreliable data in the computer, however, it is difficult to have confidence in this figure.

Current information deficiencies foreclose the possibility of appraising responsibility fairly among industrial groups and separate companies. Moreover, the HWDMS data are of little use for this purpose in that they represent what and how much a disposal facility requested permission to handle and not the actual volumes handled (see chapter 4).

Characterization

To find out what a body of data means, the information must be characterized systematically and with precision. Planning requires coordination from the start. Otherwise, inconsistencies—which can be fatal to an information system—are likely to occur. Decentralization, either within a program or between federal and state governments, can undermine these goals.

Findings of the House Government Activities and Transportation Subcommittee indicate almost no cooperation between U.S. EPA and the U.S. DOT's Bureau of Motor Carrier Safety over the regulation of hazardous wastes in transit.[22] Even the new uniform manifests that are designed to track transportation through all states and to facilitate data transfer between U.S. EPA and U.S. DOT do not require standard U.S. EPA waste identification codes. U.S. DOT's descriptions and numbers for waste identification are far less precise than U.S. EPA's; many wastes fall into the category *Hazardous Waste Not Otherwise Specified.*[23]

Congressional OTA found that detection levels under RCRA for ground water contamination are higher than those under CERCLA—sometimes as much as 1000 times higher. Thus an active facility may legally pollute ground water to a level tolerated under RCRA, but come to the attention of CERCLA for the same pollution. In essence, RCRA facilities in full compliance with regulations are not prevented from becoming uncontrolled sites that require expensive cleanup under Superfund.[24] Inconsistently defined standards make it difficult to accurately characterize a problem.

CEP has made numerous FOIA requests to the ten regional U.S. EPA offices to obtain information about the hazardous waste management industry. Despite the supposedly functional equivalence of all regional U.S. EPA databases, and the fact that our ten FOIA requests were identical, some of the U.S. EPA regions responded with complete and well-prepared information, whereas others supplied incomplete and poorly organized data. Information specialists at regional offices have met to discuss the problem of inconsistency in training and programming. Apparently, however, these meetings have not yet accomplished the goal.

A few regions, for example, were unable to program their computers to link specific wastes with the amounts handled. Consequently, if a facility listed 50 different wastes, it was impossible to determine whether 90 percent of the waste volume was accounted for by a single waste or by 20 different types of wastes. Some regions were unable to associate waste types with the process used to handle them; therefore, it was impossible to tell which wastes were incinerated, for instance, and which were landfilled.

One response included a computer printout 6 inches thick. For each facility listed, there were pages and pages of a single four-digit waste code per line, followed by pages of a single three-digit process code per line, with nothing linked and the rest of the page left completely blank.

Even information provided by just one office can be extraordinarily inconsistent. Sixty percent of the New York City TSDFs listed in the database were not properly identified as TSDFs in three printouts received during one month. Each had a different list of TSDFs. The differences could not have been due to company changes in such a short time, and in fact, the later printouts included even fewer TSDFs. If their status as TSDFs was withdrawn, on the other hand, this should have been indicated in the database.

More than 600 facilities, or 43 percent of those listed in New York City, are not characterized by the most fundamental criteria, that is, whether they are generators, transporters, or TSDFs. One of the five commercial TSDFs—those that handle other companies' wastes as a business—can be found in a commercial directory but not in the government's database! According to the database, 80 percent of the TSDFs listed in New York City have not submitted Part A of the federal permit application and the financial responsibility instruments data, the two basic requirements that identify exactly what a facility does and whether it is insured against dangerous occurrences.

CEP sent a FOIA request to U.S. EPA headquarters asking for names of owners and operators of facilities either proposed for or already on the NPL for Superfund cleanup. Unfortunately, U.S. EPA's database does not differentiate owners and operators from other "potentially responsible parties." A generator that sent a single 55-gallon drum to a site, for example, would be indistinguishable in the raw data from a party that managed the facility for 55 years. U.S. EPA headquarters advised contacting the regional offices to get an answer to the request. When we did, some EPA regions referred the query back to headquarters. Others sent related information that did not answer the request. Most simply replied that the information was unavailable. Only Region 9, the Southwestern region, was able to comply.

Enforcement

One obvious use of U.S. EPA's HWDMS should be to target enforcement activities. U.S. EPA's 1983 "Environmental Status Report for the State of New York" emphasizes concern over the poor correction rates for Class 1 violations ("those that pose direct and immediate harm or threats of harm to public health and the environment").[25]

U.S. EPA's New York data does not, however, distinguish Class 1 violations from less serious ones. One can usually determine which violators committed Class 1 violations by the enforcement action taken against them, though. Using this interpretation, U.S. EPA's database indicates that about half of the violations observed at hazardous waste management facilities in New York City are Class 1 violations, including:

- ignitable wastes stored within 50 feet of the site's boundary
- discharges of hazardous wastes
- failure to keep manifests of shipments.

U.S. EPA's HWDMS records for New York during fiscal year 1984 show that 90 percent of these Class 1 violators have not corrected their illegal activities despite U.S. EPA enforcement actions, though all had more than 6 months to do so from the time their citation was issued to the time the data printout was processed. The record for compliance among all violators—including less serious ones—is better, but hardly praiseworthy; two-thirds of all violations go uncorrected.

These shocking rates of noncompliance naturally affect the national standard. Because of inconsistencies, U.S. EPA does not use HWDMS to keep track of enforcement issues nationwide. U.S. EPA found it had to establish yet another independent database for that: the RCRA Enforcement Management System.

Unfortunately, this system cannot do the whole job either. For the most part it has information concerning major facilities, which account for less than 5 percent of the facilities recorded in the HWDMS.

CEP evaluated this enforcement system and found it also included numerous inaccuracies. Many totals that could be derived by using two separate equations were inconsistent. If, for example, major handlers are added up by state the sum equals 2944. But if they are added up by region, the total is 2863. Although a 2.8 percent difference is not enormous, it can greatly impede accurate analyses. Information for Region 2 (which includes New York State), for example, is suspect because it identifies 58 facilities as having come into compliance with regulations pursuant to EPA or state enforcement actions during 1984. But only 42 violators were identified to begin with! Region 4, on the other hand, identified 339 Class 1 violators among its major handlers, even though there are only 331 major handlers listed in the region.

The RCRA Enforcement Management System, nevertheless, contains the best available information on nationwide enforcement activities. It reveals some startling statistics on noncompliance among major handlers. More than half of all major facilities committed Class 1 violations during 1984. Even though (on average) U.S. EPA and the states took more than two enforcement actions against each violator, fewer than 60 percent corrected those violations within three months. Furthermore, the number of noncompliers appears to be growing at a rate of 74 percent each year.

Northeast's Enforcement Rates an A+. Using the RCRA Enforcement Management System it is possible to rate the performance of the regional U.S. EPA offices and, to a lesser degree, the states (see table 10-4). Despite its poor inspection performance, Region 2 gets an A+ from CEP for its enforcement record. Although its high correction rate is suspect, Region 2 does have the lowest number of Class 1 violations identified during inspections and records reviews and the lowest percentage (just over 10 percent) of major handlers with violations. Within Region 2 fewer than 1 percent of major facilities in New York have committed Class 1 violations, one of the lowest violation rates. But this could be due to inadequate information gathering. New York inspected only two-thirds of all major facilities in the state and reviewed the records of just 15 percent. New Jersey, on the other hand (along with Maryland and California), appear to have the best overall inspection and enforcement programs in the country, but not necessarily the best correction (see table 10-6).

Northwest's Enforcement Rates Poorly. In terms of enforcement Region 10 performed poorly. The number of noncompliers in the Northwest more than doubled during fiscal year 1984—even though as of August 1984 it assessed more than half of all the penalties nationwide ($194,975 out of $334,375).[26] Given the limited precision of U.S. EPA's data, it is difficult to determine for some whether the rate of noncompliance has simply increased

TABLE 10-6. *State Inspection and Enforcement of Performance*[a]

State/ Territory	A	B	C	D	E	F	G
Alabama	75	90	24	34	43	4	0
Alaska	12	3	0	0	0	0	0
Arizona	19	13	11	7	14	0	0
Arkansas	11	12	10	12	9	0	1
California	117	325	242	32	70	101	0
Colorado	37	33	13	1	0	0	0
Connecticut	196	185	68	93	62	5	0
Delaware	5	11	7	3	5	0	0
District of Columbia	0	1	0	0	0	0	0
Florida	35	90	9	22	23	1	0
Georgia	51	61	20	19	130	4	0
Guam	6	13	10	5	7	0	0
Hawaii	17	0	0	0	0	0	0
Idaho	15	9	0	0	0	0	0
Illinois	66	135	93	61	68	0	0
Indiana	112	59	99	24	32	0	0
Iowa	39	45	64	13	20	0	0
Kansas	32	28	50	6	9	0	0
Kentucky	66	104	112	30	104	5	0
Louisiana	149	107	136	78	41	0	0
Maine	40	17	0	7	8	0	0
Maryland	21	213	4	23	163	4	1
Massachusetts	107	70	0	19	16	0	0
Michigan	102	149	64	15	8	0	0
Minnesota	47	47	9	4	8	0	0
Mississippi	40	32	98	37	61	0	0
Missouri	55	29	65	30	30	3	1
Montana	14	26	8	3	12	3	4
Nebraska	25	9	44	3	6	0	0
Nevada	9	9	6	6	20	0	0
New Hampshire	49	38	6	24	21	2	0
New Jersey	208	149	307	9	69	2	0
New Mexico	21	20	14	22	18	0	0
New York	157	105	24	1	1	0	0
North Carolina	64	72	55	47	92	2	0
North Dakota	11	23	3	0	1	0	0

Columns A to G: Statistics for fiscal year 1984

A: Major facilities
B: Inspections
C: Record reviews
D: Class 1 violators
E: Enforcement actions
F: Justice department referrals
G: Justice department filings

a. All statistics apply to major facilities only.
b. These states are graded on the basis of only the first two ratios, because no Class 1 violators were found there.
c. District of Columbia was not graded because U.S. EPA has not identified any major facilities there.

State/ Territory	H	I	J	K	L	CEP State Grades
Alabama	120.00 (%)	32.00 (%)	45.33 (%)	138.23 (%)	244.90	D+
Alaska	25.00	0.00	0.00	—	25.00	D−[b]
Arizona	68.42	57.89	36.84	200.00	289.47	D+
Arkansas	109.09	90.90	109.09	83.33	174.24	D
California	277.77	206.83	27.35	534.37	991.63	A−
Colorado	89.18	35.13	2.70	0.00	121.62	D
Connecticut	94.38	34.69	47.44	72.04	153.67	D
Delaware	220.00	140.00	60.00	166.66	466.66	C
District of Columbia	—	—	—	—	0.00	[c]
Florida	257.14	25.71	62.85	109.09	329.09	C−
Georgia	119.60	39.21	37.25	705.26	826.83	B+
Guam	216.66	166.66	83.33	140.00	439.99	C
Hawaii	0.00	0.00	0.00	—	0.00	D−[b]
Idaho	60.00	0.00	0.00	—	60.00	D−[b]
Illinois	204.54	140.90	92.42	111.47	364.50	C−
Indiana	52.67	88.39	21.42	133.33	252.97	D+
Iowa	115.38	164.10	33.33	153.84	399.99	C−
Kansas	87.50	156.25	18.75	150.00	375.00	C−
Kentucky	157.57	169.69	45.45	363.33	645.15	B−
Louisiana	71.81	91.27	52.34	52.56	163.30	D
Maine	42.50	0.00	17.50	114.28	139.28	D
Maryland	1014.28	19.04	109.52	730.43	1654.24	A+
Massachusetts	65.42	0.00	17.75	84.21	131.87	D
Michigan	146.07	62.74	14.70	53.33	247.45	D+
Minnesota	100.00	19.14	8.51	200.00	310.63	C−
Mississippi	80.00	245.00	92.50	164.86	397.36	C−
Missouri	52.72	118.18	54.54	113.33	229.69	D+
Montana	185.71	57.14	21.42	633.33	854.76	B+
Nebraska	36.00	176.00	12.00	200.00	400.00	C
Nevada	100.00	66.66	66.66	333.33	433.33	C
New Hampshire	77.55	12.24	48.97	95.83	136.64	D
New Jersey	71.63	147.59	4.32	788.88	1003.79	A
New Mexico	95.23	66.66	104.76	81.81	138.96	D
New York	66.87	15.28	0.63	100.00	181.52	D
North Carolina	112.50	85.93	73.43	200.00	325.00	C−
North Dakota	209.09	27.27	0.00	—	236.36	D+[b]

Columns H to L: Ratios and grades for fiscal year 1984

H: Inspections/majors

I: Record reviews/majors

J: Class 1 violators/majors

K: Enforcement actions/Class 1 violators

L: Summation of ratios for grades (H + I + J − K)

SCALE: 0 100 200 300 400 500 600 700 800 900 1000 1100 1200

| − | D | + | − | C | + | − | B | + | − | A | + |

A = Excellent; B = Above average; C = Standard; D = Below Standard.

DERIVED FROM: U.S. Environmental Protection Agency, "Quarterly Report—Regional Data" (November, 1984).

(continued)

TABLE 10-6. *(continued)*

State/ Territory	A	B	C	D	E	F	G
Ohio	135	79	283	10	37	4	0
Oklahoma	37	30	14	8	9	0	1
Oregon	41	11	3	6	11	0	0
Pennsylvania	120	518	27	48	153	9	0
Puerto Rico	46	63	5	8	6	1	0
Rhode Island	40	16	17	13	7	0	0
South Carolina	27	24	13	8	36	2	0
South Dakota	5	3	1	0	0	0	0
Tennessee	54	158	13	42	78	14	0
Texas	205	231	227	133	62	4	1
Utah	32	47	12	15	16	0	0
Vermont	10	18	0	10	4	1	1
Virginia	28	45	24	16	26	1	0
Washington	65	24	4	4	12	0	0
West Virginia	24	30	53	10	35	0	0
Wisconsin	28	23	13	4	5	2	0
Wyoming	17	0	0	0	0	0	0
Totals	2944	3652	2384	1025	1668	174	10

Columns A to G: Statistics for fiscal year 1984
A: Major facilities
B: Inspections
C: Record reviews
D: Class 1 violators
E: Enforcement actions
F: Justice department referrals
G: Justice department filings

a. All statistics apply to major facilities only.
b. These states are graded on the basis of only the first two ratios, because no Class 1 violators were found there.
c. District of Columbia was not graded because U.S. EPA has not identified any major facilities there.

or if the variance reflects a more rigorous data-gathering program. Region 10's poor inspection performance indicates the former, however.

COMMUNICATION

South Central Tops in Communication; Southeast Is at Bottom

CEP has rated the ability of government agencies to communicate to the public by evaluating their responses to information requests. We paid particular attention to the completeness of information provided and the degree to which the information is computerized. Both are indicators of comprehensiveness and efficiency. Region 6 (the South-Central states) performed best among regional U.S. EPA offices. Region 6 responded to all but one infor-

State/ Territory	H	I	J	K	L	CEP State Grades
Ohio	58.51	209.62	7.40	410.00	670.74	B−
Oklahoma	81.08	37.83	21.62	125.00	222.29	D+
Oregon	26.82	7.31	14.63	183.33	202.84	D+
Pennsylvania	431.66	22.50	40.00	337.50	751.66	B
Puerto Rico	136.95	10.86	17.39	87.50	217.93	D+
Rhode Island	40.00	42.50	32.50	53.84	103.84	D
South Carolina	88.88	48.14	29.62	475.00	582.40	C+
South Dakota	60.00	20.00	0.00	—	80.00	D−[b]
Tennessee	292.59	24.07	77.77	219.04	457.93	C
Texas	112.68	110.73	64.87	50.37	208.91	D+
Utah	146.87	37.50	46.87	106.66	244.16	D+
Vermont	180.00	0.00	100.00	60.00	140.00	D
Virginia	160.71	85.71	57.14	168.75	358.03	C−
Washington	36.92	6.15	6.15	300.00	336.92	C−
West Virginia	125.00	220.83	41.66	350.00	654.16	B−
Wisconsin	82.14	46.42	14.28	175.00	289.28	D+
Wyoming	0.00	0.00	0.00	—	0.00	D−[b]
Averages	124.04	80.97	34.81	180.68	350.89	C−

Columns H to L: Ratios and grades for fiscal year 1984

H: Inspections/majors

I: Record reviews/majors

J: Class 1 violators/majors

K: Enforcement actions/Class 1 violators

L: Summation of ratios for grades (H + I + J − K)

SCALE: 0 100 200 300 400 500 600 700 800 900 1000 1100 1200

| − | D | + | − | C | + | − | B | + | − | A | + |

A = Excellent; B = Above average; C = Standard; D = Below Standard.

DERIVED FROM: U.S. Environmental Protection Agency, "Quarterly Report—Regional Data" (November, 1984).

mation request with computerized data. Its neighbor, Region 4 (the Southeastern states), could provide information for only half the requests (see table 10-3).

The ability of the regional offices to collect and communicate information appears to be a factor primarily of management skill. It has less relationship to the scale of the hazardous waste industry in each region. Some of the FOIA officers from regions that had below-average scores remarked that CEP information requests were too demanding or expensive to fill. Also, the regions often took 2 to 3 months to respond.

Region 2 for example called CEP to complain about a clause in our standard FOIA request letter that stated, "As provided by the Freedom of Information Act, we expect to receive a response within ten (10) working days," saying it was "too pushy." The region then took 5 months to respond to our request for facility status sheets, a standard RCRA information report, because they did not know how to retrieve it from the HWDMS.

A more efficient and sensible policy would recognize that agency information, obtained at the cost of millions of taxpayer dollars, is too valuable to underutilize by failure to release it. Region 5, which oversees the largest amount of waste generation and disposal of all the regions, appears to have adopted the more sensible approach, providing much useful and valuable data within a reasonable time frame.

CEP also sent information requests concerning air, water, and RCRA issues to appropriate agencies in 27 states. One-third of these requests were simply never answered. The agencies replied to another third of the requests stating that the information could not be provided because either adequate staff was unavailable, the information was spread out in too many different local offices to access, or a fee was required. In sum, even if the fees were all paid, the states still could not comply with half of the requests. In this regard the states performed far worse than U.S. EPA's regional offices (see table 10-7).

Good State Communication

Again, it is possible to score the relative quality of responses of the state agencies to CEP's three separate information requests. Almost none of the data requested were computerized, so this variable could not be used as an indicator of competent information management. The extensiveness of the information provided related to the number of hazardous waste facilities inquired about in each state is a useful measure of a state's ability to communicate information, however. Illinois in particular—along with Michigan, Nevada, and Texas—were able to comply most fully with CEP's requests. Idaho, Ohio, California, and New York, on the other hand, could not provide information in response to any of the requests.

As with the U.S. EPA regional offices, the size of the state programs was not a major factor in determining the quality of a state's ability to communicate information. The data suggest, however, a relationship between quality of a state's communication abilities with the ability of the corresponding U.S. EPA region. Oddly enough, the states with less federal authorization to handle their enforcement program appeared to be better able to disseminate information. In fact, the states that are fully authorized to handle the RCRA program (having shown their regulations and enforcement are "substantially equivalent" to U.S. EPA's), performed the most poorly as a group in handling information requests (see table 10-7).

Michigan, which as of this writing is not authorized to enforce any federal waste laws, for example, has one of the strictest victim's compensation laws in the country and the state communicated well. Utah's RCRA program, on the other hand, has been granted final authorization. The state did not answer one CEP request, and responded to two others by saying it had inadequate staff to provide any information. CEP inquired about only one facility in Utah.

Hotlines and Tangled Lines

U.S. EPA's RCRA/Superfund toll-free hotline (1-800-424-9346) is often busy, a fact that reveals the sizeable industrial and public demand for information. State hotlines will be increasingly useful as regulation and information fall more and more into the hands of state agencies. Each state should at least have a "waste hotline" number, even if it can only direct inquiries to the proper personnel.

Because state agencies responsible for ground water enforcement are often different from those responsible for the RCRA program, "in many cases, the roles and responsibilities of the various agencies involved are not clearly defined, which allows many critical activities to 'slip through the cracks' " in the words of U.S. EPA's "Interim Status Ground Water Monitoring Implementation Study."[27] California, New Jersey, Pennsylvania, Massachusetts, and Washington were highlighted as states with organizational structures that restricted effective implementation. Simply trying to identify the appropriate person to answer a question in these and other states can be time consuming—if not impossible.

CEP found California to pose particularly difficult problems. As U.S. EPA's ground water study points out, the California Department of Health Services (CADOHS) is the key agency in the state, but nine Regional Water Quality Control Boards oversee ground water problems. The study found that the water boards were granting unauthorized exemptions and, in general, not collecting the required RCRA data. CEP telephoned several water boards, and in turn, the boards assured the delivery of the requested material that had been gathered. It was never sent. Similarly, in response to CEP's request for air-quality information, the state Air Resources Board advised contacting the state's six Air Pollution Control Districts.

Worst of all, in response to CEP's request concerning matters that pertained strictly to RCRA, CADOHS stated that CEP should contact the "enforcement coordinator interdepartmental liaison" of 51 county district attorney's offices and county health departments. CADOHS claimed that all of the facilities CEP inquired about were undergoing litigation; therefore, the requested information was confidential and a matter for the district attorneys. A half dozen DAs were unanimous: the CADOHS claim was untrue.

SPEAKING THE SAME LANGUAGE

Why couldn't information-reporting requirements be standardized for all agencies at all levels of government? After William D. Ruckelshaus came in to clean up U.S. EPA, the agency's regional offices began to request Part A of the federal permit application from facilities that had notified they were already

TABLE 10-7. *Responses to CEP Requests*

State	EPA Size Category[a]	EPA Region	Authorization			Response to Requests			No. of Sites Inquired About	Response Quality Score[b]
			Phase I	*Phase II (As of 1/11/85)*	*Final*	*Air (7/23/84)*	*RCRA (7/24/84)*	*Water (8/30/84)*		
CEP Grade A (Excellent/ > or = 3.00)										
Illinois	E	5	5/17/82			EXT	EXT	—	10	3.69
Michigan	E	5				FEE	SUM	EXT	4	3.00
Nevada	A	9	7/19/83	7/19/83[c]		SUM	SUM	SUM	1	3.00
Texas	E	6	12/24/80	9/1/83	12/26/84	FEE	EXT	SUM	9	3.00
CEP Grade B (Above Average/ > or = 2.00)										
Wisconsin	D	5	1/15/82			SUM	—	EXT	2	2.60
Alabama[d]	D	4	2/25/81			SUM	FEE	SUM	2	2.40
New Jersey	D	2		4/6/84[c]		EXT	I M	I M	3	2.40
Oregon	E	10	7/16/81			SUM	I M	SUM	1	2.25
Kentucky	D	4		2/29/84		FEE	EXT	—	1	2.00
Louisiana	D	6	12/19/80	1/24/84		I M	I M	EXT	7	2.00

CEP Grade C (Standard/ > or = 1.00)

Arkansas	B	6	11/19/80	1/24/84	1/25/85	FEE	FEE	FEE	1	1.50
Pennsylvania	E	3	5/26/81			SUM	I M	I M	2	1.40
Connecticut	C	1	4/21/82	6/29/83		SUM	I M	—	1	1.25
Indiana	D	5				SUM	—	I M	2	1.20
Mississippi	B	4	1/7/81	8/31/82	7/27/84	SUM	—	—	1	1.00
North Carolina	D	4	12/18/80	2/7/84	12/31/84	SUM	—	—	1	1.00
Puerto Rico	B	2	10/14/82			SUM	—	—	3	1.00
Tennessee	D	4	7/16/81			SUM	—	—	2	1.00

CEP Grade D (Below Standard/ > or = 0.00)

Massachusetts	D	1	2/25/81	8/3/84		I M	I M	—	1	0.50
Utah	B	8	12/12/80		10/24/84	I M	I M	—	1	0.50
Colorado	C	8			11/2/84	FEE	—	—	2	0.40
Oklahoma	C	6	1/14/81	6/24/83	1/10/85	I M	I M	—	2	0.40
South Carolina	C	4	2/25/81	12/6/83		FEE	—	—	3	0.33
New York	E	2	12/27/83			—	I M	—	2	0.20
California	E	9	6/4/81	1/11/83[e]		DEC	DEC	DEC	14	0.18
Ohio	E	5	7/15/83			—	I M	I M	12	0.13
Idaho	B	10				—	—	—	1	0.00

NOTE 1: The twenty-one states are listed from highest to lowest quality of response.

NOTE 2: EXT (value = 4): state provided extensive response; SUM (value = 3): state provided summary response; FEE (value = 2): state would provide information with fee; IM (value = 1): state unable to respond because of inadequate manpower; DEC (value = 1): state unable to respond because of decentralization; — (value = 0): state did not respond at all.

a. established to estimate federal grants to meet state resource needs; based upon population (40%), amount of waste generation (40%), number of generators (15%), and land area (5%).

b. Number of sites inquired about added twice for every EXT and once for every SUM to sum of response values divided by number of sites inquired about added to number of requests (3).

c. State not authorized to control land disposal.

d. Interim authorization reverted to EPA on 8/1/84.

e. State not authorized to control storage/treatment in surface impoundments.

treating, storing, and disposing of hazardous wastes. An executive order put a stop to this data gathering. As part of the president's Paperwork Reduction Act, the verification process ceased midstream. What's left is a database full of incomplete information. The little it tells, few understand.

Personnel of New York City's Department of Environmental Protection telephoned CEP twice to obtain information about U.S. EPA's databases; they found it easier and faster to deal with CEP than with U.S. EPA. If governmental agencies have trouble exchanging information, how can other organizations and citizen groups be expected to obtain and utilize it? Information is a regulatory agency's most valuable resource. Squandering this precious asset is one of the greatest wastes of government spending.

APPENDIX A

Evaluating a Hazardous Waste Facility

The following is a comprehensive checklist for evaluating hazardous waste management facilities. The list is based on work by The Keystone Siting Process Group published in *The Keystone Siting Process Handbook—A New Approach to Siting Hazardous Waste Management Facilities* (Austin, Texas: Texas Department of Water Resources, 1984).

I. Economic considerations
   A. Is this facility needed?
      1. How was need established?
         a. Local or regional survey
         b. On-site or off-site needs
         c. Is it beneficial to good regional hazardous waste management?
         d. Is there a generic survey of hazardous waste needs for the area? Does the survey provide information that is compatible with this plan?
         e. Is the technology proposed an improvement over that presently available?
         f. Will this facility replace an outmoded or more hazardous one?
         g. What geographical area will it serve?
   B. Profit expectations
      1. High or low risk
      2. Longevity of facility
         a. Expansion anticipated
         b. In what time frame?
      3. Who owns the facility?
      4. Are owners financially backed by others?
      5. Who are the competitors?
   C. Facility operators
      1. Prior experience and operating record
      2. Will company that owns the facility also operate it?
      3. Who will seek the permits?
      4. How can expertise be evaluated if the operator is new to this field?
   D. Economic effects on community
      1. Possible effects on property values
      2. Who will receive any increase in tax base?
      3. How much tax revenue could be generated?

4. Will public costs rise?
   a. Police protection
   b. Fire protection
   c. Road maintenance
   d. Emergency response equipment and facilities

E. Potential for compensation to community
   1. Donated equipment community may need because of facility
   2. Fees to general revenue fund
   3. Property value guarantees
   4. Parks, etc.

F. Closure and postclosure
   1. When is closure anticipated?
   2. Who is responsible for the site after closure?
   3. What assurances will there be that the site will be closed in accordance with the plan?
   4. Financial assurances to establish ability to handle problems after closure
   5. Who certifies that site is properly closed?
   6. How are people protected from unwittingly buying land after closure?
      a. Recorded in deed
      b. What future uses are possible?

II. Function of facility

A. Storage, treatment, or disposal (processes)

B. Wastes to be handled
   1. What wastes will be handled?
      a. In what quantities
      b. Physical and chemical characteristics
         (1) Degree of hazard anticipated
         (2) What makes the waste hazardous?
   2. What wastes will not be handled? Why?
   3. Sources of waste
      a. On-site generation
      b. Off-site generation
         (1) Local
         (2) Regional
         (3) Statewide
         (4) Out of state
         (5) Out of country
      c. Consumer products from which such waste results
   4. Will nonmanifested wastes be accepted?
   5. Where will waste go if not handled at this site?

C. Does this facility fit into an integrated hazardous waste management system (reduction, recovery, recycling, sale or exchange, storage, treatment, disposal)?
   1. On site
   2. Regional

D. Is this facility part of a master plan to provide hazardous waste management?
   1. Whose plan?
   2. How does it fit into the plan?
   3. Geographical area served by plan

- E. Plans for future expansion
  1. Additional facilities
  2. Additional types of facilities
  3. Time frame anticipated

III. Technology used—general questions
- A. Why was this technology chosen?
  1. Are others available?
  2. Can wastes be recycled, sold, exchanged or treated to avoid disposing of as hazardous waste?
  3. Engineering design and operating techniques to compensate for any site deficiencies
- B. Quality assurance and control
  1. In identifying wastes
     - a. Role of generator
     - b. Role of facility
  2. Plans for lab work
  3. How are out-of-spec wastes handled?
  4. What happens to rejected wastes?
- C. Reliability of technology
  1. Past experiences with it
  2. Any serious environmental impacts
  3. How was it tested to assure long-term safety and effectiveness?
- D. Sequence of technology used from arrival of wastes to end process at facility
  1. Analysis of waste
  2. Unloading
  3. Storage
  4. Treatment
  5. Disposal
  6. Any residuals requiring further handling
  7. Monitoring
  8. Closure
  9. Postclosure

IV. Technologies to be used—specific questions
- A. Recovery or recycling
  1. How will it be accomplished?
  2. Plans for energy conversion
  3. Will supplemental fuel be needed? What type?
  4. Reliability of waste characteristics
  5. Long-term demand
- B. Storage
  1. How is waste stored?
  2. Length of time in storage
  3. Where does the waste go next?
- C. Incineration
  1. How complete will destruction be?
  2. Will supplemental fuel be needed? What type?
  3. Air quality protection
  4. Anticipated air quality impact

5. What monitoring will be done?
6. What are procedures in case of an upset?

D. Treatment
1. What type of treatment will be used?
2. What type of wastes will be treated?
3. How completely will waste be rendered nonhazardous?
4. What will happen to the treated waste?

E. Land disposal
1. Types
    a. Surface impoundment
    b. Land application or treatment
    c. Landfill (burial)
    d. Other—specify
2. Technical processes preceding land disposal
    a. Treatment, stabilization
    b. Segregation of noncompatible wastes
    c. Handling of containerized waste
3. Can waste be disposed of in another fashion?
    a. Technological limitations
    b. Cost factors
    c. Other reasons
4. Technology to protect environment
5. Closure plans
    a. Interim partial closure of each cell
    b. Final closure of full facility
6. Postclosure plans
    a. Periodic monitoring and maintenance
    b. Postclosure period
    c. Financing and cost assurance
    d. Responsibilities of facility operator, land owner, local and state units of government

F. Deep-well injection
1. Well construction
    a. Depth
    b. Casings
    c. Monitoring equipment
    d. Local faults
    e. Previous wells in area
2. Pretreatment
    a. Sediment removal
    b. Waste compatibility
3. Processes to assure environmental protection
4. Closure technology
5. Technology available if remedial action is needed

V. Site Characteristics

A. How are site characteristics determined?
1. What is done in a geotechnical investigation?
2. Other assessment techniques

B. Characteristics to be considered
   1. Site geology
   2. Hydrology
   3. Topography
   4. Soil properties
   5. Aquifer location
      a. Relationship to water table
      b. Wells presently in area
      c. Flow rate and direction of ground water flow
      d. Ground water quality
      e. Does aquifer connect with others?
      f. Aquifer recharge area
   6. Climatic conditions
      a. Normal
      b. Potential for natural disasters
   7. Is site in or near environmentally sensitive area?
      a. Wetlands
      b. Shoreline
      c. Flood-prone area
      d. Aquifer recharge zone
      e. Critical habitat of an endangered species
      f. Hurricane storm surge area
      g. Prime agricultural area
      h. Other
   8. Subsidence problems
   9. Proximity to residences, schools, etc.
   10. Evacuation routes in area
   11. Current character of surrounding area
   12. Zoning of site and areas nearby
   13. What plans currently exist for site and area?
   14. What transportation routes will be used?

C. Why was this site chosen?
   1. Were others considered?
   2. Still under consideration
   3. Why were others rejected?

VI. Environmental quality

A. Surface drainage
   1. Is the site in a flood plain?
      a. Which one? By whose standards?
         (1) How current are maps used to make determination?
      b. What is the elevation of the land?
      c. Dikes planned
         (1) Internal
         (2) Perimeter
      d. Diking required or desired
         (1) Height
         (2) How protected from erosion

(3) Design storm used
(4) Access to site over or through dikes

2. Stormwater management
   a. How will it be controlled?
   b. Treatment/discharges
   c. Effect on receiving body of water
   d. Will residuals remain (sludge management plans)?
   e. Design storm used
3. Hurricane vulnerability
   a. Is site in an area subject to storm surge?
   b. Design storm specifications
   c. Damage from wave action possible
   d. For what levels of wind speed is facility designed?

B. Ground water protection
1. Ground water resources
   a. Are aquifers used for drinking water?
      (1) Now or possibly in future
      (2) Location of known wells
   b. Other uses of ground water now or in future
   c. Proximity to surface water
2. Soil
   a. Physical characteristics, including permeability
   b. Chemical characteristics, including compatibility with wastes to be handled
3. Leachate collection
   a. How will leachate be collected?
   b. How will leachate be treated?
   c. For how long will leachate be collected? Treated?
4. Liners
   a. What is required? What is desired?
   b. What areas of the facility will be lined?
   c. Integrity of liner
      (1) Type: clay or synthetic
      (2) Thickness
      (3) How constructed
      (4) Compatibility with wastes to be handled—how tested
   d. Remedial action possible
5. Caps
   a. Same questions as liners
   b. Erosion control
   c. Prevention of water standing on site, correction of settlement
   d. Revegetation planned, postclosure maintenance
6. Deep-well injection has additional concerns
   a. Relationship of aquifers to injection zone
   b. Compatibility of waste with area geology
   c. Remedial action possible
   d. Limitations of future use of land for mining, etc.

C. Air emissions

1. What protection is afforded from which contaminants?
2. Potential for unregulated emissions
3. Odor-control plans
4. Who will be affected by emissions?
    a. Direction of prevailing winds
    b. Frequency of "bad" air conditions
5. Control of vapors at various stages of process

VII. Transportation
  A. Mode of transportation now and in future
    1. Truck
    2. Rail
    3. Barge
    4. Other possibilities
  B. Containmnent of waste during transport
    1. Type of container
      a. Bulk
      b. Drums
      c. Other
    2. Protection against leakage
      a. Compatibility of wastes with packaging
      b. Reliability record of container
    3. Labeling of containers
  C. Who is responsible for transport?
    1. If the company is responsible, what is the company's record?
    2. Training of drivers
      a. Safe driving skills
      b. Emergency response
    3. Manifest system
    4. Labeling of trucks
  D. Timing of arrivals
    1. Days
    2. Hours
  E. Routing
    1. Routes to be used
    2. Any restrictions existing
      a. Who imposed them?
      b. Who enforces them?
      c. Can penalties be assessed on offenders?
    3. Effects on area traffic
    4. Effects on area road conditions
  F. Spill response
    1. Whose responsibility?
    2. Cleanup techniques
    3. Who pays for cleanup?

VIII. Operations
  A. What actions will be taken when there are operating problems?
    1. Back-up systems planned

   2. Start-ups and shutdowns
      a. Effects on permitted emissions
      b. Frequency/longevity anticipated
B. Emergency response
   1. What is included in the contingency plan?
   2. How will fire protection be provided?
      a. On-site equipment
      b. Mutual aid agreements
      c. Alarm systems
C. Site security
   1. Controlled entry
   2. Fencing
   3. Warning signs
   4. Surveillance systems
D. Personnel
   1. Experience required
   2. Technical levels and working experience of staff
   3. Training plans
   4. Participation in quality assurance and safety programs

IX. Enforcement
A. Regulations
   1. Which apply to this facility and site?
   2. Permits needed
      a. Who grants each?
      b. Public participation opportunities
   3. Contents of permit applications
   4. Penalties for noncompliance
B. Monitoring to ensure environmental protection
   1. Techniques used
   2. Equipment available
   3. Frequency
   4. Any citizen review planned—for example, Operations Review Committee
   5. Will communities be informed when a sudden release occurs? How?
C. Who is responsible for enforcement?
   1. Owner/operator—self-reporting system
   2. Local level
      a. City
      b. County
      c. Regional or special district
   3. State level
   4. Federal level
D. Sequence and timing of possible enforcement actions
   1. Corrective measures prescribed
   2. Penalties assessed
   3. Litigation
E. What is the government's capability to ensure compliance?

APPENDIX B

Federal Hazardous Waste Databases

Various agencies of the United States government operate computerized data systems that maintain information on hazardous waste sites. Databases exist that identify locations that have been contaminated by the mismanagement of hazardous wastes, and facilities that currently generate, transport, treat, and dispose of hazardous wastes. The data range from the type and seriousness of damage caused by an uncontrolled dump, to the technologies used to process every type and volume of hazardous waste handled by an operating facility. Certain data systems record the number of inspections, violations, and corrective actions occurring at a site; others record the types and value of a site's insurance for sudden accidents or ground water contamination. And still others characterize the demographic and hydrogeological environment at a site, or even the details of a transportation accident resulting in a spill of hazardous waste.

U.S. EPA operates the following databases either at Research Triangle Park in Durham, North Carolina; at the ten regional U.S. EPA offices; or by using contractors for programs authorized under RCRA and CERCLA.

I. RCRA Databases
   A. "Hazardous Waste Data Management System" (HWDMS): HWDMS is EPA's most comprehensive hazardous waste data system, containing a wide variety of information such as:
      1. Notification: Lists all generators, transporters, treatment, storage, and disposal facilities (TSDFs) that notified EPA of their activities (approximately 50,000 entries). Includes: names, addresses, facility type, waste types handled.
      2. Part A: Lists all TSDFs granted interim status for operating (approximately 5000 to 10,000 entries). Includes: names, addresses, latitude, longitude, existence date, closure date, ownership, operator, processes, capacities, waste types, estimated quantities.
      3. Inspections (a regional database): Lists all inspections of RCRA facilities. Includes: inspection date, type, status, responsible agencies (certain regions include regulation violated).
      4. Enforcement (a regional database): Lists all enforcement actions taken against RCRA facilities. Includes: enforcement type, status, violation type, penalties.

5. Financial Instruments: Lists financial instruments implemented at all RCRA facilities pursuant to subpart G and H (sudden and non-sudden occurrences insurance, closure and postclosure assurance). Includes: instrument type, data, issuing institution, value, expiration date.
6. Major facility status sheets: Outlines adequacy of all major handlers' (approximately 10,000 entries) ground water monitoring programs and financial instruments.
7. Permit Development Status: Includes process capacity applications, dates, and comments relevant to facility management plan, exposure information, withdrawal requests, permit determinations and modifications, public notice, monitoring program, PA/SI, remedial investigations, and corrective measures.

B. "Biennial Reports": Many state governments are in the process of computerizing data reported through the manifest system that documents the actual quantities of waste generated, transported, treated, stored, and disposed. States are required to send aggregate information to EPA regional offices, which in turn send aggregate data to EPA headquarters. The "Biennial Reports" system is based on this aggregated data.

C. "RCRA Enforcement Management System": This system is based on two survey forms filled out by regional EPA offices concerning inspection and enforcement data. Although only in aggregate form, EPA claims the data is more accurate than that in HWDMS.

D. "National Survey": EPA hired Westat, Inc. to survey hazardous waste activities during 1981. The data results are computerized and more extensive than any other database regarding operating sites. The information includes: distance to nearest faults and other geological data, information regarding quantities of waste actually handled, number of workers, pay scales for employees, detailed process data, and a wide variety of other information. It is, however, only a survey and does not include as large a universe as the other EPA systems (approximately 11,000 entries).

E. "Surface Impoundment Assessment": Contains extensive information about hazardous pits, ponds, lagoons (approximately 200,000 entries).

F. "Open Dumps": Contains information on all solid waste facilities that are not in compliance with RCRA (approximately 2100 entries) including type of damage.

G. "Risk/Cost Assessment": A complex computerized model that estimates risks and costs associated with any facility for which one can define certain parameters, including ZIP code.

H. "Disposition of CERCLA Wastes Data Base": Lists the RCRA sites that have received CERCLA wastes and the CERCLA sites that sent the waste (approximately 100 RCRA entries).

II. CERCLA Databases

A. "Emergency and Remedial Response Information System" (ERRIS): Includes name, location, description of site and hazard, and status and type of inspection and corrective action (approximately 20,000 entries).

B. "Listing of Sites Scored with Hazard Ranking System": Contains hazard scores for total migration, ground water, surface water, and air migration, fire and explosion, and direct contact (approximately 1500 to 2000 entries, including all Superfund sites).

C. Preliminary Tracking System (PTS): Tracks the nature and progress of planned and ongoing remedial action projects and activities, as well as actual performance schedules and expenditures.

D. "Listing of Potentially Responsible Parties": Lists all identified parties that are suspected to have either owned, operated, shipped, carried, generated, or treated wastes disposed of at all Superfund sites.

E. "Remedial Investigations/Feasibility Studies": Tracks status of cleanup measures taken at Superfund sites (database is reportedly under development).

Other federal agencies, including the Departments of Health and Human Services, Labor, and Transportation, also maintain information systems relating to hazardous wastes. These include:

III. Department of Health and Human Services

"Agency for Toxic Substances and Disease Control": Lists all sites that state governments have closed or restricted to public use because of severe contamination by toxic substances (approximately 1500 entries).

IV. Department of Labor

A. "OSHA Inspection History Listing": Includes all inspections conducted by U.S. OSHA from 1972 to the present and identifies type of inspection, regulation violated, hazardous substance involved, injuries, deaths, penalties, and abatement.

V. Department of Transportation

A. "Hazardous Materials Information System" (HMIS): Lists all transportation incidents involving hazardous materials from 1977 to the present, identifying location, type of incident, material, shipper, carrier, amount spilled, injuries, deaths, and penalties.

B. "U.S. Coast Guard National Response Center Data System": Lists all self-reported spills of hazardous materials identifying location, cause of spill, material, carrier, amount, injuries, deaths, and remediation measures.

Notes

Chapter 3

1. Executive Scientific Panel on the Health Aspects of the Disposal of Waste Chemicals, Universities Associated for Research and Education in Pathology, Inc., "Final Report" (1985).
2. For an excellent discussion of state hazardous waste programs, see David J. Lennett and Linda E. Greer, "State Regulation of Hazardous Waste" in *Ecology Law Journal,* vol. 12, no. 2 (Berkeley, CA: University of California School of Law, 1985).
3. See Wendy Gordon and Jane Bloom, "Deeper Problems: Limits to Underground Injection As A Hazardous Waste Disposal Method" (New York: Natural Resources Defense Council, 1985) and U.S. EPA, "Report to Congress on Injection of Hazardous Waste," EPA 570/9-85-003 (May 1985).
4. See Toxic Waste Assessment Group, Governor's Office of Appropriate Technology, *Alternatives to the Land Disposal of Hazardous Wastes, An Assessment for California* (1981).
5. U.S. EPA "Prohibition on the Disposal of Bulk Liquid Hazardous Waste in Landfills—Statutory Interpretive Guidance" (draft 3/19/85).
6. U.S. EPA "Draft Guidance on Implementation of Minimum Technology Requirements of HSWA of 1984, Respecting Liners and Leachate Collection Systems; Reauthorization Statutory Interpretation #5d" (1/31/85 and 5/24/85); U.S. EPA, "Draft Minimum Technology Guidance on Double Liner Systems for Landfills and Surface Impoundments—Design, Construction, and Operation" (12/19/84 and 5/24/85); and U.S. EPA "Draft Minimum Technology Guidance on Single Liner Systems for Landfills, Surface Impoundments, and Waste Piles—Design, Construction, and Operation" (July 1982 and 5/24/85).
7. Telephone interviews with Don White and Casey Husvedt, U.S. EPA (4/3/85).
8. U.S. EPA "Permit Writers' Guidance Manual for Hazardous Waste Land Storage and Disposal Facilities. Phase I Criteria for Location Acceptability and Existing Applicable Regulations—Final Draft for Public Comment" (October 1984).
9. U.S. EPA "Permit Applicants' Guidance Manual for Exposure Information Requirements Under RCRA Section 3019—Final" (7/3/85).
10. Jack McGraw, "Reauthorization Statutory Interpretation: Guidance on Corrective Action for Continuing Releases (1/30/85).
11. William N. Hedeman, Jr., to regional administrators, "Requirements for Selecting an Off-Site Option in a Superfund Response Action" (January 28, 1983).
12. Jack McGraw to regional administrators, "Procedures for Planning and Implementing Off-site Response Actions" (May 6, 1985).
13. U.S. EPA "The Extent of the Hazardous Release Problems and Future Funding

Needs," CERCLA section 301(a)(1)(C) Study (Washington, DC: December 1984), p. 4–2.

14. U.S. EPA "Environmental Auditing Policy Statement—Revised Draft" (4/11/85).
15. See "Proposed Guidelines for Carcinogen Risk Assessment; Request for Comments," "Proposed Guidelines for Exposure Assessment; Request for Comments," "Proposed Guidelines for Mutagenicity Risk Assessment; Request for Comments," and "Proposed Guidelines for the Health Assessment of Suspect Developmental Toxicants and Request for Comments," in *Federal Register,* vol. 49 (9/23/84); James L. Regens et al., "Risk Assessment in the Policy-Making Process: Environmental Health and Safety Protection" in *PAR,* vol. 43, no. 2 (March/April 1983); and Andrew P. Sage and Elbert B. White, "Methodologies for Risk and Hazard Assessment: A Survey and Status Report" in *IEEE Transactions on Systems, Man, and Cybernetics,* vol. 10 (August 1980).
16. Jack McGraw, memo to regional administrators (2/13/85).
17. B. Goldman, J. Hulme, S. Ross, A. Tepper Marlin, and B. Wieghart, CEP meeting with S. Caldwell, M. Cook, W. Kovalick, S. Lowrance, and H. Snyder, U.S. EPA (3/7/85).
18. From fiscal year 1982 to fiscal year 1985, U.S. EPA made final permit determinations on 1471 out of a total 5037 estimated TSDFs. See Bruce R. Weddle, U.S. EPA "Summary Report on RCRA Activities" (June 1985).
19. *Hazardous Waste News* (Silver Spring, MD: Business Publishers, Inc., 4/15/85).
20. *Inside EPA* (4/5/85).

Chapter 4

1. Booz–Allen & Hamilton, Inc., *Hazardous Waste Generation and Commercial Hazardous Waste Management Capacity—An Assessment,* SW-894 (Washington, DC: U.S. EPA, 1980). See also Booz–Allen & Hamilton, Inc., "Review of Activities of Major Firms in the Commercial Hazardous Waste Mangement Industry: 1982 Update" (Washington, DC: U.S. EPA, 1983).
2. Ibid.
3. Ibid.
4. Ibid.
5. Westat, Inc., *National Survey of Hazardous Waste Generators and Treatment, Storage, and Disposal Facilities Regulated Under RCRA in 1981* (Washington, DC: U.S. EPA, 1984).
6. Within a 95 percent statistical confidence level, U.S. EPA's figure could be as high as 395 million metric tons (MMT) or as low as 132 MMT. Other estimates of annual waste generation are also within range of U.S. EPA's figure. The CBO has estimated that the 70 major waste-producing industries generated some 266 MMT during 1983 (223 MMT to 308 MMT with 95 percent confidence). CBO used a mathematical model to arrive at its estimate and employed a broader definition of hazardous waste than that established under RCRA, so its figures are not directly comparable with U.S. EPA's. Congress's OTA derived an estimate of 255 MMT to 275 MMT for 1981 from a survey conducted by the Association of State and Territorial Solid Waste Management Officials. See CBO, *Hazardous Waste Management: Recent Changes and Policy Alternatives* (1985) and OTA, *Technologies and Management Strategies for Hazardous Waste Control* (1983).
7. Telephone interview with Robert Axelrad, U.S. EPA (1/17/85).
8. Fred C. Hart Associates, *Assessment of Hazardous Waste Mismanagement Damage Case Histories* (Washington, DC: U.S. EPA, 1984).
9. *Chemical Week* (2/6/85).
10. See Chemical Waste Management, Inc., "National Accounts Bulletin" (9/18/84), *Hazardous Waste Report* (8/20/84), *New York Times* (11/1/84), and *Wall Street Journal* (5/29/84 and 9/13/84).
11. *Washington Post* (7/19/84).
12. Lee M. Thomas, "Statement Attachments" (2/22/85).

13. William Sanjour, "OTA Internal Memorandum: Where Have All the Superfund Wastes Gone?" (5/17/84).
14. Telephone interview with David Fege, U.S. EPA (4/2/85).
15. U.S. EPA "Disposition of CERCLA Wastes Data Base" (10/17/84).
16. Lee M. Thomas, "Internal Memoranda to Regional Administrators Regarding RCRA Major Facility Status Sheets" (August 1984).
17. Subcommittee on Oversight and Investigations of the Committee on Energy and Commerce of the United States House of Representatives, "Ground Water Monitoring Survey" (Washington, DC: April 1985).
18. U.S. EPA, "CERCLA Section 301 (a)(1)(c) Study: Extent of the Hazardous Release Problem and Future Funding Needs" (December 1984).
19. Ibid.
20. *Cincinnati Enquirer* (2/27/83).
21. Walter C. Barber to Alice Tepper Marlin, et al. (October 1985).
22. Karaganis, Gail & White, Ltd., "Report to the Board of Directors of Waste Management, Inc." (9/22/83).
23. U.S. EPA, "Interim Status Ground Water Monitoring Implementation Study—Phase III—Draft" (Washington, DC: U.S. EPA, 1984).
24. See Geophysics Study Forum, National Research Council, *Groundwater Contamination* (Washington, DC: National Academy Press, 1984); Office of Technology Assessment, *Technologies and Management Strategies for Hazardous Waste Control,* OTA-M-196 (1983); U.S. General Accounting Office, "Hazardous Waste Disposal Methods: Major Problems with Their Use," CED-81-21 (November 19, 1980); U.S. General Accounting Office, "Interim Report on Inspection, Enforcement, And Permitting Activities At Hazardous Waste Facilities," GAO/RCED-83-241 (September 21, 1983), U.S. General Accounting Office, "Federal and State Efforts to Protect Ground Water," GAO/RCED-84-80 (February 21, 1984); and U.S. General Accounting Office, "Inspection, Enforcement, And Permitting Activities At New Jersey and Tennessee Hazardous Waste Facilities," GAO/RCED-84-7 (June 22, 1984).
25. Philip Shabfcoff, "Toxic Leak Feared from Niagara Site" in *New York Times* (10/29/85).
26. U.S. EPA, untitled draft report on the CECOS International, Inc. Facility in Niagara Falls, New York (October, 1984).
27. U.S. EPA "Disposition of CERCLA Wastes Data Base" (10/17/84).
28. William Sanjour, "OTA Internal Memorandum: Where Have All the Superfund Wastes Gone?" (5/17/84).
29. *Hazardous Waste News* (4/30/84).
30. Michael Wines, "EPA Apparently Engaged in Toxic Shell Game, May Sidestep Policy on Toxic Dump Sites" in *Los Angeles Times* (1/11/85).

Chapter 5

1. See Committee on Institutional Considerations in Reducing the Generation of Hazardous Industrial Wastes, Environmental Studies Board, Commission on Physical Sciences, Mathematics, and Resources, National Research Council, National Academy of Sciences, *Reducing Hazardous Waste Generation, An Evaluation and a Call for Action* (Washington, DC: National Academy Press, 1985). The first in-depth study of on-site waste reduction and reuse practices has been conducted by the nonprofit organization Inform of New Jersey plants. Contact David Sarokin at Inform in New York, NY.
2. Chemical Manufacturers Association, "Five-Plant Study" (1982).
3. Office of Technology Assessment, *Technologies and Management Strategies for Hazardous Waste Control,* OTA-M-196 (1983).
4. Contact Environmental Information, Ltd. in Minneapolis, MN, or the Hazardous Waste Treatment Council in Washington, DC, for more information on the companies involved.

5. U.S. EPA, Office of Water and Waste Management, *Waste Exchanges—Background Information,* SW-887.1 (1980).
6. Ibid.
7. Laurie A. Rich, "Waste Exchanges—An Underutilized Way to Help the Bottom Line" in *Chemical Week,* (5/16/84).
8. U.S. EPA, "Surface Impoundment Assessment—National Report" (Washington, DC: December 1984).

Chapter 6

1. For examples of other release ranking systems see Environ Corp., "Review and Analysis of Hazard Ranking Schemes, Final Report. Prepared for the Studies and Methods Branch, Office of Solid Waste, Environmental Protection Agency" (May 11, 1984); Harriet L. Greenwood, Douglas J. Bach, and C. Bruce Bartley, "Evaluation of Hazardous Waste Sites: The Federal HRS Model and State-Initiated Ranking Systems" (Environmental Research Group, Inc., 1984); JRB Associates, Inc., "Methodology for Rating the Hazard Potential of Waste Disposal Sites" (McLean, VA: JRB Associates, 1980); and "Site Assessment System (SAS) For the Michigan Priority Ranking System Under the Michigan Environmental Response Act" (Act 307, P.A. 1982). Other states with ranking systems include: Alaska, California, Louisiana, Maryland, Rhode Island, and Washington.
2. The average cost of a RI/FS is $800,000. See U.S. EPA, "The Extent of the Hazardous Release Problems and Future Funding Needs," CERCLA section 301(a)(1)(c) Study (Washington, DC: December 1984), pp. 4–5, and U.S. EPA, "National Oil and Hazardous Substances Pollution Contingency Plan; Proposed Rule (40 CFR Part 300)" (Washington, DC: *Federal Register,* February 12, 1985). See also CH2M Hill, Inc., and JRB Associates, Inc., "Guidance Documents for Remedial Investigation and Feasibility Studies" (Washington, DC: U.S. EPA, October 1984) and U.S. EPA, *Modeling Remedial Actions at Uncontrolled Hazardous Waste Sites* (Cincinnati, OH: U.S. EPA, April 1985).
3. Lee M. Thomas, statement before the Subcommittee on Commerce, Transportation, and Tourism of the House Committee on Energy and Commerce (March 7, 1985).
4. 42 USC 9604 section 104 (c)(1)(iii).
5. Harry Seraydarian, memo to Lee M. Thomas, "Hazard Ranking System Score for BKK Sanitary Landfill in West Covina, California" (June 28, 1984).
6. Releases of methane gas would not be considered an observed release to the air using HRS; rather, such releases would be factored as fire and explosion hazards.
7. See U.S. EPA, "The Extent of the Hazardous Release Problems and Future Funding Needs," CERCLA section 301(a)(1)(c) Study (Washington, DC: December 1984).
8. U.S. EPA, "Comments on the Hazard Ranking System and the National Priorities List" in *Federal Register,* volume 47, number 137 (July 16, 1982). See also the excellent discussion in Harris et al., "Comparative Risk Assessment: Tools for Remedial Action Planning" in *Hazardous Waste,* volume 1, number 1, (1984) pp. 19–33.
9. 40 CFR part 300, appendix A.
10. Ibid.
11. Ibid.
12. Telephone interview with Steve Caldwell, U.S. EPA (5/18/85).
13. Martin T. Katzman, "Chemical Catastrophes: Regulating Environmental Risk Through Pollution Liability Insurance. A Report to the Huebner Foundation for Insurance Education" (July 1984), pp. 5–20.
14. See *Wall Street Journal* (3/19/85); and statements of William O. Bailey, president, Aetna Life Insurance and George K. Bernstein on behalf of the American Insurance Association before the Committee on Environment and Public Works of the United States Senate (4/3/85).
15. For an excellent discussion of HAZAN and HAZOP, see Trevor A. Kletz, "Eliminating Potential Process Hazards" in

Chemical Engineering (4/1/85), pp. 48–68.
16. Ibid.
17. ICF, Inc., *The RCRA Risk-Cost Analysis Model Phase III Report* (Washington, DC: U.S. EPA, March 1, 1984). For an excellent review of the many other available analytical models, see Environ Corp., "Review and Analysis of Hazard Ranking Schemes, Final Report. Prepared for the Studies and Methods Branch, Office of Solid Waste, Environmental Protection Agency" (May 11, 1984). Also see Princeton University Water Resources Program, *Ground Water Contamination From Hazardous Wastes* (Englewood Cliffs, NJ: Prentice Hall, 1984); Jeffrey M. Bass, "A Comparison of Conjoint Analysis and Decision Analysis," (Cambridge, MA: Arthur D. Little, Inc., July 9, 1982); and J. M. Bass to B. Goldman (5/21/84).
18. Eric Maláes, OSW, memorandum to the Environmental Engineering Committee of the Science Advisory Board, "Use of OSW's RCRA Risk-Cost Analysis (WET) Model" (7/12/84).
19. *Environmental Reporter* (10/1/82), p. 743.
20. Ibid.
21. *Inside EPA* (10/1/82), p. 11 and Robert A. Michaels and Benjamin A. Goldman, "RCRA Must be Strengthened" (New York: CEP, 3/15/84).
22. ICF, Inc., *The RCRA Risk-Cost Analysis Model Phase III Report* (Washington, DC: U.S. EPA, 3/1/84), pp. 1–16.
23. Telephone interview with Susan Bromm, U.S. EPA (5/21/85) and U.S. EPA Science Advisory Board Environmental Engineering Committee, transcript of hearing (5/15/84).
24. Telephone interview with Eric Maláes (4/9/85).
25. Anne Sprightley Ryan, Massachusetts Hazardous Waste Facility Site Safety Council, "Approaches to Hazardous Waste Facility Siting in the United States" (September 1984). For examples of siting approaches, see Clark–McGlennon Associates, Inc., "A Decision Guide for Siting Acceptable Hazardous Waste Facilities in New England" (Boston, MA: New England Regional Commission, November 1980); New Jersey Department of Environmental Protection, "Basis and Background Document for New Jersey Major Facility Commercial Hazardous Waste Facility Siting Criteria" (February 12, 1983); and The Keystone Siting Process Group, "The Keystone Siting Process Handbook—a New Approach to Siting Hazardous Waste Management Facilities" (Austin, TX: Texas Department of Water Resources, January 1984).
26. Environmental Resource Management, Inc., "Technical Criteria for Identification and Screening of Sites for Hazardous Waste Facilities" (West Chester, PA: Delaware River Basin Commission and New Jersey Department of Environmental Protection, August 1981).

Chapter 7

1. Congressional Budget Office, *Hazardous Waste Management: Recent Change and Policy Alternatives* (1985), p. 28.
2. Ibid., p. 5.
3. Environmental Information, Ltd., *Industrial and Hazardous Waste Management Firms, 1985* (Minneapolis, MN, 1984) and U.S. EPA, "Summary Report on RCRA Activities" (May 1985).
4. The parameters that are ratios have been rounded to the nearest tenth, and the rankings are based on the rounded tenth.
5. U.S. EPA Regional Hazardous Waste Data Management System (HWDMS) printout received in response to CEP's FOIA request of 10/22/84. See appendix B.
6. U.S. EPA Regional (HWDMS printout received in response to CEP's FOIA request of 7/9/84. See appendix B.
7. U.S. EPA, "National Priorities List" (1985) and telephone interviews with Mike Nalpinski, U.S. EPA Region 3; Mary Villarreal and Tim O'Mara, U.S. EPA Region 5; Alice Furst, U.S. EPA Region 7; Randy Landin, U.S. EPA Region 8; John Hazen and Frank Richardson, New Jersey Department of Environmental Protection; and Jim Burkett-Bauer, Wisconsin

Department of Natural Resources (July 29–30, 1985).

8. Citizens Clearing House for Hazardous Wastes, Environmental Action Foundation and National Campaign Against Toxic Hazards.
9. U.S. EPA Regional HWDMS printout received in response to CEP's FOIA request of 10/22/84. See appendix B.
10. Ibid.
11. U.S. Department of Labor, "OSHA Inspection History Listing" (July 20, 1984). See appendix B.
12. Ibid.
13. Department of Transportation, "Hazardous Materials Information System" (7/11/84) and U.S. Coast Guard to Benjamin A. Goldman (10/24/84). See appendix B.
14. U.S. EPA, "Major Facility Status Sheets" received in response to CEP's FOIA request of 2/27/85.
15. Ibid. and Subcommittee on Oversight and Investigations, Committee on Energy and Commerce, U.S. Congress, "Ground Water Monitoring Survey" (Washington, DC: Government Printing Office, April 1985).
16. U.S. EPA and state documents and company responses to CEP inquiries.
17. U.S. EPA Regional HWDMS printout received in response to CEP's FOIA request of 7/9/84 and Environmental Information, Ltd. (1985).
18. Ibid.
19. CEP search using LEXIS, a product of Mead Data Central Corp.
20. See appendix B and chapter 7 footnotes 9, 11, 14, 16, 18, and 19.
21. Estimate of AMEC's 1984 hazardous waste revenue based on "Comparison of 1984 to 1983," p. 17 in American Ecology, Inc., "10-K Report to the Securities and Exchange Commission" (1984).
22. AMEC, "10-K Report," financial statement, note 11, p. 34.
23. U.S. EPA "National Priorities List Proposed Update #2," (October 1984).
24. Telephone interview with Joseph Boyle, U.S. EPA Region 5 (11/15/85).
25. Marvin Resnikoff, "U.S. Radioactive Landfill Experience," paper presented at the Annual Conference of the Institute of British Geographers, (Leeds, England: January 7–10, 1985).
26. Resnikoff, p. 6.
27. J. J. Scoville, president, U.S. Ecology, letter to B. Goldman (10/24/85) and "Agreement between Commonwealth of Kentucky Executive Department for Finance and Administration and Nuclear Engineering Company, Inc." (May 22, 1978).
28. AMEC, "10-K Report," p. 10.
29. Ibid., p. 1.
30. Ibid., part II, item 7, p. 16; p. 17; and financial statement, note 10.
31. Ibid., pp. 2, 9–10, and 19.
32. Ibid., p. 11.
33. Ibid., p. 16.
34. Ibid., financial statement, note 10, p. 32.
35. Resnikoff, p. 6.
36. Scoville to Goldman (10/25/85).
37. Ibid.
38. AMEC, "10-K Report," financial statement note 4 and p. 19.
39. Ibid., p. 13.
40. Telephone interview with Joseph Boyle, U.S. EPA Region 5 (11/15/85).
41. AMEC, "10-K Report," p. 14.
42. Scoville to Goldman (10/25/85).
43. "The 100 Largest Diversified Service Companies," in *Fortune* (June 10, 1985), pp. 176–179.
44. Standard & Poor's, "Standard NYSE Stock Reports—Browning–Ferris, Inc." (June 7, 1985).
45. *Moody's Ratings* (Summer 1985).
46. BFI, "10-K Report to the Securities and Exchange Commission" (1984) p. 11.
47. Ibid., p. 8.
48. Ibid., p. 3.
49. CECOS, "Fact Sheets" #4, #6, and #7.
50. BFI, "Annual Report," (1984).
51. Ibid., p. 3.
52. CECOS, "Fact Sheet #10."
53. CECOS, "Contact Company Bulletin."
54. BFI, "10-K Report," p. 3.
55. U.S. EPA, "Hazardous Waste Data Management System" (1984).
56. BFI, "10-K Report," pp. 2–7.
57. Ibid., p. 11.
58. Ibid., p. 10.
59. Ernest Neal, vice president governmental affairs, CECOS International, to B. Goldman (11/1/85).

60. U.S. EPA "National Priorities List Update #1" (September 1984).
61. CEP search using LEXIS, a product of Mead Data Central Corporation.
62. State of Louisiana, Department of Environmental Quality, Environmental Control Commission, "Compliance Order and Schedule issued on Browning–Ferris Industries, Chemical Services, Inc." (2/29/84).
63. *Baton Rouge State Times* (5/12/83).
64. Neal to Goldman (11/1/85) quotation by CECOS from Louisiana Supreme Court decision (April 1984).
65. Kay Hahn, "Investment Research—Environmental Systems Company" (Chicago: The Chicago Corp., March 6, 1985).
66. Jeffrey Klein, "Company Follow Up, Environmental Systems Company" (New York: Kidder, Peabody & Co., March 13, 1985).
67. ENSCO, "10-K Report to the Securities and Exchange Commission" (1984).
68. ENSCO, "RCRA, Part A Application" (12/31/84).
69. State of Minnesota, District Court, First Judicial District, *The County of Scott* v. *Pollution Controls, Inc. and Minnesota Pollution Control Agency,* Notice of Motion (10/9/75) and *State of Minnesota and Minnesota Pollution Control Agency* v. *Pollution Controls, Inc.,* Order (7/29/82).
70. ENSCO, "RCRA, Part A Application" (12/31/84).
71. Telephone interview with Charles L.Robertson, ENSCO (11/11/85).
72. ENSCO, "RCRA, Part A Application" (12/31/84).
73. ENSCO, "10-K Report."
74. Charles Robertson.
75. Interview with Dr. John Brugger, U.S. EPA Region 2 (July 1985) and telephone interview with Robert Pojasek (8/29/85).
76. ENSCO, 10-K Report."
77. Ibid.
78. Jeffrey Klein.
79. Kay Hahn.
80. ENSCO, "10-K Report."
81. Kay Hahn.
82. Charles Robertson.
83. *Hazardous Waste News* (7/1/85).
84. *New York Times* (July 11, 1985).
85. IT, "10-K Report to the Securities and Exchange Commission" (1985).
86. Environmental Information Ltd., *Industrial and Hazardous Waste Management Firms* (Minneapolis: 1985).
87. IT, "10-K Report."
88. *Engineering News-Record* (October 18, 1984).
89. Douglas Augenthaler, "Environmental Control, Equity Research: International Technology Corp." (New York: E. F. Hutton & Co., Inc., August 3, 1984).
90. Patrick McVeigh, "Insight: Equity Briefs" (Boston: Franklin Research and Development, July 1985).
91. Douglas Augenthaler.
92. Toxics Assessment Group, "Nowhere to Go: The Universal Failure of Class I Hazardous Waste Dump Sites in California" (The Environmental Defense Fund: 1985).
93. David L. Bauer, vice president, Environmental Affairs, to B. Goldman (10/31/85).
94. IT, "10-K Report."
95. Supreme Court of Louisiana, *Save Ourselves* v. *The Louisiana Environmental Control Commission and the Louisiana Department of Natural Resources,* No. 83-C-1480 (5/14/84).
96. Commission on Ethics for Public Employees, "Opinion on IT Corporation" (April 22, 1983).
97. IT, "10-K Report."
98. Telephone interview with Donald Frost, ESI (8/12/85).
99. "The 100 Largest Diversified Service Companies," *Fortune* (6/10/85).
100. Environmental Information Ltd., *Industrial and Hazardous Waste Management Firms* (Minneapolis: 1985).
101. CEP estimate based on telephone interview with Donald Frost (8/9/85).
102. *Wall Street Journal* (7/12/85).
103. Environ Corp., "Report Prepared for the City of Oregon OH, on Fondessy Enterprises" (Princeton, NJ: 1984).
104. *Hazardous Waste News* (7/1/85).
105. Ohio Environmental Protection Agency, "Director's Final Findings and Orders In the Matter of Fondessy Enterprises, Inc." (1/10/85).
106. Clement Associates, Inc., "The Potential

for Gradual Environmental Impairment Arising from the Operations of Envirosafe Services of Idaho, Inc., Grandview, Idaho Site B—Final Report Prepared for Conversion Systems" (November 7, 1984).

107. IU, "Annual Report" (1984).
108. IU, "10-K Report to the Securities and Exchange Commission" (1984).
109. Donald Frost.
110. *Directory of Corporate Affiliations* (Skokie, IL: National Register Publishing Co., Inc., 1983).
111. Donald Frost (8/12/85).
112. IU, "10-K Report."
113. Ibid.
114. Ibid.
115. Clement Associates, Inc., p. 17.
116. Environ Corp., "Report prepared for the City of Oregon, OH, on Fondessy Enterprises."
117. RES, Inc., "10-K Report to the Securities and Exchange Commission" (1984).
118. Ibid.
119. John Quarles, Morgan, Lewis & Bockius, counselors at law to Alice Tepper Marlin (10/28/85).
120. RES, "Part A Application EPA I.D. Number TX DO55141378" (August 25, 1981).
121. Quarles to Tepper Marlin (10/28/85).
122. Fred C. Hart, "Damage Incident Summary Form VI-74" in "Assessment of Hazardous Waste Management Damage Case Histories" (Washington, DC: U.S. EPA, 1984).
123. U.S. EPA, "HWDMS Master Facility Listing" (12/19/83) and Quarles to Tepper Marlin (10/28/85).
124. *Wall Street Journal* (8/26/85).
125. Quarles to Tepper Marlin (10/28/85).
126. Superior Court of New Jersey, Law Division, Gloucester County, *Rollins Environmental Services (NJ), Inc.*, v. *Township of Logan,* 199 NJ Super 70 (1985). Appeal pending.
127. Quarles to Tepper Marlin (10/28/85).
128. *Wall Street Journal* (8/7/85).
129. U.S. EPA "ID Number LAD 000778514 Interim Status Permit Application" (8/6/82).
130. Jeffrey Klein, "Company Close-up on Rollins Environmental Services, Inc." (New York: Kidder, Peabody & Co., Inc., 5/13/85).
131. Telephone interview with Thomas Blank, RES (7/29/85).
132. Telephone interview with Robert Pojasek (8/29/85).
133. Thomas Blank and Quarles to Tepper Marlin (10/28/85).
134. Jeffrey Klein (5/13/85), and "Company Comment on Rollins Environmental Services, Inc." (New York: Kidder, Peabody & Co., Inc. 7/25/85 and 8/8/85).
135. USPCI, "10-K Report to the Securities and Exchange Commission" (1984).
136. Ibid.
137. Ibid.
138. Telephone interview with Ken Raymond, OK DOH (8/18/85).
139. USPCI, "Annual Report" (1984).
140. Response to CEP corporate questionnaire. Ken Jackson, vice president, operations, to B. Goldman (3/12/85).
141. USPCI, "10-K Report."
142. Ibid.
143. Ibid.
144. Standard & Poor's Corp., "Standard ASE Stock Reports—Beard Oil" (7/12/84).
145. USPCI, "10-K Report."
146. WMI, "10-K Report to the Securities and Exchange Commission" (1984).
147. Nancy Shute, "The Selling of Waste Management," *The Amicus Journal* (Summer 1985) and Walter C. Barber, vice president of environmental management, CWM to CEP (10/24/85).
148. Moody's Industrial Manual (New York: Moody's Investors Service, 1985) and "10-K Report."
149. U.S. EPA, "Hazardous Waste Data Management Systems" (1984).
150. U.S. EPA, "National Priorities List, Update #1" (September 1984) and "National Priorities List Proposed Update #2" (October 1984).
151. "Alabama: The Nation's Dumping Ground" in *Montgomery Advertiser* (June 1984); CWMI, "National Accounts Bulletin" (5/24/84); and Barber to CEP.
152. WMI, "10-K Report."
153. *Moody's*; WMI, "10-K Report"; and Barber to CEP.
154. Barber to CEP.

155. Standard & Poor's, "Standard NYSE Stock Report—Waste Management" (5/13/85); "The 100 Largest Diversified Service Companies," *Fortune* (6/10/85), and *Moody's Ratings* (Summer 1985).
156. WMI, "10-K Report."
157. *Hazardous Waste News* (4/15/85) p. 120; *Wall Street Journal* (3/5/85); and Barber to CEP.
158. *Chemical Week* (7/31/85) and Barber to CEP.
159. *Chicago Business* (9/19/83).
160. *New York Times* (3/21/83).
161. Chemical Waste Management, Inc., "News Releases" (3/23/83).
162. Karaganis, Gail, & White, Ltd., "Report to the Board of Directors of Waste Management, Inc." (9/22/83).
163. *Wall Street Journal* (10/4/83).
164. WMI, "10-K Report."
165. *Montgomery Advertiser,* (June 1984).
166. *Wall Street Transcript* (11/19/84).
167. Barber to CEP.
168. *New York Guardian* (11/28/84).

Chapter 8

1. B. Goldman, J. Hulme, S. Ross, A. Tepper Marlin, and B. Wieghart, CEP, meeting with S. Caldwell, M. Cook, W. Kovalic, S. Lowrance, and H. Snyder, U.S. EPA (3/7/85).
2. 40 CFR part 300, appendix A.
3. U.S. EPA Science Advisory Board, "Draft Report on Incineration of Hazardous Liquid Waste," (Washington, DC: December 1984).
4. "Comments" in 40 CFR part 300.
5. In an attempt to establish a facility in Devers, Texas, Envirosafe Services of Texas, Inc. (ESTI) may be the first firm to be denied an operating permit on the basis of potential impact to air quality. The hearing examiner's report concludes: "It is the recommendation of this Examiner that ESTI's request for an exemption be denied on the basis that the proposed hazardous waste management facility will make a significant contribution of air contaminants to the environment. It is further the recommendation of this Examiner that there is *insufficient* evidence in the record that the proposed facility will not cause a nuisance." See Hearing Examiner's Report for the Contested Case Hearing Number 173, In the Matter of Envirosafe Services of Texas, Inc. (1985).
6. 40 CFR part 300, appendix A.
7. Telephone interview with Steve Caldwell, U.S. EPA (5/18/85).
8. 40 CFR part 300, appendix A.
9. Telephone interviews with Raymond Crawford, fire chief, Wendover, NV (for USPCI UT); Idaho state fire marshal; Mr. Gleason, Louisiana state fire marshal; El Dorado, AR, fire marshal; Dave Rinker, fire chief, Niagara Township, NY; and Thomas Roach, chief, fire prevention, Chicago (1/2/85).

Chapter 9

1. For a complete description of how to use U.S. EPA's Hazard Ranking System, see "Appendix A—Uncontrolled Hazardous Waste Site Ranking System: A Users Manual" in *Federal Register,* vol. 50, no. 29 (1985), pp. 5910–5927.
2. See *Wall Street Journal* (8/7/85 and 8/26/85) and *Hazardous Waste News* (8/12/85).
3. CEP site visit and interview with L. H. Shepherd, USPCI Grassy Mountain facility plant manager (4/26/85).
4. Ibid. and telephone interview with Bill Sinclair, Utah Department of Solid and Hazardous Waste (11/26/84).
5. U.S. Pollution Control, "Grassy Mountain Facility RCRA Part B Application," volume I (1984), p. 2.
6. Ibid., pp. 165–167.
7. Bill Sinclair (11/26/84).
8. "Part B," volume I, pp. 165–167.
9. Ibid.
10. Ibid., p. 85.
11. Robert Benton, U.S. Department of Interior to Steve Mason, USPCI, (5/29/84) and James L. Dykman, Utah Department of Cultural and Economic Development, Division of State History to Steve Mason, USPCI (5/25/84).

12. U.S. EPA "Facility Status Sheet—USPCI Grassy Mountain" (1/9/85).
13. Before the Utah Solid and Hazardous Waste Committee, USPCI, "Notice of Violation and Compliance Order" (6/7/84).
14. Before the Utah Solid and Hazardous Waste Committee, USPCI, "Notice of Violation" (9/21/84).
15. Before the Utah Solid and Hazardous Waste Committee, USPCI, "Notice of Violation" (10/15/84).
16. Before the Utah Solid and Hazardous Waste Committee, USPCI, "Notice of Violation" (11/13/84). Other enforcement actions including compliance orders (3008) issued in 1981 and 1983 and a warning letter (3007) issued in 1983 are recorded in U.S. EPA, "Hazardous Waste Data Management System;" however, CEP was unable to find information about these actions.
17. Ken Jackson, USPCI to Dale Parker, USHWC (6/21/84).
18. CEP site visit.
19. Ibid.
20. Dale Parker, USHWC to Harry Hansen, USPCI (10/15/84).
21. USPCI, "RCRA Part A Application" (1984).
22. "CEP Hazardous Waste Management Study General Corporate Questionnaire" (completed by USPCI in 1984).
23. USPCI, "Part A." In addition, there is conflicting information regarding whether this facility has received wastes removed from Superfund sites. EPA's "Disposition of CERCLA Waste Database" indicates that as of October 1984 the facility had not received such materials; however, the Subcommittee on Oversight and Investigations of the House Energy and Commerce Committee's "Ground Water Monitoring Survey" (April 1985) indicates that the facility has received CERCLA wastes.
24. CEP site visit.
25. Ibid.
26. Telephone interview with June Wickham (1984).
27. Fred C. Hart & Associates, "Environmental Assessment of SCA Chemical Services Company Hazardous Waste Management Facility, Chicago/Lake Calumet" in SCA Chemical Services, "RCRA Part B Application for Chicago Incinerator" volume I, appendix IIA (1983), pp. IV–22.
28. Ibid., p. IV–919.
29. IEPA, *The Southeast Chicago Study: An Assessment of Environmental Pollution and Public Health Impacts* (draft) (Springfield, IL: IEPA, April 1984), figure 1.1.
30. Telephone interview with Roger Selburg, Chicago Public Water Supply (12/10/84).
31. "Environmental Assessment," "Part B," pp. IV–9.
32. Ibid., pp. IV–10.
33. Ibid., "Part B," pp. IV–2, 5.
34. Ibid., pp. IV–9 and Robert Bartholomew, CWM, "Comments on CEP HRS Package" (1984).
35. *The Southeast Chicago Study,* figure 4.8.
36. "Environmental Assessment," "Part B," pp. IV–9.
37. U.S. Geological Survey, "Lake Calumet, IL–IN Quadrangle" (1977) and Robert Bartholomew.
38. Ibid.
39. "Part B," pp. B-6.
40. Ibid., "Environmental Assessment," pp. IV–14 and appendix D.
41. U.S. EPA, RCRA inspection (9/20/80) and IEPA, inspections (1/20/84, 3/16/84, and 5/22/84).
42. "Part B," "Environmental Assessment" pp. II–2.
43. Robert Bartholomew and telephone interview with Clifford Gould, IEPA (12/7/84).
44. Clifford Gould.
45. U.S. EPA, "Facility Status Sheet for SCA Chemical Services, Chicago" (1/20/84) and telephone interview with Juana Rogo, U.S. EPA (12/6/84).
46. U.S. EPA, in the matter of SCA Chemical Services, Inc., Chicago, Illinois, docket #V-W-81-R-24, "Complaint and Findings of Violation" (12/23/80).
47. U.S. EPA in the matter of SCA Chemical Services, Inc., Chicago, Illinois, docket #V-W-81-R-24, "Consent Agreement and Final Order" (3/30/81).
48. Ibid.
49. Robert Bartholomew.

50. Ibid.
51. Ibid.
52. U.S. OSHA, "Inspection History Listing" (March 1984).
53. "Facility Status Sheet" and telephone interview with John Bayer, CWMI (11/5/85).
54. Robert Bartholomew.
55. U.S. EPA, "Disposition of CERCLA Wastes Data Base" (October 1984).
56. CEP site visit and interview with Donald Brady, plant manager (8/17/84).
57. Walter C. Barber, CWM, to Alice Tepper Marlin, CEP (4/30/85).
58. Robert Bartholomew.
59. Robert Ginsburg, "Comments on the United States Environmental Protection Agency Region V Proposed Approval of the SCA Chemical Services Incinerator To Dispose of Liquid and Non-Liquid Polychlorinated Biphenyls (PCBs)" (Chicago: Citizens For A Better Environment, 1983).
60. Telephone interview with Jim Foley, Avalon Trails Improvement Association (8/15/84).
61. *The Southeast Chicago Study,* figure 4.1.
62. See Robert Ginsburg, CBE and Mary Ellen Montes, I-ACT, to William Ruckelshaus, U.S. EPA "Citizen Petition for an Investigation and Rulemaking Action" (7/13/84); U.S. EPA "Investigation and Rulemaking Action Concerning Southeast Chicago; Denial of Citizens' Petition" in *Federal Register* (10/31/84); and U.S. EPA "Investigation and Rulemaking Action Concerning Southeast Chicago; Denial of Citizens' Petition—Support Document" (10/19/84).
63. Jim Foley.
64. Walk, Haydel & Associates, "CWM Lake Charles Facility RCRA Part B Permit Application" revision 0, volume I, figure 18, (1983).
65. 26 buildings × 3.8 residents as directed in HRS. U.S. Geological Survey, "Black Lake Quadrangle" (1975) and "Geographical Factors" in "Part B," volume I, p. 1.
66. Sam Vincent reportedly has two wells screened at 50 feet, one at 200 feet, two at 500 feet, and one at 700 feet approximately 4000 feet southeast of the facility. Telephone interview with Marvin Hargar (11/8/84). The 27-foot well is owned by Ortega and is used for watering livestock. Woodward–Clyde Associates, "Water Well Survey" (11/5/81) in "Part B."
67. "Introduction," "Part B," vol. I, p. 5.
68. "Water Well Survey," "Part B" and Louisiana Department of Health and Human Services, "Data Summary of Water Wells Within 3-Mile Radius of Chemical Waste Management, Carlyss" in "Report on Phase II Drinking Water Surveillance Program" (12/29/82).
69. CWM, comments on CEP data sheets (3/14/85).
70. "Site Geology," "Introduction," "Part B," vol. I, p. 4.
71. Ibid., p. 2.
72. Ibid., pp. 14–15.
73. Ibid.
74. "Site Hydrology," "Part B, p. 1."
75. LADEQ, "Emergency Response Record" and "General Inspection Reports" (2/28/84, 2/29/84, 3/2/84, and 3/5/84).
76. Telephone interview with Victor J. Montelaro, LADEQ (11/7/84).
77. Ibid.
78. Kenneth H. Anderson, CWM, to Gerald D. Healy, Louisiana Department of Natural Resources (LADNR) (7/12/84).
79. Ibid.
80. Ibid., W. V. Walls, CWM, to Gerald D. Healy, LADNR (3/18/83) and Anderson to Healy (2/15/84).
81. U.S. EPA "Facility Status Sheet for CWMI Lake Charles Facility" (11/19/84).
82. LADEQ, in the matter of Chemical Waste Management, Inc. Lake Charles facility. Proceedings under the Louisiana Environmental Quality Act LA R.S. 30:1051 et seq. "Compliance Order." (8/30/84) superseding LADEQ, in the matter of Chemical Waste Management, Inc. Lake Charles facility. Proceedings under the Louisiana Environmental Quality Act LA. R.S. 30:1051 et seq. "Compliance Order" (6/28/84).
83. Walter C. Barber, CWM, to Alice Tepper Marlin, CEP (4/30/85), p. 2.
84. CEP site visit and interview with Norb Bolda, former CWM LA plant manager (8/7/84).

85. Interview with Walter C. Barber and Robert Bartholomew, CWM (3/14/85).
86. CWM response to CEP questionnaire (1984).
87. "Facility Status Sheet."
88. "Operations Plan," "Part B," pp. 2 and 6.
89. CWM response to CEP questionnaire and CWM comments on CEP data sheets.
90. "Introduction," "Part B," p. 4.
91. CWM response to CEP questionnaire.
92. Interview with Leonard Knapp, Calcasieu district attorney and Peggy Franklin et al. from CLEAN (8/7/84).
93. Spetrix Laboratory, "Spetrix Reports" (1/6/82, 9/15/82, 10/7, 20, and 26/82).
94. Interviews with Marvin Hargar (8/7/84, 10/19/84 and 11/8/84).
95. CWM response to CEP questionnaire and interview with Walter C. Barber and Robert Bartholomew, CWM (3/14/85).
96. Telephone interview with Dr. Batti, U.S. EPA Office of Safe Drinking Water (November 1984).
97. U.S. Geological Survey, "Tonawanda West and Niagara Falls Quadrangles" (1980) and Hana Umlaufl Lane, ed., *The World Almanac and Book of Facts 1983,* (New York: Newspaper Enterprise Association, Inc., 1983), p. 236.
98. NYSDEC, "Appendices" in *Toxic Substances in the New York State Environment: Technical Report* (1979), pp. 2–73.
99. Telephone interview with Bob Speed, NYSDEC, Buffalo (11/16/84).
100. Telephone interview with John McMann, NYSDEC regional engineer (11/16/84).
101. NYSDEC, *Toxic Substances in New York State.*
102. William Myers, U.S. EPA untitled internal draft report (1984).
103. U.S. EPA draft report and Wehran Engineering Co., "Document 16" (1980) in CECOS, "RCRA Part B Application" (11/3/83), p. 36.
104. "Part B," "Document 6," pp. 2–1, 2–3, and 2–4; "Document 15," pp. 13–16 and 19–20; and "Document 16," pp. 26 and 30.
105. "Part B, Document 6," p. 2–4, and "Document 15," pp. 19 and 20.
106. U.S. EPA Regional II, in the matter of CECOS, Niagara Falls, New York. Index number II RCRA-85-3013-50201. "Administrative Order on Consent" (2/22/85).
107. "Tonawanda West and Niagara Falls Quadrangles."
108. Telephone interview with D. Firth, NYSDEC, Buffalo, Surface Water Branch (11/27/84) and telephone interview with Frank Graber, NYSDEC, Buffalo (11/28/84).
109. Telephone interview with Paul Eisman, NYSDEC (11/16/84).
110. NYSDEC, "Consent Order" (8/10/81).
111. NYSDEC, "Consent Order" (4/8/82).
112. NYSDEC, "Inspection Report" (11/15/82).
113. RECRA Research, "Ground Water Monitoring Report" (5/26/83) in "Part B Addition" (September 1984), pp. 11g–11h.
114. U.S. EPA Region 2, "Inspection Report" (5/5/83).
115. NYSDEC to CECOS (2/15/84).
116. NYSDEC, "Weekly Summary of Daily Inspections" (12/17/83, 12/21–28/83, and 1/9–16/84).
117. NYSDEC, "Inspection Report" (1/26/84–2/1/84).
118. Roger Schlaf, CECOS to Carol Jones, NYSDEC, Memorandum: "High Effluent Phenols—Request of Variance from Niagara Falls" (12/16/83).
119. CEP site visit (10/26/84).
120. U.S. EPA, untitled draft report.
121. U.S. EPA Region 2, "Administrative Order on Consent."
122. Ibid., telephone interview with Faith Halter, U.S. EPA Region 2 (7/8/85) and Subcommittee on Oversight and Investigations of the Committee on Energy and Commerce of the U.S. House of Representatives, "Ground Water Monitoring Survey" (April 1985).
123. U.S. EPA Region 2, "Consent Agreement" (12/22/78).
124. U.S. EPA Region 2, "Compliance Complaint and Notice of Opportunity for a Hearing" (5/15/80).
125. NYSDEC, "Consent Order" (8/10/81).
126. U.S. EPA Region 2, "Final Administrative Order" (1981).

127. U.S. EPA Region 2, "Compliance Complaint" (March 11, 1982).
128. NYSDEC, "Consent Order" (4/8/82).
129. U.S. EPA Region 2, "Hazardous Waste Data Management System" (1985).
130. Ibid.
131. Edward Belmore, NYSDEC, to John Dugan, CECOS (11/28/83).
132. Ibid.
133. *Hazardous Waste News* (Silver Spring, MD: Business Publishers, Inc., 11/5/84), p. 351 and *Toxic Material's News* (11/7/84), p. 352.
134. U.S. EPA Region 2, "Administrative Order on Consent."
135. U.S. EPA Region 2, in the matter of CECOS, Niagara Falls, New York. Index number II RCRA-85-3008-0202. "Consent Agreement and Consent Order" (2/22/85).
136. Interview with Thomas L. Moran, formerly of CECOS during CEP site visit (10/26/84).
137. Interview with Frank Grabar, NYSDEC (10/25/84).
138. Ibid. and RCRA Research, "Assessment of Ground Water Monitoring Program at CECOS" (September 1984), p. 1.
139. Telephone interview with Ted Gable, U.S. EPA Region 2 (7/8/85).
140. Telephone interview with Earnest Neal, CECOS (11/5/85).
141. U.S. OSHA, "Inspection History Listing" (March 1984).
142. NYSDEC, "Weekly Summary of Daily Inspections" (12/8–17/83), p. 1.
143. U.S. EPA Region 2, "Hazardous Waste Data Management System" (1984).
144. CECOS, "RCRA Part A Application" (12/13/83).
145. Environmental Information, Ltd., *Industrial and Hazardous Waste Management Firms, 1985* (Minneapolis, MN: 1984), pp. 316–317.
146. Thomas L. Moran, CECOS corporate communications vice president, "Press Release" (10/29/84) and interview with Michael J. Hanchak, corporate manager of research and development (10/26/84).
147. U.S. EPA "Disposition of CERCLA Waste Data Base" (October 1984).
148. RECRA Research, "Part B," "Environmental Impact Statement" (1979), p. 2-1 and interview with Frank Grabar.
149. Interview with Michael J. Hanchak, CECOS (10/26/84).
150. Philip Shabecoff, "Toxic Leak Feared From Niagara Site" in *New York Times* (10/30/84), p. 1.
151. Interview with Tony Bassili (10/25/84).
152. Interview with Margaret Guiliani (10/25/84).
153. Telephone interviews with Bernice Stella, Tony Lapin, and Anita Fama (7/1/85).
154. Hana Umlauf Lane, ed., *The World Almanac & Book of Facts* (New York: Newspaper Enterprises Associates, 1983).
155. U.S. Geological Survey, "El Dorado East" and "El Dorado West" Quadrangles (1981).
156. Arkansas Department of Pollution Control and Ecology (ADPC&E), "Arkansas Water Supply Inventory," pp. 313–314 and ADPC&E, "Ground Water Use in Arkansas" (1980).
157. Developers International Services Corp. (DISC), "Final Report for an Environmental Study of the ENSCO Facility El Dorado, Arkansas" (December 1983).
158. Dr. George D. Combs, ENSCO, "Ground Water Problems at ENSCO's El Dorado Plant," (1/19/84), p. 8.
159. Ibid.
160. Telephone interview with Mark Witherspoon, ADPC&E (11/20/84).
161. AMI Engineering, "Part B," "Geological Profile," (1/3/83).
162. DISC, pp. 1–6.
163. ADPC&E, "Consent Administrative Order LIS 84-016" (7/26/84), p. 4.
164. Combs, pp. 9–11.
165. DISC.
166. Combs.
167. U.S. EPA Region 6, "Litigation Report," p. 1.
168. Energy Systems Co., "RCRA Part B Application, Revision 1" (2/23/84), p. B-1.
169. Telephone interview with Cecil Harrel, ADPC&E (11/16/84).
170. Centaur Associates, *Siting of Hazardous Waste Management Facilities and Public Opposition* (EPA: November 1979), p. 282.

171. "Litigation Report," pp. 5 and 10.
172. "Consent Decree."
173. Combs, pp. 3–4.
174. Ibid.
175. Ibid., pp. 6–8.
176. Allyn M. Davis, EPA Air & Waste Management Division director to Dr. George D. Combs (6/20/83).
177. Holly Anderson, EPA inspector, "Compliance Monitoring Inspection Report" (12/19/83) and Jack B. Danielson, ENSCO to Benjamin A. Goldman, CEP (4/11/85).
178. U.S. Coast Guard.
179. Combs, p. 10–12.
180. U.S. Coast Guard.
181. Ibid.
182. ADPC&E, "Consent Administrative Order LIS 84-016" (7/26/84).
183. U.S. EPA Region 6, "Record of Communication" (1/3/84).
184. U.S. EPA, "Facility Status Sheet" (12/5/84).
185. *United States of America* v. *Energy Systems Company,* "Consent Decree," (1/28/81). See *U.S.* v. *Energy Systems Company,* "Notice of Defendant's Satisfaction of Consent Decree—Civil Action No. WDA-81-1006" (6/19/85).
186. U.S. EPA Region 6, "Litigation Report," pp. 5 and 10 and ENSCO, "RCRA Part A Application, Revision 2" (12/31/84), p. A-S.5.
187. ENSCO, "Supplement to Part A."
188. William H. Taylor, U.S. EPA Enforcement Section chief to Chery Terai, ADPC&E Compliance and Technical Assistance Branch chief (3/20/84).
189. Holly Anderson, U.S. EPA inspector, "Compliance Monitoring Inspection Report," (12/19/83).
190. ADPC&E, "Consent Administrative Order LIS 84-016" (7/26/84), p. 4.
191. Telephone interview with Richard Quinn, ADPC&E (11/20/84) and telephone interview with Cecil Harrel, ADPC&E (11/16/84).
192. Danielson to Goldman.
193. U.S. OSHA, "Inspection History Listing" (1984).
194. "Facility Status Sheet."
195. "Part A."
196. U.S. EPA "Disposition of CERCLA Wastes Data Base" (10/17/84).
197. Centaur Associates, pp. 280–291.
198. Adlene Harrison, U.S. EPA Region 6 administrator to Melvyn Bell, ENSCO (1/26/81).
199. Telephone interview with Rachel McKinney (7/15/85).
200. Telephone interview with J. Vestal (7/15/85).
201. 220 buildings × 3.8 residents per building as directed by HRS. U.S. Geological Survey, "Scotlandville, TX Quadrangle" (1975).
202. Louisiana Department of Health and Human Services, "Data Summary of Potable Wells within 3-mile Radius of Petro Processors (Old and New Sites) and Rollins Environmental Services" in *Report on Phase II Drinking Water Surveillance Program* (12/29/82).
203. Ibid.
204. Ibid.
205. Geraghty and Miller, Inc., "Exhibit 5b" (1981) in Rollins Environmental Services of Louisiana, "RCRA Part B Permit Application," volume II (8/15/83) pp. 2-25–2-27.
206. Ibid., p. 2-65 and John Quarles, retained by RES (LA) to Alice Tepper Marlin, CEP (7/10/85), p. 4.
207. Geraghty and Miller, "Part B, Addendum," "EPA Deficiency No. 27—Response," p. 30-1.
208. "Exhibit 5b," p. 2-16.
209. U.S. EPA, in the matter of RES (LA), Baton Rouge, LA, Administrative Order, docket number RCRA-3013-08-84, "Order Requiring Submission and Implementation of Proposal for Monitoring, Testing, Analysis, and Reporting" (10/12/84).
210. Ibid.
211. "Scotlandville Quadrangle."
212. U.S. EPA, docket number RCRA-3013-08-84.
213. "Part B," volume I, p. 1-20-1.
214. Cox, Walker and Associates, *Project #48* (6/4/76 through present) and E. R. Cox, Jr., P.E., to Lynn E. Williams, parish attorney (10/24/83).
215. U.S. EPA, docket number RCRA-3013-08-

84 and Quarles to Tepper Marlin (7/10/85), p. 2.
216. Confidential telephone interview with U.S. EPA inspector (11/5/84) and Quarles to Tepper Marlin (7/10/85), p. 6.
217. Eustis Engineering Co., "Summary of Visitation to site of Rollins Environmental Services Baton Rouge LA" (7/21/83) in "Part B," volume IV, pp. 3-3-4–3-3-8.
218. "Reports 134 & 135" (8/11/83 and 2/13/84) in *Project #48.*
219. U.S. EPA "Facility Status Sheet for Rollins Environmental Services (LA) Baton Rouge" (9/27/84).
220. Cox to Williams (10/24/83).
221. Ibid.
222. "DNR Refers Two Odor Complaints to Attorney General for Action" in *State Times* (Baton Rouge, LA: 1/21/83).
223. LADEQ, "Notice of Violation to LAD010395127" (5/11/84) and Patricia Norton, LADEQ, to Wiley Irby, RES (LA) (5/22/84).
224. U.S. EPA in the matter of RES (LA) Baton Rouge, LA, docket number RCRA VI-441-H, "Compliance Order and Notice of Opportunity for Hearing" (10/12/84).
225. Ann Banks, U.S. EPA Region 6 to Lowell F. Martin, retained by RES (LA) (10/4/85).
226. U.S. EPA, docket number RCRA-3013-08-84.
227. Telephone interview with Gary Aydell and Ken Fladderman, Louisiana Water Pollution Control Division could not provide a list of RES (LA) discharge violations.
228. Telephone interview with Randy Cox, P.E. (September 1984).
229. Quarles to Tepper Marlin (7/10/85), p. 2.
230. U.S. EPA, docket number RCRA-3013-08-84.
231. CEP site visit to RES (LA); interview with Wiley Irby, RES (LA) (8/6/84); interview with John Quarles (2/20/85); and Quarles to Tepper Marlin (7/10/85), p. 3.
232. Quarles to Tepper Marlin (7/10/85), p. 2, and enclosures.
233. U.S. EPA, docket number RCRA-3013-08-84.
234. Telephone interview with Steven Schwartz, U.S. EPA Region 6 hydrologist (10/22/84).
235. Quarles to Tepper Marlin (7/10/85), pp. 3 and 6.
236. Pamela Phillips, U.S. EPA Region 7 to Lowell F. Martin, rctained by RES (LA) (10/23/85).
237. U.S. OSHA, "Inspection History Listing" (April 1984).
238. "Facility Status Sheet" (9/27/84).
239. RES (LA), "RCRA Part A Application" (7/17/85).
240. Environmental Information, Ltd., *Industrial and Hazardous Waste Management Firms, 1985* (Minnesota, 1984).
241. U.S. EPA "Disposition of CERCLA Waste Data Base" (October 1984).
242. RES (LA), "RCRA Part A Application" (1984).
243. Environmental Information, Ltd.
244. "*Mary McCastle et al.* v. *Rollins Environmental Services of Louisiana, Inc., et al.,* no. 83-C-2621" in *456 Southern Reporter, 2nd Series* (Louisiana, 1984), p. 612–615.
245. Interview with Mary McCastle and Steve Irving (8/6/84).
246. Rand McNally, *Universal World Atlas* (1982).
247. U.S. Geological Survey, "Driscoll East Quadrangle" (1975).
248. Telephone interview with Paul Lewis, Texas Department of Water Resources (TDWR) Hydrogeologist (11/1/84) and Texas Ecologists, "RCRA Part B Application," volume IVa (1983), p. 8.
249. Paul Lewis.
250. "Attachment K. Location Map, Texas Department of Water Quality Household Survey" (1979) in "Part B."
251. Paul Lewis.
252. "Paul B," volume IVa, pp. 9, 12, and 13.
253. U.S. Geological Survey, "Ground Water Resources of Nueces and San Patricio Counties, Texas" (Texas Department of Water Resources, 1982).
254. "Part B," volume IVa, pp. 8–10.
255. Ibid.
256. Ibid.
257. U.S. Geological Survey, "Driscoll East Quadrangle."
258. Paul Lewis.
259. "There is no doubt. . . . that the facility has a history of odor problems and that

better management practices should be utilized at this facility." Texas Department of Water Resources, "TECO—Robstown, Texas."

260. Ibid.
261. J. L. Grant, Law Engineering Testing Co., to B. R. Carter, Nuclear Engineering Co. (10/6/80) and "Dump Linings Are Questioned," *Corpus Christi Caller* (6/6/80), pp. 1 and 12.
262. TDWR, "TECO."
263. Paul Sweeney, "Poison Acres" in *Texas Observer* (4/11/80), p. 6.
264. Research and Special Programs Administration, U.S. Department of Transportation, "Selected Hazardous Materials Incidents" (7/10/84).
265. TACB, "Complaint Investigations on Account NE0126S Since 1/1/82" (12/12/84) and Sabino Gomez, TACB to Ben Goldman, CEP (12/14/84).
266. Paul Kutchinski, TDWR to Bill Jones, TECO (2/14/84).
267. TDWR, "Major Facilities Status Sheet for Texas Ecologists Inc." (2/14/84).
268. TDWR, "TECO."
269. Paul Lewis.
270. R. Fleming, TDWR to H. Pruett, TDWR "Interoffice Memorandum" (1/20/84).
271. Paul Lewis.
272. *Texas Observer* (4/11/80), p. 6.
273. Telephone interview with Tom Palmer, TACB (11/2/84).
274. CEP site visit to TECO TX and interview with Bill Jones, TECO TX plant manager (4/29/85).
275. Ibid.
276. "Facilities Status Sheet."
277. TECO TX, "RCRA Part A Application" (1982).
278. "Part B," volume III, pp. 4–5.
279. Telephone interview with William Rhea, director U.S. EPA Region 6 (10/31/84).
280. *Texas Observer* (4/11/80), p. 7 and Kevin Carmondy, "Shamrock Denies EPA Charge" in *Beaumont, Texas Enterprise* (10/23/84).
281. U.S. EPA "Disposition of CERCLA Waste Data Base" (October 1984) and William Rhea.
282. Bill Jones.
283. Ibid.
284. Ibid.
285. *Texas Observer* (4/11/80), p. 5.
286. Telephone interview with Kenneth Ahlrich (10/18/84) and personal interview with Kenneth Ahlrich (4/29/85).
287. Telephone interview with Mrs. Douglas Salyer (11/2/84).
288. Bryant–Park and Associates, Inc., "J & J Disposal Site, Geological and Soils Report, Benicia, California" (1974) in IT, "RCRA Part B Application for Benicia Facility," volume II (1983), p. 1.
289. Personal interview with Mustafa Radman, EPA Region 9 Permit Writer for IT Benicia (July 1984).
290. San Francisco Bay Region Water Quality Control Board (SFBRWQCB), "Waste Discharge Requirements for IT Corporation Class I Disposal Site Benicia, Solano County. Order #81-31" (1981).
291. Morris A. Balderman, IT to David Bauer, IT, "Response to EPA Compliance Order (Docket #RCRA-09-84-0040)" and attached "Hydrogeological Investigation" (1984) in Toxics Assessment Group, "Nowhere To Go, The Universal Failure of Class I Hazardous Waste Dump Sites in California" (The Environmental Defense Fund: 1985), p. I-34.
292. David Bauer, IT to Benjamin A. Goldman (10/31/85).
293. Telephone interview with Mustafa Radman, U.S. EPA (11/4/84) and ibid.
294. Bryant–Park & Associates in "Part B," p. 5.
295. Balderman to Bauer in "Nowhere To Go," p. I-39.
296. Bryant–Park & Associates in "Part B," p. 13.
297. Ibid., p. 2. IT states this dip does not characterize site.
298. "Part B," volume I, p. 150 and Balderman to Bauer in "Nowhere To Go," pp. I-40–I-41.
299. "Part B," volume I, p. 198.
300. Kit Custis, State Water Resources Control Board to Harold Singer, SFBRWQCB, "Additional recommendations and questions on monitoring wells at IT Corporation, Benicia Class I Waste Disposal Site" (2/25/85) in "Nowhere To Go," p. I-38.
301. Interview with Barbara Acker, IT Corp. (7/24/84).

302. J. H. Kleinfelder & Associates, "IT Corporation Benicia Slope Stability Site Location Map" in "Part B."
303. Harding, Miller, Lawson & Associates, "Geologic Investigation J & J Disposal Company Liquid Waste Disposal Site, Solano County, California" (1971) in "Part B," p. 1.
304. SFBRWQCB, "Waste Discharge Requirements. Order #81-31," p. 2.
305. U.S. Geological Survey, "Vine Hill Quadrangle" (1980).
306. SFBRWQCB, "Clean-up and Abatement Order No. 85-003" (1/16/85).
307. Woodward–Clyde Consultants, "Geotechnical Integrity of Pond Nos. 12 and 13, IT Benicia Site" (1981), p. 3 in "Part B."
308. Ibid., pp. 9–10.
309. SFBRWQCB (1/16/85).
310. U.S. EPA in the matter of IT Corporation (Benicia facility) docket #RCRA-09-84-0040. "Determination of Violation, Compliance Order, and Notice of Right to Request A Hearing" (9/27/84).
311. SFBRWQB (1/16/85).
312. U.S. EPA (9/27/84).
313. U.S. EPA, "Inspection Report" (3/9 and 12/84), p.2.
314. U.S. EPA (9/27/84).
315. John T. Schofield, IT to Benjamin A. Goldman, CEP (3/21/85).
316. SFBRWQCB (1/16/85).
317. Ibid.
318. Custis to Singer (2/25/85) in "Nowhere To Go," pp. I-37–I-38 and Bauer to Goldman (10/31/85).
319. U.S. EPA, "Facility Status Sheet for IT Corp. Benicia."
320. Bruce H. Wolf, SFBRWQCB, "Status of Regional Board Involvement [IT Benicia]" (11/6/84) in "Nowhere to Go," p. I-36.
321. Harry Seraydarian, U.S. EPA to Richard Wilcoxon, CADOHS, "Notice Letter" (5/23/84).
322. U.S. EPA (9/27/84).
323. U.S. EPA, "IT Corporation, Benicia, Completeness Check" (10/4/84).
324. SFBRWQCB (1/16/85).
325. *State of California Department of Health Services* v. *IT Corp., "Consent Decree" (2/21/85).*
326. CEP site visit of IT, Benicia (7/24/84).
327. U.S. EPA, "Facility Status Sheet for IT Corp. Benicia" (3/12/84).
328. IT, "RCRA Part A Application" (1980).
329. U.S. EPA, "Disposition of CERCLA Waste Data Base" (October 1984) and telephone conversation with Jody Sparks and John Wesnousky, CADOHS (5/28/85) in "Nowhere To Go," p. I-34.
330. "Part A" and Wolf (11/6/84) in Ibid.
331. Sandy Avol and Steve Simanonok, U.S. EPA to Bob Mandel, U.S. EPA (2/25/84) in Ibid.
332. "Part A."
333. Bauer to Goldman (10/31/85).
334. ESII ID, "RCRA Part B Application, Revision 1" (May 17, 1984), p. B-12.
335. Clement Associates, Inc., "The Potential for Gradual Environmental Impairment Arising from the Operations of Envirosafe Services of Idaho, Inc., Grandview, Idaho Site B—Final Report Prepared for Conversion Systems" (November 7, 1984), p. 6.
336. U.S. Geological Survey, "Castle Butte Quadrangle" and "Jack Ass Butte Quadrangle" and Hana Umlauf Lane, ed., *The World Almanac & Book of Facts* (New York: Newspaper Enterprises Associates, 1983).
337. "Part B," p. B-12.
338. Telephone interview with Jack Sceva, U.S. EPA Region 10 geologist (12/3/84).
339. Donald J. Frost, ESI to Benjamin A. Goldman, CEP (March 20, 1985).
340. "Part B," pp. D.5-8–D.5-10, D.15-28; telephone interview with Patrick Stoll, senior hazardous materials specialist, Idaho Health and Welfare Department, Division of the Environment (12/2/84); Frost to Goldman, p. 2; and Clement Associates, pp. 21–22.
341. "Part B," p. B-16, Sceva, U.S. Geological Survey, and Clement Associates, p. 32.
342. Clement Associates, p. 6.
343. U.S. EPA, "Damage Incident Summary Form."
344. *U.S. EPA* v. *Envirosafe Services of Idaho, Inc.,* "Findings of Fact, Agreed Compliance Order, and Agreed Order Requiring Submission and Implementation of Proposal for Sampling, Analysis, Monitoring and Reporting" (11/30/83).

345. Ibid., p. 9.
346. *U.S. EPA* v. *Envirosafe Services of Idaho, Inc.,* "Consent Agreement and Final Order" (10/23/84), p. 2.
347. Charles Scott, Owyhee County Safety Officer, "ESI Weekly Inspection" (March 26, 1984).
348. Charles Scott, "ESI Weekly Inspection" (June 5, 1984).
349. Consent Agreement and Final Order" (10/23/84), p. 7.
350. Ibid., p. 2.
351. U.S. EPA "Status Sheet for ESII Site B" (1984).
352. U.S. EPA "Notice of Violation and Warning" (9/8/83).
353. "Compliance Order" (11/30/83).
354. Clement Associates, p. 20. EPA may have also issued a TSCA enforcement action to ESII in 1981; however, CEP was unable to get information about this action.
355. *U.S. EPA* v. *Envirosafe Services of Idaho, Inc.,* "complaint, Compliance Order and Notice of Opportunity for Hearing" (8/24/84).
356. "Consent Agreement and Final Order" (10/23/84).
357. Clement Associates, p. 20.
358. Ernest B. Barnes, U.S. EPA regional administrator to Congressman L. E. Craig (10/23/84).
359. Arnon E. Garonzick, ESI to Benjamin A. Goldman, CEP (3/7/85).
360. "CEP Hazardous Waste Management Facility Questionnaire for ESII, Grandview, ID" (Filled out by Larry Haack, ESI).
361. Scott, "ESI Weekly Inspection" (4/5/84).
362. *Idaho Statesman* (4/5/85), p.1.
363. U.S. EPA, "Hazardous Waste Data Management System."
364. ESII, "Part A" (1984); Frost to Goldman, p. 1; and Clement Associates, pp. 15–16.
365. U.S. EPA "Disposition of CERCLA Wastes Data Base" (10/17/84).
366. Clement Associates, p. 6.
367. CEP facility questionnaire.
368. "Part B."
369. Personal interview with Nick and Betty Nettleton (4/25/85).
370. CECOS, "RCRA Part B Application," "Site Characteristics." (1983).
371. "Part B," "Folder #4," "Document #1" p. 6.
372. Ibid., p. 4.
373. *New York Times* (11/14/84).
374. Cheryl Belmont, Office of the District Board of Health of Clermont County to Ben Goldman, CEP (12/11/84).
375. Ohio–Kentucky–Indiana Regional Council of Governments, "CER's Hazardous Waste Disposal Facility. A Report to OKI's Executive Committee" (December 1980), p. 6.
376. "Part B," "Facility Description."
377. *Clermont Sun* (7/4/84).
378. Telephone interview with Bill Strachan, OEPA full-time site inspector (12/11/84).
379. OKI, p. 6.
380. "Part B," "Folder #4" pp. D.1-11, D.1-34–D.1-35.
381. U.S. Geological Survey, "Williamsburg, OH Quadrangle" (1971) "Newtonsville, OH" (1967).
382. Mary Bauer, CECOS to Valdus Adamkus, U.S. EPA Region 5 administrator (9/8/82).
383. OEPA to Bridget Wieghart, CEP (12/19/84), p. 2.
384. Bill Strachan, OEPA on-site inspector, quoted in *Clermont Sun* (12/14/83).
385. Telephone interview with Thomas Winston, OEPA (12/11/84); *New York Times* (11/14/84); and *Clermont Sun* (10/17/84).
386. Mrs. Raymond K. White to Ohio State Board of Health (6/13/84).
387. *Clermont Sun* (10/17/84).
388. *New York Times* (11/16 and 27/84).
389. OEPA, in the matter of CECOS, "Director's Final Findings and Orders" (5/9/85); telephone interview with Donald Marshall, OEPA (6/14/85); and *New York Times* (5/10/85).
390. Ibid.
391. Warning letter from W. H. Miner, OEPA to Alan Orth, CECOS (3/28/83).
392. *Clermont Sun* (2/15/84).
393. OEPA to Bridget Weighart, CEP (12/19/84).
394. *Clermont Sun* (9/19/84).
395. U.S. EPA Region 5, "Complaint and Finding of Violations and Order" (9/24/84).
396. OEPA, in the matter of CECOS "Direc-

tor's Emergency Final Findings and Orders" (11/9/84) and *New York Times* (11/26/84).

397. OEPA, in the matter of CECOS, "Director's Final Orders" (11/26/84).
398. *New York Times* (11/28/84).
399. OEPA, "Director's Final Findings and Orders" (5/9/85) and telephone interview with Don Marshall, OEPA (6/14/85).
400. *Wall Street Journal* (6/3/85).
401. CEP site visit and interview with Mary Bauer, formerly of CECOS (8/13/84).
402. U.S. OSHA, "Inspection History Listing" (March 1984).
403. Donald S. Marshall, OEPA to Richard O. Toftner, CECOS (7/12/82).
404. Donald S. Marshall, OEPA to Gregory Melia, CECOS (5/21/82).
405. *Courier* (10/5/84).
406. "Part B."
407. CEP site visit and CECOS "RCRA Part A Application" (1980).
408. U.S. EPA "Disposition of CERCLA Waste Data Base" (October 1984).
409. CEP site visit.
410. Interviews with William Stewart and Bill and May Marshall (8/11/84); George Pattison, Clermont County prosecuting attorney to Clermont County Commissioners (6/10/83); and *Clermont Environmental Reclamation Company* v. *Lois Hancock and Frank E. Parker*, "Memorandum and Judgement Entry" (3/29/84).

Chapter 10

1. Telephone interview with Fred Bruin, U.S. DOT (4/11/84).
2. U.S. EPA, "Hazardous Waste Data Management Systems" (New York: U.S. EPA Region 2, November 1984).
3. Westat, *National Survey of Hazardous Generators and Treatment, Storage, and Disposal Facilities Regulated Under RCRA in 1981* (Washington, DC: U.S. EPA, 1984), p. 135.
4. U.S. EPA "Assessing the Releases and Costs Associated with Truck Transport of Hazardous Waste" (June 1984).
5. Telephone interview with Bill Wilson, U.S. DOT (April 1984).
6. U.S. EPA Science Advisory Board, "Draft Report on Incineration of Hazardous Liquid Waste," (Washington, DC: December 1984) and *New York Times* (1/17/85).
7. Telephone interview with Casey Husveldt (4/3/85) and *Hazardous Waste News* (8/26/85).
8. "A Ground Water Protection Strategy for the Environmental Protection Agency" (Washington, DC: U.S. EPA, 1984), p. 4.
9. A. T. Kearney, Inc., "Strategic Waste Management Considerations in Siting Industrial Facilities," p. 21.
10. OTA, "Superfund Strategies—Summary" (1985). See also Charles B. Reingaum and Walter T. Hang, "Toxic Wastelands: Environmental and Public Health Implications of Six New York City Municipal Landfills" (New York: NYPIRG, 1983).
11. Telephone interview with Kent Stoddard (July 1984).
12. Assembly Office of Research, "Toxic Ponds: Antiquated Methods and Unacceptable Dangers—Executive Summary" (Sacramento: California Legislature, February 1984).
13. Office of Drinking Water, "Surface Impoundment Assessment National Report" (Washington, DC: U.S. EPA 1983), pp. 4–6 and 60; see also National Research Council, Geophysics Study Committee, *Groundwater Contamination* (Washington, DC: National Academy Press, 1984), p. 4.
14. "Interim Status Ground Water Monitoring Implementation Study—Phase III—Draft" (Washington, DC: 1984), p. 1.
15. Ibid.
16. Robert L. Duprey, Director, Air and Waste Management Division, Region 8, "Letter: Re: FOIA Request RIN-0329-84" (12/27/84).
17. Lee M. THomas, "Memorandum: Definition of 'Major Handlers' of Hazardous Waste" (July 11, 1983).
18. See Dr. Ruth Ruttenberg's testimony to Congress on behalf of CEP in Committee on Government Operation, U.S. Congress, "OSHA's Failure to protect the

Health and Safety of Workers at Hazardous Waste Sites" (Washington, DC: U.S. Government Printing Office, 1985).
19. Dr. George Chu, California Division of Occupational Safety and Health to Benjamin A. Goldman (10/5/84).
20. NYSDEC, Division of Solid and Hazardous Waste, "1982 New York State Generators of Hazardous Waste Annual Report, Volume II" (December 1983).
21. *New York Times* (8/31/83).
22. Government Activities and Transportation Subcommittee, Committee on Government Operations, U.S. House of Representatives, "Improving the Effectiveness of the Bureau of Motor Carrier Safety and its Enforcement of Hazardous Materials Regulations" (Washington, DC: U.S. Government Printing Office, 1983).
23. "Uniform Hazardous Waste Manifest, Joint EPA/DOT Rule" in *Federal Register,* vol. 49, no. 55 (3/20/84).
24. Industry, Technology, and Employment Program, Office of Technology Assessment, U.S. Congress, "Groundwater Protection Standards for Waste Land Disposal Facilities: Will They Prevent More Superfund Sites?" (4/6/84). See also Office of Technology Assessment, *Protecting the Nation's Groundwater From Contamination* (1985).
25. U.S. EPA, "Environmental Status Report for the State of New York" (New York: U.S. EPA Region 2, 1983).
26. U.S. EPA "EPA RCRA 3008 Administrative Civil Complaints/Compliance Orders—FY 84" (8/10/84).
27. U.S. EPA "Interim Status Ground Water Implementation Study—Draft."

Glossary

The following words and expressions are defined as used in this book.

Absorbant: a material that absorbs liquid through its entire bulk (not merely at its surface).

Acre-foot: the volume that would cover 1 acre to a depth of 1 foot, equal to 43,560 cubic feet.

Activated carbon: carbon particles made extra porous by heat treatment.

Activated sludge: organic sludge with special cultures of microorganisms added to aid waste decomposition.

Air release: see *Release*.

Aqueous: dissolved in water.

Aquiclude: a geological formation of impermeable clay or bedrock that may impede water migration.

Aquifer: an underground water-bearing geological formation.

Aquitard: a semipermeable geological formation that reduces water migration.

Assessment monitoring: monitoring to assess the extent and constituency of ground water contamination.

Assessment report: written description of results of assessment monitoring.

Berm: a ridge constructed of clay or other material designed to contain wastes and prevent runon and runoff.

Bioaccumulative: substances that accumulate in living organisms without being metabolized or excreted.

Bioassay: a test to determine the potency of a substance to affect the biological activity of a living organism.

Biodegradation: chemical decomposition through biological activity.

Boiler: a pressure vessel that produces vapor or heated liquid after the application of heat.

Cancer: one of various diseases characterized by a malignant neoplasm or tumor and proliferation of epithelial cells in various parts of the body. Cancer spreads into adjacent tissue, with consequent progressive degeneration which is often fatal. There are many types of cancer and—possibly—many causes.

Carcinogenic: substance shown to cause cancer either in humans or laboratory animals.

Carrier: a vehicle, or its owner or operator, that carries hazardous wastes during transport.

Cell, landfill: a section of a landfill that is segregated from the rest by a berm or other means of containment.

Class A explosive: materials that are highly explosive including dynamite, nitroglycerine, picric acid, lead azide, mercury fulminate, black powder, blasting caps, and detonating primers.

Class B explosive: materials that are flammable and include such products as propellant explosives and some special fireworks.

Class I violation: a violation that results in a release or serious threat of release of hazardous waste to the environment, or involves the failure to assure that ground water will be protected, that proper closure and postclosure activities will be undertaken, or that hazardous wastes will be destined for and delivered to permitted or interim-status facilities.

Class II violation: any violation of RCRA requirements that does not meet the criteria listed above for Class I violations.

Closure: the owner or operator of a hazardous waste facility must close the facility in a manner that minimizes the need for further maintenance, and controls, minimizes or eliminates, to the extent necessary to prevent threats to human health and the environment, postclosure escape of hazardous wastes. A written closure plan must be submitted to and approved by appropriate regulatory agencies.

Commercial: (CEP definition) a company or facility that manages hazardous waste as a business, receiving more than 50 percent of its wastes from off-site generators.

CERCLA (Comprehensive Environmental Response Compensation and Liability Act of 1980): the federal law that governs the clean-up of uncontrolled hazardous waste sites. Popularly known as the Superfund law.

CERCLA waste recipient: an operating hazardous waste facility that receives wastes removed from a Superfund site.

Confined liquid: a liquid contained within a vessel, usually in a 55-gallon drum.

Containment: method or technology that prevents migration of hazardous waste into the environment.

Destruction efficiency: the percentage of a substance destroyed during thermal treatment.

Detection monitoring: monitoring to detect the presence of hazardous constituents in ground water.

Disposal: the final method used to manage wastes—often, long-term storage underground.

Environmental Impairment Liability: the category of insurance coverage for sudden and non-sudden occurrences that pose risks to third parties and the environment.

Exposure pathways: the environmental routes through which living organisms can be exposed to hazardous waste.

Feedstock: a raw material used during production processes.

Financial instrument: one of nine alternative insurance and assurance mechanisms required for certain RCRA facilities, including: trust fund, financial guarantee bond, performance bond, letter of credit, insurance, financial test, corporate guarantee, state-required mechanism, and state guarantee.

Flash point: the lowest temperature at which a liquid will give off enough flammable vapor at or near its surface so that a spark or flame will ignite it in air.

French drain: an underground drainage system that uses a series of pipes, covered trenches, or permeable channels to direct leachate to a collection point.

Generators: companies or facilities that produce hazardous wastes.

Ground water: water found underground.

Ground water monitoring: monitoring of ground water to detect and assess possible contamination.

Halogenated: organic (carbon-containing) substances that have bromine, chlorine, fluorine, or iodine atoms in their molecular structure. Chlorinated hydrocarbons are the most important commercially.

Hazard Ranking System: the model used to determine inclusion of a waste site on U.S. EPA's National Priorities List for Superfund cleanups.

Heavy metal: any metal with a specific gravity greater than 4 (that is, any metal more than four times as dense as water).

Hydraulic gradient: the direction of ground water flow.

Hydrocarbon: any of a class of compounds containing only hydrogen and carbon, such as an alkane, methane, alkene, ethylene, alkyne, acetylene, or aromatic compound.

Hydrogeology: the science of the properties of water confined within geological formations.

Hydrology: the science that deals with the occurrence, circulation, distribution, and properties of water.

Incompatible: two or more substances that cause fire, explosion, generation of flammable or toxic gases, or other violent reactions when mixed.

Inorganic: containing no carbon.

Interim status: any hazardous waste management facility that began operation prior to November 19, 1980 and is applying for a RCRA permit.

Land emplacement: any method used to dispose of wastes in the ground, including landfill, surface impoundment, and underground injection well.

Landfill: a technology used for burial of solid and hazardous wastes that attempts to isolate the wastes from the surrounding environment.

Leachate: liquid (often rainwater or other precipitation) that may be contaminated by hazardous components after migrating underground from a land emplacement unit into a collection system or the surrounding environment. Not to be confused with a leak.

Leak: migration of liquid wastes outside of containment structures.

Liner: a synthetic or earthen material built under and around a landfill, surface impoundment, or waste pile to reduce leachate formation or migration.

Major facility: state-designated sites that include all facilities subject to ground water-monitoring and/or protection requirements; all incinerators, up to 10 percent of remaining treatment, storage, and disposal facilities; and up to 3 percent of generators and transporters.

Manifest: federal and state form that indicates generator, quantity, and type of waste and its destination during transportation.

Midnight dumping: the disposal of hazardous wastes by illegal methods.

Migration route: the environmental media, that is, air, water, or land, through which hazardous wastes can be released.

Mutagenic: causing alterations to the structure of genetic material of living organisms.

National Priorities List: the list of hazardous waste sites targeted for cleanup action under CERCLA.

Net working capital: a firm's current assets less current liabilities. The result shows the relative liquidity of capital.

Non-methane hydrocarbon: a hydrocarbon compound that is generally larger than methane (one carbon, four hydrogen atoms alone).

Non-sudden occurrence: a gradual, accidental event arising from the operations of a hazardous waste facility that may cause bodily injury or property damage to a third party.

Off site: located away from the site of waste generation.

On site: hazardous waste that originates from the site.

Organic: containing carbon.

Part A: the first part of a hazardous waste management facility's permit application. Part A outlines general characteristics of the facility's location, waste types, quantities, process types, and capacities.

Part B: the second part of a hazardous waste management facility's permit application. Part B outlines in great detail exactly how the operator will comply with all applicable RCRA requirements.

Permeability: the degree to which a liquid (usually water) can pass through a substance or mass (10^{-7} centimeters per second equals approximately 1 inch per year; 10^{-6} centimeters per second equals 1 foot per year; 10^{-5} centimeters per second equals 10 feet per year). Note: Some solvents can move quickly through clays that are "impermeable" to water.

Permit: EPA and authorized states issue a permit to a hazardous waste management facility once it has shown that it meets and will continue to meet the set of conditions required.

pH: a measure of the acidity or alkalinity of a solution on a scale of 0 to 14 (low is acidic, high is alkaline or caustic, 7 is neutral).

Photochemically reactive: reacts with other chemicals when exposed to sunlight to produce smog or other atmospheric pollutants.

Pickling liquor: corrosive liquid (usually acid) used for removing scale and oxides from metals.

PCB: polychlorinated biphenyl.

Postclosure: 30-year period after a RCRA facility has been closed during which proper monitoring and reporting must be conducted.

Reactive: a substance that is normally unstable and readily undergoes violent change; or reacts violently with water; or generates toxic gases, vapors or fumes in a quantity sufficient to present danger to human health or the environment when mixed with water.

Release: the event of hazardous waste dispersing from an uncontrolled or improperly controlled site via air, water, or land.

Remedial action: those cleanup measures consistent with a permanent remedy taken instead of or in addition to removal.

Removal Action: the cleanup or removal of released hazardous substances from the environment to prevent, minimize, or totally mitigate damage.

Residence time: the length of time during which a hazardous waste is subject to elevated temperatures during incineration.

RCRA (Resource Conservation and Recovery Act): the federal law passed in 1976 to regulate management of hazardous waste.

RCRA Hazardous Waste: a waste that exhibits measurable characteristics that may cause or significantly contribute to an increase in deaths; serious irreversible or incapacitating illness; or pose a substantial present or potential hazard to human health or the environment when it is improperly treated, stored, transported, disposed of, or otherwise managed.

RCRA solid waste: garbage, refuse, or sludge irrespective of whether it is discarded, used, reused, recycled, reclaimed, stored, or accumulated for any of these purposes.

Runoff: a liquid that migrates over an uncontained, or improperly contained, area.

Runon: a liquid that migrates onto an uncontained, or improperly contained area.

Scrubber: a device that purifies gases (usually those emitted from an incinerator) by contacting the gases with a liquid spray or dripping surface.

Shell game: a sleight-of-hand game employing walnut shells or the like, under which a pea or similar object is moved around. Also, any deception, swindle, or fraud.

Shipper: the generator of waste that is transported.

Sludge: a waste that appears semisolid and is often composed of 90 percent or more water (by weight).

Slurry wall: an underground vertical wall made of relatively impermeable material that significantly retards leachate and ground water migration.

Solvent: a liquid capable of dissolving another substance.

Specific conductance: the capacity of material to conduct electrically charged particles.

Sudden occurrence: a sudden, accidental event arising from the operations of a hazardous waste facility that may cause bodily injury or property damage to a third party.

Superfund: the colloquial name applied to the federal fund established under CERCLA.

Surface impoundment: pits, ponds, and lagoons used for treating, storing, or disposing of liquid hazardous wastes (and sometimes for drying sludges).

Surface water: any body of water that is found above ground; canals, creeks, lakes, ponds, rivers, streams, etc.

Surficial: originally being on the surface of the earth.

10-K Report: the annual financial statement submitted to the Securities and Exchange Commission by all publicly owned U.S. companies.

10-Q Report: the quarterly financial statement submitted to the Securities and Exchange Commission by all publicly owned U.S. companies.

Teratogenic: causing injury to a developing fetus.

Test burn: a trial incineration of hazardous wastes that tests an incinerator's destruction efficiency.

Total Dissolved Solids (TDS): a measurement of all solids that are dissolved in water, waste water, or leachate, usually stated in milligrams per liter (parts per million).

Total Organic Carbon (TOC): a preliminary step in overall analysis of a water sample. TOC is measured by breaking organically bound carbon compounds into carbon dioxide and measuring the carbon dioxide level; usually given in milligrams per liter (or parts per million).

Total Organic Halogen (TOX): a measurement used to estimate total quantity of halogenated organic material in water, waste water, or leachate. TOX represents total amount of all halogens combined as opposed to individual component compounds. TOX does not give information about the structure or nature of the organic compounds involved, nor does it detect fluorinated compounds. See *Halogenated.*

Toxic: capable of producing injury, illness, or damage to living organisms through ingestion, inhalation, or absorption through any body surface.

Toxic Substances Control Act (TSCA): the federal law enacted in 1976 to regulate commerce and protect human health and the environment by requiring testing and necessary use restrictions on certain chemical substances at the point of manufacture.

Toxicology: the science of poisons, their effects, antidotes, and detection.

Transporter: company that transports hazardous wastes.

Unconfined liquid: a liquid that is not contained in a vessel.

Uncontrolled: a waste site without adequate measures to prevent wastes from migrating off site, or from being released through fire, explosion, and direct contact.

Unqualified auditor's opinion: opinion of a certified public accountant that a firm's financial statement is presented fairly and according to standard accounting practices. The statement need not be adjusted for existing contingencies that may have material impact on the firm's economic viability.

Unsaturated zone: the geological layer below the earth's surface and above the uppermost seasonal level of the water table.

Volatile: evaporating readily at normal temperatures and pressures.

Wetlands: fragile environments that are regularly inundated or saturated with water. Wetlands are susceptible to severe impact from pollution. Marshes, coastal wetlands, bayous, and swamps are examples of wetlands.

Index

Also Available from Island Press

An Environmental Agenda for the Future,
by Leaders of America's Foremost Environmental Organizations.
$5.95, paper.

The chief executive officers of the ten major environmental and conservation organizations launched a joint venture to examine what goals the environmental movement should pursue now and on into the 21st century. This book presents policy recommendations to effect changes needed to bring about a healthier, safer living experience. Issues discussed include: nuclear issues, human population growth, energy strategies, toxic and pollution control, and urban environments.

> *". . . a substantive book addressing the most serious questions about the future of our resources."* —JOHN CHAFEE, SENATOR, ENVIRONMENTAL & PUBLIC WORKS COMMITTEE

> *"While I am not in agreement with many of the positions the authors take, I believe this book can be the basis for constructive dialogue with industry representatives seeking solutions to environmental problems."*
> —LOUIS FERNANDEZ, CHAIRMAN OF THE BOARD, MONSANTO COMPANY

Water in the West, by The Western Network.
Vol. I: What Indian Water Means to the West. $15.00, paper.
Vol. II: Water for the Energy Market. $15.00, paper.
Vol. III: Western Water Flows to the Cities. $25.00, paper.

An essential reference tool for water managers, public officials, farmers, attorneys, industry officials, and students and professors attempting to understand the competing pressures on our most important natural resource—water. This three-volume series provides an in-depth analysis of the effects of energy development, Indian rights, and urban growth on other water users.

Community Open Spaces,
by Mark Francis, Lisa Cashdan, Lynn Paxson. $24.95, cloth.

Over the past decade thousands of community gardens and parks have been developed on vacant neighborhood land in America's major cities. *Community Open Spaces* documents this movement in the U.S. and Europe, explaining how planners, public officials, and local residents can work in their own community to successfully develop open space.

Private Options: Tools and Concepts for Land Conservation,
by Montana Land Reliance and Land Trust Exchange. $25.00, paper.

Techniques and strategies for saving the family farm are presented by 30 experts. *Private Options* details the proceedings of a national conference and brings together, for the first time, the experience and advice of land conservation experts from all over the nation.

The Conservation Easement in California,
by Thomas S. Barrett and Putnam Livermore
for The Trust for Public Land. $44.95, cloth; $24.95, paper.

The authoritative legal handbook on conservation easements. This book examines the California law as a model for the nation. It emphasizes the effectiveness and flexibility of the California code. Also covered are the historical and legal backgrounds of easement technology, the state and federal tax implications, and solutions to the most difficult drafting problems.

Building an Ark:
Tools for the Preservation of Natural Diversity
through Land Protection,
by Phillip M. Hoose. Illustrations. $12.00, paper.

The author, The Nature Conservancy's national protection planner, presents a comprehensive plan to identify and protect each state's natural ecological diversity, and shows how plant and animal species can be saved from destruction without penalty to the landowner. Case studies augment this blueprint for conservation.

Land-Saving Action,
edited by Russell L. Brenneman and Sarah M. Bates.
$39.95, cloth; $24.95, paper.

The definitive guide for conservation practitioners. A written symposium by the 29 leading experts in land conservation. This book presents in detail land-saving tools and techniques that have been perfected by individuals and organizations across the nation. This is the first time such information has been available in one volume.